Environmental Medicine — Beginnings and Bibliographies of Clinical Ecology

Theron G. Randolph, M.D.

Published by:
Clinical Ecology Publications, Inc.
109 West Olive Street
Fort Collins, CO 80524
(303) 482-6001

Environmental Medicine — Beginnings and Bibliogaphies of Clinical Ecology
by Theron G. Randolph, M.D.

©1987 Clinical Ecology Publications, Inc.

Library of Congress Cataloging-in-Publication Data

Randolph, Theron G.
    Environmental medicine: Beginnings and bibliographies of clinical ecology

    Includes bibliographies and index.
    1. Environmentally induced diseases—History.
2. Environmentally induced diseases—Bibliography.
I. Title. [DNLM: 1. Ecology—bibliography. 2. Ecology—
history. 3. Environmental Health—bibliography.
4. Environmental Health—history. WA 11.1 R194e]
RB152.R36   1987   616.98'009   87-26790
ISBN 0-943771-00-5
ISBN 0-943771-01-3 (soft)

All rights reserved. No part of this publication may be reproduced or transmitted in any manner, electronic or mechanical, including photocopying or recording on any information storage or retrieval system, whether for internal, personal or commercial use, except for brief quotations in published reviews, without express written permission from Clinical Ecology Publications, Inc.

Cover design by John Sorbie.

Printed in the United States of America by Citizen Printing, Fort Collins, Colorado.

ISBN: 0-943771-00-5 (cloth)
       0-943771-01-3 (paperback)

Dedication

To my wife, Tudy, whose consistent help has enabled me to write this book.

Acknowledgments

To L. Grace Nelson, my manuscript secretary, to Robert M. Lamont, who collected the bibliographies, to John C. Kern, for several helpful suggestions, to Lawrence D. Dickey, M.D., who helped in numerous ways with this publication, and to the many contributors to the Human Ecology Research Foundation over the past thirty-five years.

Preface

There are two related themes in this book — an introductory autobiographical account and the main thesis — the emergence of clinical ecology as a major environmental focus in medicine.

As far as this autobiographical sketch is concerned, my life has been likened to that of Semmelweis.[1] Although I was complimented by this comparison at the time, this failed to register fully until reading accounts of the life of Semmelweis more recently.[2,3]

Semmelweis' description of how childbed fever spreads[4] entitles him to be called an early spokesman of what came to be called the germ theory of disease at a time when this interpretation was widely contested in medicine. I have been referred to as a spokesman for the roles of common foods and environmental chemicals in the induction and perpetuation of many poorly understood physical and mental illnesses in specifically susceptible persons. This interpretation, despite its demonstrability, is also widely debated currently. In short, both Semmelweis and I were impressed by the apparent significance of our clinical observations. We both felt strongly that a wider application of our concepts and techniques to demonstrate these relationships in the practice of medicine was mandatory.

The second theme of this book and its major consideration concerns the emergence of clinical ecology as an alternative interpretation applicable in the management of many chronic illnesses. This brings up the subject of creativity. Creativity in science leading to productive scientific revolutions is initiated by individuals, rather than by a committee of peers. Scientific revolutions in medicine are usually started by individual physicians as a result of observing and treating their patients rather than emerging from board rooms of medical schools. This is precisely how the subject of clinical ecology developed.

As will be described in detail later, the subject of allergy, first described as altered reactivity at the turn of the century, developed on the heels of infectious diseases. Under these circumstances it is not surprising that these two environmental thrusts with medical applications should both be interpreted in terms of immunologic mechanisms. However, Arthur F. Coca, founder and editor of the Journal of Immunology, objected strenuously to this immunologic interpretation of allergy as accepted by medical academia. Nevertheless, this deductively derived and limited interpretation of allergy has been taught as fact in the curriculum of medical schools as well as in most postgraduate courses and training programs sponsored by national allergy societies.

Coincident with acceptance of the immunologic interpretation of allergy, observations of a half-dozen practicing allergists (including Duke, Rowe and later Coca and Rinkel) laid the groundwork for what came to be called clinical ecology, by reporting their inductively derived observations of interrelations between specific environmental exposures and chronic illnesses of specifically susceptible patients. It was subsequently learned that Hare in Australia had made similar observations starting two decades earlier. I noticed these interrelationships beginning in the mid-1940s. Inasmuch as localized allergic syndromes had already been described, I focused my attention on manifestations resulting from the impingement of specific foods and environmental chemicals on susceptible persons presenting as systemic illnesses. These personal-exogenous interrelationships were interpreted in terms of specific adaptation as they manifested either as stimulatory or withdrawal reactions, depending on the patients' degree of individual susceptibility and factors of exposure.

The point to emphasize here is that the original observations of these pioneers which led to the present concepts and techniques of clinical ecology were all made inductively, in that they were based on thousands of detailed clinical observations of chronically ill patients in the course of their responses to given environmental exposures. This accumulated knowledge led to hypotheses which were then confirmed and extended as they were applied more widely in the practice of medicine. Workable concepts and techniques from this accumulated evidence of causal interrelationships led to the development of the diagnostico-therapeutic technique of comprehensive environmental control in a hospital unit. Many of the major clinical interpretations in this book are based on the application of this methodology in thousands of cases. This technique of isolating a patient from inadvertent exogenous exposures as a baseline for observing the clinical effects associated with single re-exposures to previously avoided materials turned out to be the ultimate diagnostico-therapeutic technique in this new approach to this old field. In my clinical experience, this regimen has served as the standard of reference with which the results of less rigidly controlled diagnostic and therapeutic techniques must be compared. More recently, the additional role of endogenously stored lipid-soluble toxic environmental agents is also being evaluated.

The fact that I had worked closely with Coca, Rowe and Rinkel and that I am familiar with the contributions of other pioneers in clinical ecology enables me to write this account of the beginnings and bibliographies of this aspect of environmental medicine.

1. Shambaugh GE Jr: Ahead of Their Time, Arch. Otolaryngol. 79: 118, 1964
2. Thompson M: The Cry and the Covenant, New York, Doubleday & Co., 1949
3. Bendmer E: Semmelweis, Lone Rager Against Puerperal Fever, Hospital Practice, February 5, 1987
4. Semmelweis I: The Etiology, Concept and Prophylaxis of Childbed Fever 1861

Introduction

In General

How clinical ecology started as an environmentally focused specialty of medicine has not been recorded in detail. My friends and associates tell me that I should do this, inasmuch as I am the only living person who had worked closely with Arthur F. Coca, Albert H. Rowe and Herbert J. Rinkel — the three pioneer clinical observers most responsible for this development.

Although I had reminisced about my relationships with Coca, Rowe and Rinkel a few times, these recollections had not been recorded. I have been prompted to report these anecdotes by Dr. and Mrs. Francis J. Waickman of Cuyahoga Falls, Ohio who heard one of these accounts at the home of Dr. and Mrs. William H. Wilson of Denver, Colorado in 1974.

More recently, Dr. Ronald Finn of Liverpool, England and Dr. Don Jewett of San Francisco asked me questions about the beginnings of specific adaptation, addiction and the rotary diversified diet. But I had difficulty in answering their questions as isolated events. Also during recent years I have been asked to supply writers about my part in this development, but this was also difficult without integrating my findings with detailed observations of others.

I also needed this account of my background, acquaintances and attainments for Francis Silver V who was nominating me for an environmental-ecology award. Not having completed this account by the 50th anniversary of my graduation from medical school in 1983, I then aimed for my 80th birthday, July 8, 1986. But missing this publication date, I am now trying to finish this for the 20th Annual Meeting of the Society for Clinical Ecology, recently renamed The American Academy of Environmental Medicine — the Discipline of Clinical Ecology, scheduled for October 27-30, 1986.

After several changes in format, it was finally decided to write an autobiographical account, bringing in the relationships of others as I had encountered them. It also seemed advisable to include the bibliographies of Coca,

Rowe, Rinkel and others, as well as my own updated list of publications. References will be made to my publications by number as listed in the Appendix, starting on page 349.

This account is prepared for both physicians and laymen. Since most physicians are already somewhat knowledgeable in how environmentally focused medicine differs from conventional medical care, as well as differences of opinion presently existing in this area, nonmedical persons would understand this book better if they were familiarized with these controversies. It is recommended, therefore, that laymen read the article prepared for the public on this score, presented as the first item of the Appendix, page 281, before reading this book.

Geographical Setting

The geographical setting of this autobiographical account begins in rural southern Michigan, a few miles off the direct highway between Chicago and Detroit. More specifically, this sketch starts in Hillsdale County, Michigan, continues at Ann Arbor, Michigan, from which the setting changes to Milwaukee and Chicago, except for a two-year sojourn in the Boston, Massachusetts area.

Timing

The dates of this account cover a period of 80 years. During this span of a single lifetime more profound changes occurred in our way of life, as a result of the application of the industrial revolution, than transpired during any other similar period of time. Although I grew up in a relatively self-sufficient rural agricultural society, I ended up 80 years later in being primarily concerned with the health hazards associated with an industrial urban environment. For instance, the living conditions of my parents and my younger years are so different from those in which my five-year-old grandson is being raised as to suggest a different world. Indeed it is a different world. Whereas my parents earned their living as general and stock farmers, my grandson's father is a technician and his mother is president of an electronics manufacturing firm.

Medical Background

This autobiographical sketch is especially concerned with the medical implications and applications of the industrial revolution. Previous considerations in respect to geography, timing, occupational and economic background are merely introductory to the medical aspects of this account.

It seems that I am reasonably well qualified to write this report inasmuch as my own life has been a microcosm of this profound change in American life, as this transition has affected the health and behavior of individuals. Personal qualifications also include a better than average knowledge of history, an active interest in reading and writing and, especially, more concern with the underlying whys and wherefores of medical problems than exhibited by most physicians. These interests led to considerations of causative environmental factors in medical syndromes at a time when these factors were poorly understood and appreciated. This environmental-medical interest led to the development of techniques to demonstrate the inciting and perpetuating roles of the commonplace environment in many chronic illnesses, especially those occurring in highly susceptible individuals.

As will be pointed out later, many of these environmental-medical relationships were suspected, demonstrated and reported before any consensus either within the medical profession or at the public level that these influences were operating to change the lives of selected individuals. Neither did these environmental-medical interrelationships occur evenly across the board. Initially, there were only a few highly susceptible persons engulfed in this environmental maelstrom. But with time, the scope, intensity and depth of this whirlpool involving the human environment engulfed an ever-increasing number of people.

In fact, so rapidly are people becoming susceptible to some of the circumstances of their intake and surroundings that such environmentally related illnesses have now apparently passed the majority mark. Moreover, all degrees of specific environmental susceptibility seem to be involved. As one drifts to the edge of this ever-expanding whirlpool, such environmentally related medical problems are more annoying than impairing, more intermittent than continuous. But with time, increasing individual susceptibility and cumulative specific exposures, a vulnerable person tends to be drawn increasingly deeper into these environmental problems of the modern world. Sooner or later, and despite all efforts to remain physically active and mentally alert, many victims tend to become so chronically ill that they sink into impaired productivity and, sometimes, functional disability.

Likening the progressive development of chronic physical and/or mental illnesses from unrecognized environmental exposures to which suspected persons are highly susceptible, to drifting gradually into a whirlpool characterized by advancing illness and disability, is appropriate for several reasons. When the cumulative impact of given environmental materials as enhanced by individual susceptibility remains unrecognized, victims of this subtle process are usually unable to protect themselves. Because of the essentially addictive nature of this involvement with one's environment, involved persons tend to resort to the responsible exposure(s) as often as necessary and in an amount sufficient so as to remain relatively stimulated. Such dose-exposure stimulatory responses of specifically addicted persons vary greatly as this process develops. Examples from chronic food allergy will be described first as this is the area where these maladaptive clinical observations were first observed. Later the additional impingement of environmental chemical exposures on susceptible persons will be considered.

Overall Environmental-Medical Interrelationships

Where this account of environmental-medical interrelationships fits into what has already been described needs to be stated to serve as background for this presentation. The first major thrust of the environment on the course of modern medicine was the germ theory of disease. The concept of infectious disease came to this country precipitously in 1880, although its full implications were not widely applied for at least another generation or two.

The second great thrust of the environment on modern medicine was what later came to be called the allergic interpretation of illness, in which given environmental exposures impinged on the health of susceptible persons. Individual susceptibility to pollens and animal danders manifesting in hay fever and asthma was described by the Englishmen, Bostock and Elliotson, by 1830. (Bostock J. Of the Catarrhus Aestivus or Summer Catarrh. Med Chirurg, Trans., London 14:437, 1828; Elliotson J. Clinical Lectures. Lancet 2:370, 1830) Salter, of London, extended this list of environmental exposures precipitating bouts of asthma in selected persons to include — in addition to pollens and danders — foods, drinks, dusts, feathers, fogs, smokes and fumes, in 1862. (Salter HH. *On Asthma; Its Pathology and Treatment*. Philadelphia: Blanchard & Lea, 1864)

The next important development occurred in the experimental animal laboratory. While attending an international symposium on stress in Monaco in 1980 and visiting the Museum of Oceanography, I was astounded to find a replica of a turn of the century laboratory inside the front door. The story of this laboratory is as follows: About 1900, Prince Albert of Monaco arranged for Charles Richet, a physiologist, to accompany him on his yacht during a cruise of the Mediterranean to study the problem presented by the Portuguese man-of-war. It seems that the Prince liked to swim in the Mediterranean but was prevented from doing so by the stinging tentacles of the man-of-war. In the course of his studies of this problem, Richet injected guinea pigs with dog serum in the absence of any immediate or delayed reaction. But when such an injection was repeated three weeks later, these guinea pigs developed what Richet called acute anaphylaxis (meaning without protection) and promptly died. Upon returning to Monaco, Richet demonstrated this phenomenon to his friend, Portier. Prince Albert set up Richet in a laboratory to continue his physiological observations, a reproduction of which from the Musee Oceanographique is illustrated in Figure 1.

After the Frenchmen, Richet and Portier, reported their findings in the medical literature, (Richet C. and Portier P. De L'Action Anaphylactique de Certains Venins. Compt Rend Soc de Biol 54:170, 1902) von Pirquet, an Austrian pediatrician, referred to this phenomenon as allergy in 1906, describing it as altered reactivity occurring with time. (von Pirquet C. Allergie. Munchen med Wehnschr 53:1457, 1906) One by one, rhinitis, asthma and various other localized manifestations were described and accepted as allergies within the open-ended descriptive definition of allergy.

Introduction

Subsequent developments within this field, a major part of which has since been called clinical ecology, will be presented in the course of this book.

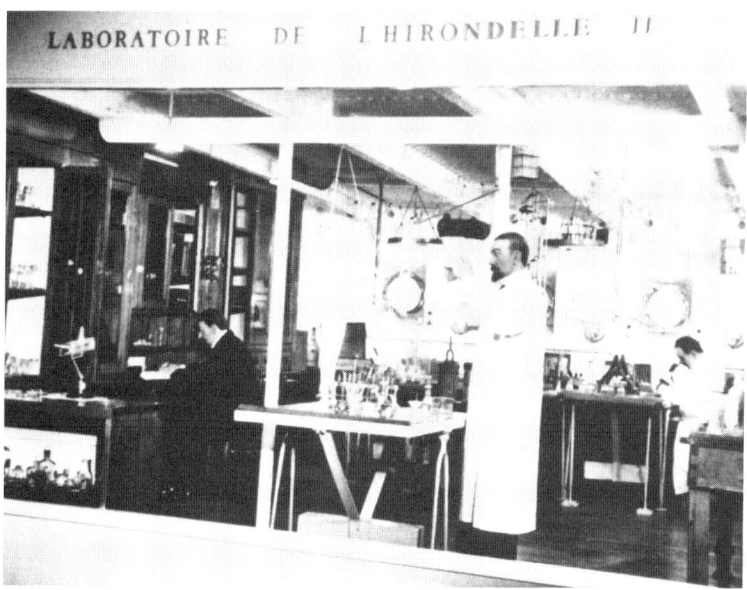

Figure 1. A replica of the research laboratory of Charles Richet in the Museum of Oceanography, Monaco.

Contents

Preface vii

Introduction ix

Table of Illustrations xvii

Chapter I 1906 - 1943 1
Early Life and Medical Training

Chapter II 1944 - 1950 21
Private Practice — Chicago, Illinois

Chapter III 1950 61
Further Anecdotes — Coca, Rowe and Rinkel

Chapter IV 1951 - 1953 71
Other Environmental Exposures

Chapter V 1954 - 1956 99
Changes in Medical Points of View and Terminology

Chapter VI 1957 - 1959 117
A Period of Transition

Chapter VII 1960 131
The Environmental-Medical Watershed

Chapter VIII 1961 - 1964 141
The Beginning of an Ecological-Medical Focus

Chapter IX 1965 - 1970 157
Era of Ecological-Medical Enlightenment
Favorable Governmental Environmental Protection

Chapter X 1971 - 1976 177
Continued Ecological-Medical Enlightenment
Environmental Protection and the National Legislative Scene

Chapter XI 1977 - 1980 201
Continued Ecological-Medical Enlightenment
Continued Deterioration of the Environment
Decreasing Regulation of Environmental Hazards
Environmental-Industrial Stalemate

Chapter XII 1981 - 1987 209
Continued Ecological-Medical Enlightenment
Continued Deterioration of the Environment
Increasing Resistance From Academic Medicine
Decreasing Regulation of Environmental Hazards
Increasing Politicization of Environmental Issues
Increasing Governmental Control of Medicine

Chapter XIII 229
Environmental Aspects of Clinical Ecology Updated

Chapter XIV 241
Adapted (Stimulatory) and Maladapted (Withdrawal) Responses Updated

Summary 259

Appendix A 281
"Environmental Illness" by Steve McNamara

Appendix B 295
Training Fellowships in Clinical Ecology for Physicians
Training Fellowships in Clinical Ecology for Nurses

Appendix C 299
Comments on the Position Paper on Clinical Ecology
by the American Academy of Allergy and Immunology,
by Doris J. Rapp, M.D.

Appendix D 309
Bibliographies of Clinical Ecologists
    Arthur F. Coca, M.D. 311
    Hal M. Davison, M.D. 317
    William W. Duke, M.D. 319
    French K. Hansel, M.D. 325
    Francis W. E. Hare, M.D. 329
    Herbert J. Rinkel, M.D. 333
    Albert H. Rowe, M.D. 337
    Warren T. Vaughan, M.D. 343
    Michael Zeller, M.D. 347
    Theron G. Randolph, M.D. 349

Index 359

Illustrations

Page	Figure	
xiii	1	A replica of the research laboratory of Charles Richet in the Museum of Oceanogaphy, Monaco
2	2	Father and Mother in front of boyhood home, Jerome, Michigan
4	3	Theron G. Randolph at ages 2, 9 and 16
10	4	Theron G. Randolph at ages 22, 30 and 47
12	5	Simpson Memorial Institute, Ann Arbor, Michigan
12	6	University Hospital, Ann Arbor, Michigan
24	7	Symptoms in food allergy
25	8	Types of food allergy based on the response to specific avoidance
26	9	Types of food allergy based on the extent of involvement of the diet
65	10	Albert Rowe, Helen C. Hayden, William G. Crook, Hilda M. Hensel, Theron G. Randolph, Mrs. Schafer, Walter L. Schafer, Janet M. (Tudy) Randolph, George S. Frauenberger and Helen Frauenberger
75	11	Low pressure center to the west, bringing industrial air pollution into Chicago area
75	12	Low pressure center to the east, bringing cool, less contaminated air to Chicago area
80	13	Francis Hare, M.D. as a young and older man
89	14	Arthur F. Coca, M.D.
94	15	Albert H. Rowe, M.D.
95	16	Albert Rowe wearing his medal
111	17	Side 1 of brochure, *Specific Adaptive Illness*
112	18	Side 2 of brochure, *Specific Adaptive Illness*
128	19	Scientific exhibit in Edinburgh, Scotland with Dr. Richard Mackarness of Kew, England and Dr. Theron G. Randolph

Page	Figure	
129	20	Levels of reaction in stimulatory-withdrawal phases
148	21	Herbert J. Rinkel, M.D.
152	22	Herbert J. Rinkel, M.D.
162	23	The Jonathan Forman gold medal, front and back
162	24	Theron G. Randolph, M.D.
174	25	Bodily centered versus ecologically oriented medical care
185	26, 27	Food ingredients entering the production of alcoholic beverages
187	28	Analytic versus biologic dietetics
193	29	Manifestations — Levels of reaction in ecologic mental illness
194	30	Clinical levels of specific reactions in ecologic mental illness
211	31	ABCDEFs of clinical ecology
214	32	American College of Allergists Award of Merit
248	33	Intermittent and sustained stimulatory and withdrawal levels of physical and mental reactions
251	34	The addiction pyramid
269	35	Adaptation as observed by clinicians
311		Arthur Fernandez Coca
317		Hal McCluny Davison
319		William Waddell Duke
325		French Keller Hansel
329		Francis Washington Everard Hare
333		Herbert John Rinkel
337		Albert Holmes Rowe
343		Warren Taylor Vaughan
347		Michael Zeller

Chapter I — 1906-1943

Early Life and Medical Training

Figure 2. Father and Mother in front of boyhood home, Jerome, Michigan.

Boyhood and Adolescence

I was born at Jerome, Michigan in 1906, the third of six children. My father was Fred Emerson Randolph 1867-1937. My mother was Rena Bagley Kempton 1877-1968.

My first exposure to the so-called chemical environment occurred when I was a boy of 10 years. My home was in the midportion of the lower tier of counties in the Southern Peninsula of Michigan. My father was a cattle feeder. He bought several carloads of feeder cattle in the fall, carried them through the winter and then drove them to our ranch in Gladwin County, Michigan — north and west of Saginaw, Michigan. My job was to drive the horse and buggy behind the herd of cattle. This took us very near the Dow Chemical Company plant at Midland, Michigan. The wind must have been very favorable, judging from my father's derogatory comments. The stench from this place provided my father with a topic for a spirited soliloquy for the next several miles. Needless to say, I was also impressed unfavorably.

My next major contact, as a farm boy, was with infectious disease. This occurred the following year at the age of 11 years. My father had gone to the Chicago Stockyards and bought four carloads of cattle to feed in our barns for the coming winter. I helped drive the cattle home from the railhead. But the very next day we were placed under police enforced quarantine. A raging epidemic of hoof-and-mouth disease had been diagnosed at the Chicago Stockyards the day before, this virus infection being extremely contagious for animals. This quarantine, starting at about Thanksgiving time, was enforced through the winter. All pigeons were killed for several miles around. We were told to dig a large hole (about 15 feet wide, 100 feet long and 8 feet in depth). As soon as the symptoms of slobbering and lameness had developed in some of the cattle, all of our cattle, hogs, sheep, cats and dogs were herded into this hole by the state police and shot down. Additional police in hip boots added lime and walked over the carcasses, slashing the bodies with swords

Figure 3.

Theron G. Randolph at ages 2, 9 and 16.

Early Life and Medical Training

to preclude attempts to salvage any of the hides. All were then covered. Our barns were fumigated and whitewashed. Each time we returned from the barn we waded through a disinfecting pool. Although I don't recall its identity, the odor was unpleasant.

Toward the spring of this long confined winter my father sold the wool crop. He was not concerned with breaking the quarantine by delivering the wool to the buyer in a small town 30 miles away. While my father was closing the sale, I went to the local barber shop — looking like a shaggy dog for not having had a haircut for months. The barber recognized a new kid and asked my name and where I lived. After telling him, he identified me as a member of the hoof-and-mouth disease quarantined family in the next county. When I told my father about this later, his only comment was that there were times when it was not necessary to supply strangers with everything they might wish to know.

The following winter when I was 12 years of age, mother turned in an amazing performance in the great influenza epidemic of 1918. My brother 16, sister 18, father and I, as well as the hired girl and the hired man all had influenza simultaneously. Mother not only isolated my younger sister six, and brother three, to the kitchen, she took care of six influenza victims in other parts of our big house (without an indoor bath), but neither she nor the younger children developed influenza! Perhaps some of this know-how had rubbed off from her brother, Rockwell M. Kempton of Saginaw, Michigan who had graduated from medical school in 1918 and at the time was in the midst of an infectious disease pediatric residency. I recall having listened intently to his accounts of happenings in medical school and early years of post-graduate training.

My father, on the other hand, held very relaxed ideas about matters of hygiene. The earliest major confrontation which I had with my father occurred the following year. After the cow I was milking stepped in my milk bucket, I quickly dumped the milk in the gutter, rinsed the pail repeatedly with running water and continued milking. When father came by a few minutes later and saw the spilled milk, he criticized me sharply for having dumped it. This was enough for me. After that experience I refused to drink milk again. Despite later having developed many food allergies, sensitivity to milk and beef never occurred.

My father also believed in having boys do man-sized jobs around the farm. I didn't object to this. One day during a scheduled copulation between horses, he dismissed me summarily to the house, but allowed my brother to stay on and help with the arrangements. Although I resented this, there was nothing I could do about it at the time. However, this was the last time he ever treated me so condescendingly, as I got back at him during the next few weeks. On several occasions when he assigned me a man's job and responsibility, I did it, but grumblingly commenting that I was either too small, too young or otherwise too insufficiently experienced for this or that assignment. My father got the message. At the time of each of these grumbling protests he would look at me sharply but would not say anything. This transition from boyhood to manhood, as far as my father and I were concerned, went more smoothly after this.

My mother not only defended me against my father in this milk bucket episode, but also agreed that I should not drink cow's milk. Not being convinced that this milk pollution had been controlled, mother saw to it that only cooked milk in the form of custards, cream sauces, milk toast and various other cooked milk dishes were served. I don't recall seeing any raw cow's milk on the table after this. It was also about this time that mother was shocked to learn that all of us in district school drank from a common cup dipped into a common bucket. She took care of this nonhygienic oversight in short order.

Living on a relatively self-sufficient and isolated farm in southern Michigan and participating in all farm activities (my father saw to that), taught me the importance of the environment. Indeed, prior to going away to college, I lived very close to the land. For instance, before we had a car, truck or tractor, I remember driving a team of horses with a wagonload of apples to the cider mill, and wheat to the flour mill, bringing back cider to make vinegar and flour to make bread for the following winter. Because of shortages during the first World War, I helped my mother make soap and candles. We also butchered our own hogs, cattle and sheep each fall. My older brother and I helped our father in the yearly task of butchering and castrating the young male pigs over the aggressive howling din of the mature hogs. Consequently, I knew something about anatomy long before going to medical school.

At the age of 17, our high school basketball team played in another small town about 50 miles away. In the course of allotting the players to homes for the night, I recognized the name of the director of our earlier hoof-and-mouth disease quarantine restrictions and requested to stay there. During dinner that evening, I inquired of the man of the house if he had directed the quarantine arrangements for the hoof-and-mouth disease epidemic several years earlier. Obviously surprised at this inquiry from a high school boy, he answered affirmatively. Remembering the time, place and circumstances, he launched into a tribute to my mother which I still remember. He went on to say that of all the people with whom he had dealt in enforcing quarantine restrictions in the state, my mother impressed him the most by asking more intelligent, pertinent and down to earth questions than any other person he had encountered. I remembered this dining table conversation seven years earlier in which my mother cross-examined this state official, during the entire meal, although I did not think much about it at the time. Indeed, this was the organized way of mother's inquiries and how we had come to expect her to act under such circumstances.

A few other points about my father and mother might be mentioned. Whereas my mother saw to it that we had at least some exposure to Sunday school, my father thought that this was less important than taking his boys on some type of excursion at exactly the same time. My father also held to the general idea that laws and other regulations which interfered with personal freedoms and desires existed to be bent from time to time. For instance, and in contrast to mother's objections, he saw nothing wrong in hunting rabbits with a ferret, despite this being illegal. Moreover, Sunday morning during winter months was an excellent time for such an activity. Sunday morning during the early spring and before the leaves emerged was also a good time to tramp the woods looking for bee trees and picking wild flowers. Crops on various parts of the farm were inspected at this time, fences were repaired or fishing trips were sometimes arranged.

My parents also held different ideas about the various changes in personal hygiene which were developing in rural areas at this time. Although we were being taught in school to brush our teeth regularly and, as a former school teacher, mother went along with this, father never accepted nor applied this new fangled idea.

At this time in a rural small town setting one generally consulted a dentist for a toothache, rather than for prophylactic care. Whereas my older brother always ran away on the days of dental appointments (with absence of dental anesthesia), I went along and simply hung on to the arms of the dentist's chair. But by the time I started college, I had cavities in all teeth with multiple amalgam fillings in all of them except for porcelain fillings in incisors. I happened to be waiting on tables the day that a new dentist arrived in town. I became his first patient. He replaced all my leaking amalgam fillings with gold inlays. To this day I have lost only one tooth, in contrast to the record of my brother, four years older, who became edentulous by the age of 40. According to current views, the absence of mercury amalgam fillings may have been partially responsible for my excellent health record.

But to illustrate matters of generally accepted personal hygiene in rural surroundings at the time, all members of the family followed the practice of Saturday night baths — whether or not these were needed. The routine of daily baths (with perhaps the hazards of water-logging) had not yet arrived — neither had the eight hour day on the farm; indeed, it still hasn't.

I had not realized until now, when writing this, that the germ theory of disease had been at stake in my family during my childhood. Allow me to explain these circumstances. Father was born in 1867. His rural schooling occurred between 1873 and 1885. The germ theory of disease, which had developed gradually in Europe over several decades, arrived suddenly on this side of the Atlantic in 1880. Prior to that time Americans had been preoccupied with our Civil War, including the turmoil preceding it, the trauma of hostilities, carpetbagging and other social and political adjustments following it. But after the demonstration of Koch's famous postulates in Germany in 1880, one American institution after another accepted the germ theory of disease as factual and began teaching this concept of illness. (Allen P. Etiologic Theory in America Prior to the Civil War. J History of Medicine 2:489, 1947)

Even the medical profession was notoriously retarded in accepting and applying this new information. Acceptance by the public occurred even more slowly. I doubt if my father ever did grasp and apply its full significance. The cow stepping in the milk bucket episode indicates that this was not a part of his thinking and decision making at the time that this confrontation occurred.

On the contrary, my mother (born in 1877) had undoubtedly been taught the importance of personal cleanliness, sanitation and other measures of infection control throughout her schooling starting in 1883. Moreover, she had taught school, being responsible for teaching these basic concepts and techniques to grade students for several years prior to marrying my father in 1899.

Despite my father's deficiencies in respect to making changes in his lifestyle in keeping with concepts of infection and health, he had many commendable points. He never drank or smoked. He worked diligently and long hours. He did have a few other deficiencies, including complete color blindness. Although he was fond of flowers, his favorites were those of striking shapes and aromas rather than colors. In contrast to mother who had jet black hair and dark brown eyes, father was almost an albino with light blue eyes and sandy light red curly hair. He always wore dark shirts during summer months and never went out without wearing a broad rimmed hat because of an extreme tendency for sunburning. Father was also an astute student of farm products, practices and markets, as far as their economic implications were concerned. He was also original and quick to innovate in growing new or unusual crops. In contrast to other farmers in the region, he was the first to try growing rape, sunflowers and buckwheat in our area. In being appalled at ambient air pollution in 1917, my father was nearly a half-century ahead of his time.

In high school I drove a horse and buggy five miles to the North Adams, Michigan school daily. Although I graduated as valedictorian of my class, this was no great distinction in a class of 14.

At about the age of 18, I helped my father load a frozen haystack bottom which entailed much pulling and wrenching motions. After finishing, I took my gun and walked about a half mile to hunt rabbits. But by the time I got there, I had developed severe left-sided chest pain and progressive shortness of breath. Although I started back home immediately, it was necessary to lie down in the snow several times, splinting my left chest before reaching home. I was taken to the local doctor immediately who made a diagnosis of pleurisy and taped my left chest. This was done in the absence of a chest X ray, as such a facility was 25 miles distant. I recovered gradually and was soon playing basketball again but remained subject to recurrences of acute chest pain intermittently for the next several years.

My great-grandfather gave $100 to Hillsdale College when this institution was founded in 1844. He made this gift in the name of his grandson, my father, with the understanding that his children would have free tuition at this institution. When my older sister presented this sheepskin in registering, she was granted a $3 decrease in tuition — the 1844 equivalent in tuition. Since

Early Life and Medical Training

my four grandparents, my father and about 50 other family members had attended this college, we children grew up with the assumption that all of us would attend college.

But my father declared bankruptcy a few weeks before I enrolled in college in 1925. Fortunately, I obtained work in the local theatre earning about $1,000 per year for the following four years which more than met all of my expenses.

During my beginning biology course, I chose the subject of adaptation for my term paper. This was a curious choice in view of later developments but emphasized the fact that from the beginning of classes in science I seemed to be more interested in synthesis, interactions and interrelationships than in analysis. In the beginning course in psychology I was assigned the subject of behaviorism which, although recently popularized by John B. Watson, was still regarded as extremely controversial. I taught the psychology class that day, defending behavioristic psychology against all comers, to the amusement and enjoyment of the professor sitting in the back of the room. Having been a member of the debating teams through high school and college, defending both sides of various issues at different times, I was prepared to cope with their defense of introspection and to champion making a "science" out of psychology. This debate experience also helped with future confrontations.

Not having planned on attending medical school earlier, it was necessary to spend an extra year in college to work off the physics requirement for entrance to medical school. My major interests in college had been in biology and psychology. Two things occurred during my senior year in college which turned out to be very helpful.

In order to placate my mother, I enrolled in several education courses in my senior year in preparation for a teaching career (but which I had no intention of ever doing). Being financially independent by means of having a steady job, I made most of my own decisions. These education courses were so dull that I dropped all of them after 10 days and enrolled in the commercial department, learning how to type. Also, contrary to advice at the time, I concentrated on speed at the expense of accuracy — the latter came later. By the end of this course I was an accomplished typist and had my own typewriter. This was unusual for students in 1928.

I also was the first student assistant in the Department of Psychology in my senior year in college, being in charge of the psychology laboratory. As far as is known this was the first psychology laboratory course ever offered for credit. Since my freshman psychology course, I had had a special interest in attempting to make psychology into more of a science. However, I regretfully concluded, by the end of this year of attempting to teach psychology for science credit, that probably psychology was not yet ready for a course in "science."

Having developed an interest in abnormal psychology, I applied for admission to the University of Michigan Medical School to become a psychiatrist. My application may have been facilitated by the fact that our local doctor,

1906-1943

Figure 4.
Theron G. Randolph at ages 22, 30 and 47.

who was the only physician member of the Board of Regents of the University of Michigan, wrote me a letter on regents' stationery to the effect that I was a rare young man. Not knowing whether he thought that I was half-baked or a genius, I attached it to my application anyway.

This reminds me of one of Herb Rinkel's favorite comments of the definition of a genius: "A genius is merely a local boy away from home."

Medical School — 1929-1933

The late George Frauenberger, who was the original audio-visual expert of the Society for Clinical Ecology and second recipient of the Jonathan Forman Award, was in the same class. Some years later, I asked George if he had ever seen this fellow, Larry Dickey before and he said, "No." In the meantime, I had been looking at Larry quizzically from the time of the second meeting of the Society for Clinical Ecology at the Flying Carpet Motel in Des Plaines, Illinois, vaguely realizing that I had seen him somewhere. Finally, I recognized his identity as an instructor in gross anatomy from the other side of the anatomy laboratory.

Still thinking that I would eventually be a psychiatrist, I watched the performance of this department with more than passing interest, but with increasing dismay. It was soon apparent that the department head — a year or so before he was overtaken by the mandatory retirement age — was unable to communicate with most other staff members. Moreover, no member of this department, irrespective of age, could measure anything! Of course, I was in a good position to make this criticism, having become an expert in measuring psychological and psychiatric behavior the previous year! Consequently, I escaped from psychiatry before getting in. Nevertheless, this early experience in making the transition from introspection to behaviorism was later to stand me in good stead — little did I know at that time what the future might offer.

At this time the medical curriculum was sharply divided between the preclinical first two years and the last two clinical years. My first two patients were disastrous and prophetic, respectively. The first patient, an aged cardiac male, was propped up in a semi-reclining position in bed to be examined. I propped him up further in order to listen to the back of his chest, while my associate listened to his heart. I commented, "I don't hear anything anymore, do you?" He said, "No." Whereupon we called the intern who pronounced our first patient dead. A student nurse added to the indignity of this disaster by charting: "Patient died while being examined by junior medical students." Although I was not amused by this experience, nor so the instructor, Thomas Findley, later Professor of Medicine at Tulane University Medical School, at least each time I have seen him since he starts to smile before we have a chance to shake hands.

The second patient, a 35-year-old housewife, was what we were later told was a profound hypochondriac. Indeed, there was no end to her history, which included between 18 and 20 different complaints. It must have taken me 40 minutes to recite this patient's history to our instructor, who to my abject astonishment became progressively amused and apparently disin-

Figure 5. Simpson Memorial Institute, Ann Arbor, Michigan.

Figure 6. University Hospital, Ann Arbor, Michigan.

Early Life and Medical Training

terested as I related this long story. The crowning insult, as far as I was concerned, was his crack at the conclusion of my history: "The more numerous a patient's complaints, the less significance of any of them." I argued with him — as if to question his medical omnipotence — which only seemed to amuse him further. For many years, I still harbored a dim view of this doctor's medical ability.

During my senior year in medical school I lived and ate with the interns and residents, having been appointed the extern at the Simpson Memorial Institute — an institution for research in blood diseases. My job was to take detailed histories on all new patients (the census averaged about 16). I also made rounds with attending physicians. The detailed history taking routine turned out to be of exceptional value to me. The major research activity in the laboratory at this time was identification of the active protective principle responsible for the beneficial effects of liver in the treatment of pernicious anemia. Although not directly involved in this analytically oriented clinical research, I was interested in keeping up with this activity. In looking back at this unsuccessful quest, I am quite sure that I must have witnessed the phenomenon of throwing the "baby" (the active principle for which they were looking) out with the wash water on several occasions. I did have something to do in testing these analytical fractions, as these tests were done on relapsed pernicious anemia patients under my medical supervision.

One of the patients seen at this time, before the advent of antibiotics and long before any knowledge of immune suppression, was a 30-year-old man hospitalized for treatment of aplastic anemia. Despite anything which could be done, I watched day by day a small gum infection progress to involve the entire side of his head — a massive noma infection which killed him.

During this senior year in medical school (1933), I attended my first national allergy meeting in Atlantic City, New Jersey. This provided the opportunity of at least seeing and hearing pioneers in this field who were active at that time. I recall William W. Duke, whom I don't remember ever seeing again. Marion Sulzberger spoke at length on the need to differentiate the different types of eczema, namely, the clinical distinctions between infantile or atopic eczema and contact dermatitis as had been emphasized by European dermatologists.

The American Association of Immunologists happened to be meeting in Atlantic City at the same time. By a stroke of more luck than planning, I heard the presidential address of Arthur F. Coca, which was apparently the last gasp of his objections to the redefinition of allergy in terms of immunology. Coca, who was dean of the American immunologists at the time, cited amongst his objections the probability that there must be several, rather than only one, mechanisms responsible for the wide clinical manifestations of allergy. In this presentation he referred to the phenomenon of aspirin sensitivity and other examples not explained by immunologic mechanisms.

In writing this it occurred to me that perhaps this presidential address had been published in the Journal of Immunology. To my disappointment, I found that the policy of printing presidential addresses was first initiated in 1934. Those interested in an account of the redefinition of allergy from descriptive to immunologic terms are referred to page five of Urbach's book, *Allergy,* (Urbach E, Gottlieb PM. New York: Grune & Stratton, 1943) which quotes both Coca and Sulzberger as objecting to the redefinition of the descriptive designation of allergy to the presently accepted definition in terms of antigen-antibody-mediator mechanisms.

Internship and Residency Training — 1933-1937

As an intern rotating through orthopedic surgery at the University Hospital in Ann Arbor, Michigan, one of my jobs was to replace the maggots in the osteomyelitis wounds and to keep them there. Maggots worked around the clock but were prone to take leave of their restricted premises at night, to the dismay of the involved patient. One of my jobs was to go to the patient's rescue and, immediately corralling the wayward maggots, reconfine them to their workplace — necrotic osteomyelitic tissue.

I owe my life to an incident as a resident on the medical service a few months later. It seems that the interns and nurses on this floor had planned a picnic. Although I had arranged to be off duty that evening, a half-hour before departure a patient with lobar pneumonia was admitted to my service. Although sulfonamide drugs had just been described, supplies remained scarce and pneumonia patients had to be typed immediately in order to start specific vaccine therapy. This meant staying up all night. Early the next morning the partygoers, or part of them, returned. Two of the interns — including my roommate, the owner and driver of the car — were killed outright. No doubt I would have been in the front seat of this car had it not been for the pneumonia patient.

As I went on in my residency training and instructorship in internal medicine at the University of Michigan Medical School, it became increasingly apparent that internal medicine was too broad a field to be mastered by one person. In looking about for some special interest within this field, allergy was chosen as it seemed to me the largest and least understood of any of the so-called subspecialties of internal medicine. Consequently, I began spending my spare time in the allergy clinic and in reading about allergy and immunology. I even took Coca, Walzer and Thommens' book, *Asthma and Hay Fever in Theory and Practice,* with me on my vacation. My first medical article on allergic headache was written with Dr. John Sheldon during this period and was published in 1935.[1] The last two of the four postgraduate years I was able to spend increasing time in the allergy clinic. I was in charge of this department each August when Dr. Sheldon was on vacation.

There is a point where my medical history altered my medical career, which should be described. As stated earlier, I had been diagnosed as having pleurisy in high school following a chest injury during strenuous physical exertion. With repeated unexplained recurrences, a probable diagnosis of

Early Life and Medical Training

pulmonary tuberculosis was made but in the absence of identifying the tubercle bacillus. A left temporary phrenic nerve paralysis was done and I was placed at bed rest for a period of four months.

Having just completed my medical training in nutrition and metabolism, with Dr. Louis H. Newberg, it occurred to me that there should be a mathematical relationship between diabetic diets possessing a constant level of available glucose and variable calories, as well as constant calories and variable available glucose. While confined to bed, I worked out these relationships in 100 diabetic diets, presenting them on two sides of a circular disk. The details of each diet could be read through slots in the outer disks. This device was later sold to the E. R. Squibb Company, published as the Squibb Diabetic Diet Calculator.[4,6,7] This was distributed, gratis, to all internists, pediatricians and physicians dealing with diabetes.

Fellowship in Allergy and Immunology — 1937-1939

Although I had intended to enter the practice of internal medicine with a special interest in clinical allergy at the termination of my instructorship in internal medicine, this did not seem feasible with a probable diagnosis of tuberculosis. Consequently, in late 1936, I looked where special training in this relatively new field of allergy could be obtained. There was only one such medical school affiliated course available. This was a Fellowship at the Massachusetts General Hospital and the Harvard Medical School under the direction of Francis M. Rackemann. I applied in person, obtained the position and lived in the Boston area from July 1937 through July 1939. I was only the third person to be trained in this position.

I worked part-time in a laboratory on the fourth floor of the Bullfinch Building, around the corner from the Ether Dome where ether anesthesia had first been administered. It was also where the withdrawing of blood from a vein by means of a syringe and needle had first been demonstrated in this country. I was told this by Dr. Newberg of the faculty of the University of Michigan Medical School. Shortly after Dr. Newberg had graduated from medical school he visited in Europe where he had observed this procedure. He brought back the necessary equipment and made this demonstration approximately 80 years ago.

Considerable time was spent the first year in making blood determinations of histamine in asthma and in eosinophilia. A suggestive increased level of blood histamine over normal was found during paroxysms of asthma as compared with intervals between attacks and with normals, but normal levels of blood histamine were found in several cases of high blood eosinophilia.[3]

Early in this period, Dr. Rackemann told me about an interesting article in Nature by Hans Selye a year or so previously. This was Selye's original description of what later became known as the general adaptation syndrome. I reread this several times, inasmuch as it seemed to be important. However, I was unable to apply this concept to what I was seeing clinically. In reporting these observations to Dr. Rackemann a few days later, he agreed, saying that he could not see how it fitted in with clinical views, either. (There will be more on this subject 17 years later, as it seems to take a long time for some points to sink in!) However, I learned many things about allergy generally during this training period. My major research interest was concerned with allergy to molds, which was published in 1938.[2] I also studied the role of histamine in allergy, which led to a publication in 1941.[3]

One of the most valuable experiences of this time was working closely with Louis Dienes in immunology, who was running a series of experiments in inducing sensitivity in animals by means of injections in the presence of Freund's adjuvant.

Private Practice, Milwaukee, Wisconsin — 1939-1942

I started practice in Milwaukee in mid-1939 in association with Theodore L. Squier, an orthodox allergist but who had an active interest in allergy to foods and drugs. Dr. Squier and his associate Dr. Fred Madison, also an internist, were involved in studying purpura and other blood diseases on the basis of demonstrable reactions to foods and drugs. They reported some of the earliest clinical observations in both areas.

My first year in Milwaukee, I founded the Allergy Clinic at the Milwaukee Children's Hospital and became a member of the staffs of Marquette University Medical School and Columbia Hospital.

During my three years in practice with Dr. Squier we were diagnosing food allergy by feeding patients in the office, in lieu of breakfast or lunch. We followed the views of Rowe that it was the commonly eaten foods (such as wheat, milk or egg) which were the most troublesome. Despite patients not having fasted specifically prior to these test meals, acute and convincing postingestion reactions often occurred; these were sometimes associated with a postingestion leukopenia. Since these experimental food ingestion tests were performed in the office daily, I estimate that I must have observed at least 5,000 such ingestion tests in the absence of a preparatory period of specific avoidance. It was here that I learned the basic nature of food allergy and details of diagnosing and treating food sensitive patients.

Both in this office and at the Children's Hospital, inhalant allergy was handled traditionally by performing both cutaneous and intracutaneous tests. One child experienced a very severe reaction following a battery of intracutaneous food tests, becoming unconscious, temporarily, apparently from mustard. I wrote on the seasonal incidence of fungous allergy at this time.[5,8]

Milwaukee was an interesting place to live during events leading up to our participation in World War II because of the dominant German descended population. I made the mistake of commenting socially one evening that there was less reason for students to study German than formerly because German science had become less important under Hitler. Although the statement was true, this temporarily polarized the conversation in the room.

It was while here that Janet Sibley of Chicago and I were married. Our first son, Jonathan, was born in 1942.

It was also here that my intermittent spontaneous pneumothorax was correctly diagnosed for the first time. In view of this chest problem and to clarify my draft status, I volunteered for service in the U.S. Army. Being rejected, I was classified as 4-F. As there were few well trained available young physicians in this field, I was offered and accepted the position as Chief of the Allergy Clinic at the University of Michigan Medical School in mid-1942 after Dr. John Sheldon left for service.

Allergy Clinic, University of Michigan Medical School — 1942-1944

In the intervening years since hearing Dr. Coca criticize the immunologic interpretation of allergy, I had been reading widely on the subjects of allergy, immunology and their interrelationships. Having become increasingly critical of the immunologic interpretation of clinical allergy and seeing little evidence of how this applied to food allergy and knowing that I would be responsible for teaching these subjects to medical students, I was left in a quandary of how to do this. Although I was aware of the controversy existing for several years between Dr. Robert Doerr of Switzerland and Dr. Coca, I was unable to find a clear statement of Dr. Coca's thesis. Suspecting that this might be found in his long presentation, "Hypersensitiveness," published in 1920, I could not find this as it had been published in *Tice's Practice of Medicine* — a looseleaf publication in which earlier chapters had been replaced by later updated material.

The only documentation I could find on this redefinition of allergy was in Urbach's book, *Allergy*. He stated that European allergists, under the leadership of Robert Doerr, prevailed upon their American colleagues to accept the immunologic interpretation of allergy, although Coca never fully accepted this, as judged by his writings. Convinced at the time of these lectures that the immunologic interpretation of allergy had not been substantiated, I taught the medical students the immunologic theory of allergy. I hoped that this point of view might stimulate interest in medical students, as well as in myself, to seek out other interpretations of this phenomenon. At least I sold myself and began looking more earnestly for other more workable explanations.

One of the places to look for information regarding the mechanisms of allergy was the function of eosinophils long known to be associated with the pathology of allergy, but not well integrated into the immunologic theory of its mechanism. The hangup seemed to be in the techniques of enumerating eosinophils, as these relatively large fragile cells either tended to be ruptured or rolled off the edges of the blood smear. In trying to find a counting chamber process, the problem was to find a technique which would eliminate red blood cells but would carry a stain for the differential identification of white blood cells. In looking up the subject of hemolysis, I ran across the report that

50 percent propylene glycol was a hemolytic agent. So I simply made a white blood cell count using equal parts propylene glycol and water. Intending to watch for hemolysis under these circumstances, I was called to the telephone for a few minutes. Upon returning, there were no red blood cells in the counting chamber field. At this point, I was interrupted by a longer telephone call. Upon returning, the erythrocytes had returned! Assuming that the red cells had taken on the same density as the surrounding media, and had therefore become temporarily nonrefractile only to return when some of the water had evaporated, I demonstrated this interpretation by looking at such mixtures through a dark field microscope. Simply adding phloxine (closely related to eosin) to this mixture produced the first highly reliable technique for enumerating eosinophils. These results, including a statistical comparison of the stained film and the counting chamber techniques of enumerating eosinophils, were published as the first of a series of articles entitled "Blood Studies in Allergy."[14,22,31]

The medical technician hired for this blood work, Elizabeth Gibson, had been working in the University Health Service where the chief medical problem associated with hematologic abnormalities had been infectious mononucleosis. After a few weeks she commented that, to her amazement, she was seeing about the same number of atypical lymphocytes in the peripheral blood smears of chronic allergic patients as seen in infectious mononucleosis. I was fascinated by her comments as I had been studying allergy patients with the complaint of chronic fatigue but whose heterophil antibody determinations were consistently negative. We reported these findings in 1943.[10,15,31]

In continuing these observations of eosinophils in acute allergic reactions it was demonstrated, for the first time apparently, that a sharp decrease of eosinophils occurs during acute reactions to allergenic drugs, only to rebound to exceptionally high levels of eosinophilia after such acute clinical reactions subsided.[17,21]

Less striking findings were reported in food allergy. The common occurrence of low levels of white blood corpuscles in food and drug allergy, as had been published by others, was also reported.[29] I was especially interested in the relatively low levels of lymphocytes and the fact that the sicker the patient, often the lower the lymphocyte count. Thinking that the function of the lymphocyte must be important in all of this, I attempted to review this area. Although my cubicle at the medical library was cluttered with partially read journal articles and books bearing on the subjects of lymphocytes, allergy and immunology, I simply did not have time, in view of the manpower shortages during the War, to pursue this subject. There later turned out to be good reasons to have been upset and disappointed the day I dismantled this library cubicle.

It was difficult carrying through with an active department with approximately one-half the desired medical staff, especially in trying to fill the shoes of the former chief, John Sheldon, who had been known for his amiability to

Early Life and Medical Training

the point that his colleagues often imposed on him. I had come under criticism one day when unable to work up a patient immediately who had been referred by Dr. Peet, Chief of the Department of Neurological Surgery. Before this difference of opinion settled down, I told Dr. Peet of the case of a student nurse having intractable headaches upon whom he had operated, unsuccessfully, twice in order to locate the brain focus of the headaches. Losing two-thirds of her time in Nursing Training School, her nurses' training had since been terminated. I had seen her on an outpatient basis, having found her violently sensitive to milk and a few other foods. With the avoidance of milk and beef products her headaches came under control and she had been readmitted to complete her nursing education. I happened to be in the nursing station of the floor in which she was working one day when I noticed that she was ataxic and that one eyelid was drooping. Immediately, she grabbed her head and fell to the floor. Although it is rare to be felled by an allergic headache, this can occur.

Upon regaining consciousness, she denied any breaks in her diet. But in checking with the kitchen I learned that the substitute cook that noon had *not* removed her carrots before they had been buttered, in contrast to the practice of the regular chef. In order to demonstrate what suggestion might or might not do, I scheduled her for an electroencephalogram after milk under double-blind conditions. I had added sufficient Amphogel to simulate the turbidity of a drop of milk in a half glass of water. Of these two glasses she happened to select the sham feeding initially. After waiting two hours without reaction, she drank the second glass with the understanding that sometimes two doses are necessary. She reacted within three minutes with such pain and activity that we were unable to obtain an electroencephalogram tracing!

In taking over the allergy clinic at the University Hospital, I simply continued with already installed routines and techniques which were similar with the ones used in Milwaukee except for the approaches in food allergy. But I soon became disturbed with the clinical results from performing multiple skin tests with food extracts, both by the scratch and intradermal routes, and interpreting the results in terms of +, ++, +++ and ++++ reactions. Instructing patients on the home use of elimination diets, in order to confirm or not confirm the skin test results, prompted the frequent comment: "If the skin tests with foods are unreliable, why do you do them?" I finally got up the courage to discontinue their use; this came about as follows.

Residents from the Departments of Internal Medicine, Pediatrics and Otolaryngology rotated through the allergy clinic a month each. I called a meeting of our staff and recently instructed physicians from other departments to discuss this subject. All agreed that we had learned more about food allergy since paying little heed to skin tests and relying on ingestion tests and elimination diets, than previously. In retrospect, these comments were of interest, inasmuch as feeding tests employed were without preliminary specific avoidance and the elimination diet which we had devised contained corn as one of its basic foods!

1906-1943

I suggested that in view of this consensus we dump the skin test extracts for foods down the sink. There was no objection, so we did. This was done without permission, with the understanding that we would not say anything about it. It might be pointed out that the allergy clinic at that time was virtually autonomous in that there was no effective direction from the head of the Department of Internal Medicine, of which it was a subsidiary. At a subsequent meeting, it was agreed that discarding the use of skin tests with food extracts — mechanically performed and literally interpreted — depending on clinical observations instead, had been beneficial.

Although Rinkel had described the phenomenon of masked food allergy and the desirability of avoiding a food for at least four days prior to its test ingestion starting in 1936, (see Appendix, page 334) this concept and technique had received little publicity until 1944, when he presented his views at an American College of Physicians Instructional Course in Omaha, Nebraska. Upon first learning of this in that year, I immediately put it to test, having had the previous clinical experience of feeding patients under observation in the absence of fasting several thousand times. In comparing these two techniques, I published an appraisal of Rinkel's individual food ingestion technique in 1946,[29] confirming both the increased accuracy and sharper clinical responses when commonly eaten foods are avoided completely for several days prior to test re-exposures.

Chapter II — 1944-1950

Private Practice — Chicago, Illinois

Reasons For Selecting Chicago

Chicago was chosen as a location to practice for several reasons. It was a transportation hub and major medical center in which there was very little interest in food allergy — my major investigative concern at the time. I brought my technician from Ann Arbor and opened an office on Michigan Avenue. The only physicians I knew in Chicago were Michael Zeller and two doctors in practice from the University of Michigan Medical School. Being naive, I did not appreciate that not having attended medical school here nor having any of my training in Chicago was to be a major disadvantage. Neither was I acquainted with the city, having only been here briefly on two previous occasions.

Michael Zeller arranged for me to make night calls for a surgeon. This not only provided considerable income but also taught me the layout of principal streets. At the end of three months I was making expenses and discontinued night calls. Times were advantageous inasmuch as the majority of younger physicians were in the armed forces.

Upon starting practice in Chicago, I applied for and obtained staff privileges at Northwestern University Medical School and its subsidiary, Wesley Memorial Hospital. Coming from another medical school facility, certification by the American Board of Internal Medicine, membership in all national allergy societies and especially previous membership in the Chicago Allergy Society and the Central Society for Clinical Research were helpful. I was later subcertified in allergy by the Conjunct Board of Allergy and Immunology.

Soon after arrival, I announced that I was interested in food allergy, but did not perform skin tests for its detection. This announcement turned out to be both good and bad. Although it attracted attention and referrals, it also antagonized many other allergists — most of whom were performing skin tests with food extracts routinely and interpreting them as +, ++, +++ and ++++ in the relative absence of clinical evaluations.

I also began the practice of recording medical histories and progress notes directly on the typewriter, without any offhand interpretation of their significance. This routine turned out to be so helpful in dealing with alleged hypochondriacs that I adopted the technique of practicing "poker faced medicine." More specifically, this meant not evincing either assent or dissent, approval or disapproval, agreement or disagreement, as indicated either by word of mouth or expression in respect to all statements and comments by patients. This policy encouraged patients to report factually and to be recorded in the relative absence of the historian's preconceptions. Additionally, selected research cases were studied in detail in the absence of economic consideration. The combination of these practices provided a data base not otherwise available — data independent of snap judgements at the time collected. These unique practices led to the description of several systemic manifestations of allergy, either poorly appreciated or not previously described.

Food Allergy

During the first few years of practice in Chicago I was preoccupied by becoming as knowledgeable as possible in regard to the basic nature of food allergy. The earlier clinical observations of Coca, Rinkel and Rowe (see their bibliographies in the Appendix, pages 311, 333 and 337) were studied in detail, confirmed and sometimes extended. I talked and exhibited on this subject as often as possible because of the widespread interest in it. My knowledge of food allergy at this time is indicated by Figures 7, 8, and 9, and early publications on this subject.[36,38,46]

Figure 7. Symptoms in food allergy.

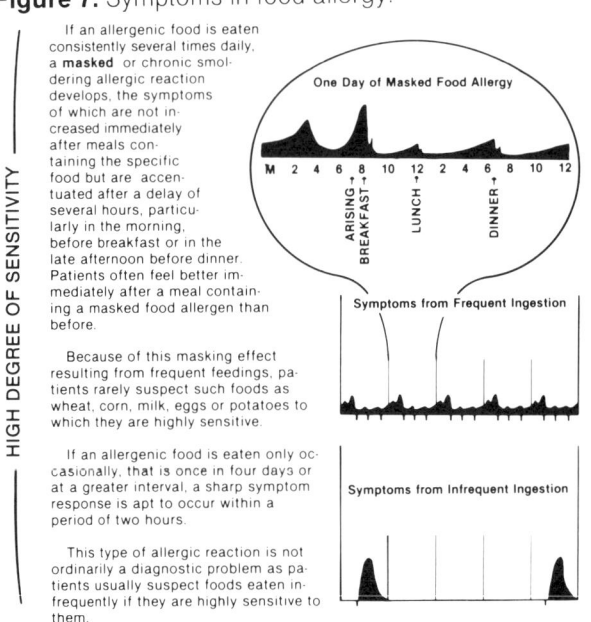

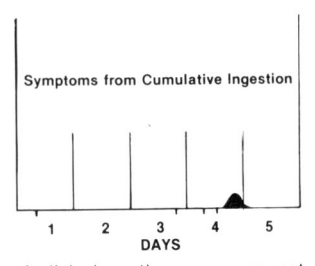

If an allergenic food is eaten consistently several times daily, a **masked** or chronic smoldering allergic reaction develops, the symptoms of which are not increased immediately after meals containing the specific food but are accentuated after a delay of several hours, particularly in the morning, before breakfast or in the late afternoon before dinner. Patients often feel better immediately after a meal containing a masked food allergen than before.

Because of this masking effect resulting from frequent feedings, patients rarely suspect such foods as wheat, corn, milk, eggs or potatoes to which they are highly sensitive.

If an allergenic food is eaten only occasionally, that is once in four days or at a greater interval, a sharp symptom response is apt to occur within a period of two hours.

This type of allergic reaction is not ordinarily a diagnostic problem as patients usually suspect foods eaten infrequently if they are highly sensitive to them.

HIGH DEGREE OF SENSITIVITY

A clinical reaction may or may not develop after eating a food to which a relatively low degree of sensitivity exists. Symptoms are more apt to occur if the food is eaten:

1) Cumulatively (several successive meals);
2) In a massive dose;
3) When the 'allergic load' is high, as in the hay fever case during the pollen season;
4) In the presence of precipitating factors such as infection, menses, etc.

LOW DEGREE OF SENSITIVITY

Private Practice — Chicago, Illinois

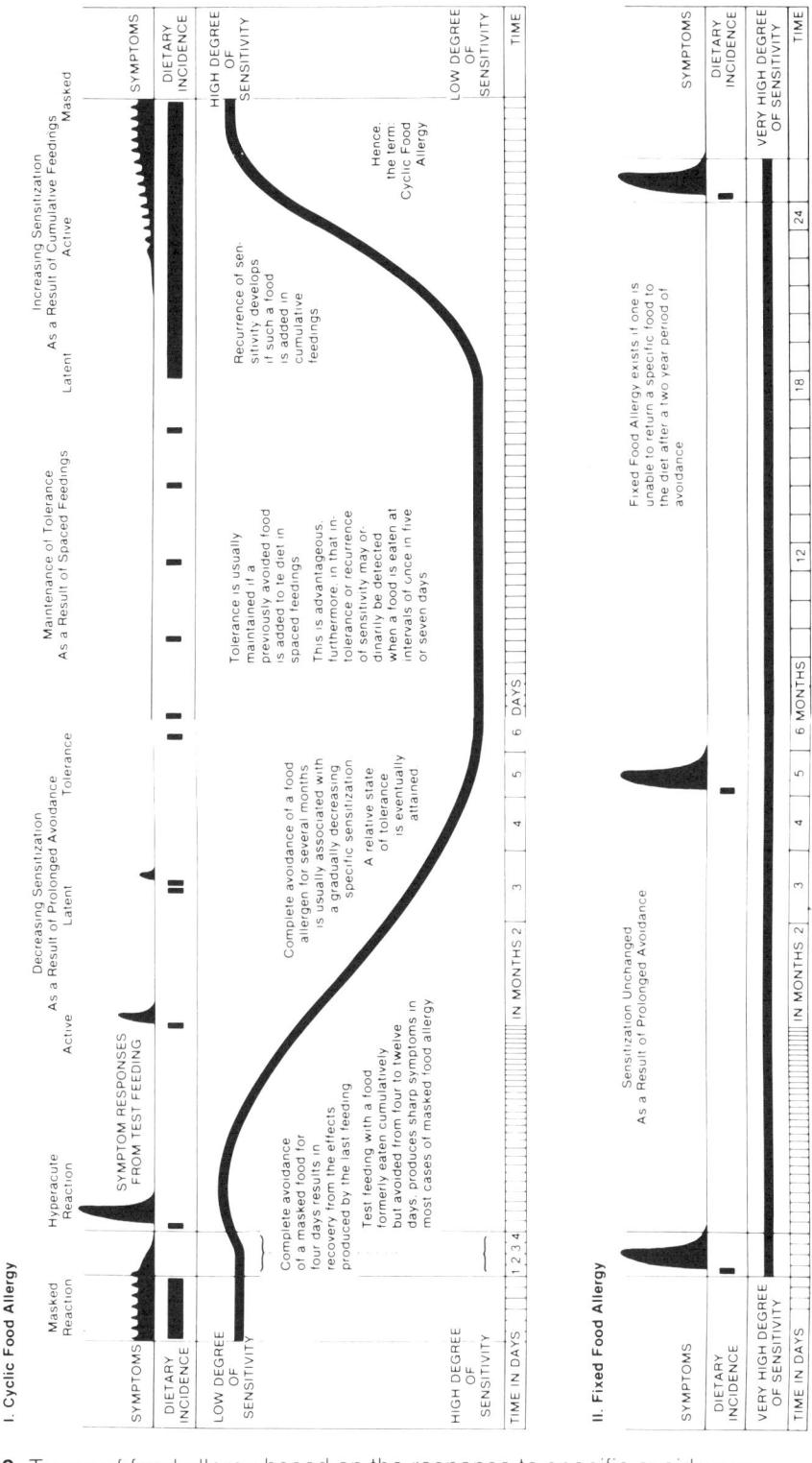

Figure 8. Types of food allergy based on the response to specific avoidance.

1944-1950

Types of Food Allergy Based on the Extent of Involvement of the Diet

I. Narrow Base of Sensitivity, High Inherent Tolerance for Foods

A narrow base of sensitivity exists if a patient reacts with symptoms on individual food tests with only one or two of the first four or five major allergenic foods listed below.

Restricted diagnostic diets may be employed in this type of case and the results of specific avoidance are usually good. Not only may the offending food be readily avoided but sensitivity is not apt to spread to other major or substituted foods.

II. Wide Base of Sensitivity, Low Inherent Tolerance for Foods

A wide base of sensitivity exists if one reacts with symptoms on individual food tests with more than half of the tests performed.

This type of case should not be placed on a highly restricted diet because of the ease with which sensitivity may spread to substituted foods. Some patients with an exceedingly low inherent tolerance for foods appear able to become sensitized to any article of the diet if it is used frequently and consistently. Patients with such potentialities are best controlled by eating a wide variety of tolerated foods, rotating them in the diet and avoiding the frequent use of foods botanically related to known allergic offenders.

The Order of Incidence of Foods Causing Masked or Cumulative Allergic Reactions

The incidence of sensitivity to corn, wheat, milk and eggs was determined from a survey of 200 consecutive cases studied for food allergy by direct methods, i.e. by means of individual food tests with those four foods and subsequent clinical follow up. The relative incidence of sensitivity to other foods is estimated from clinical experience.

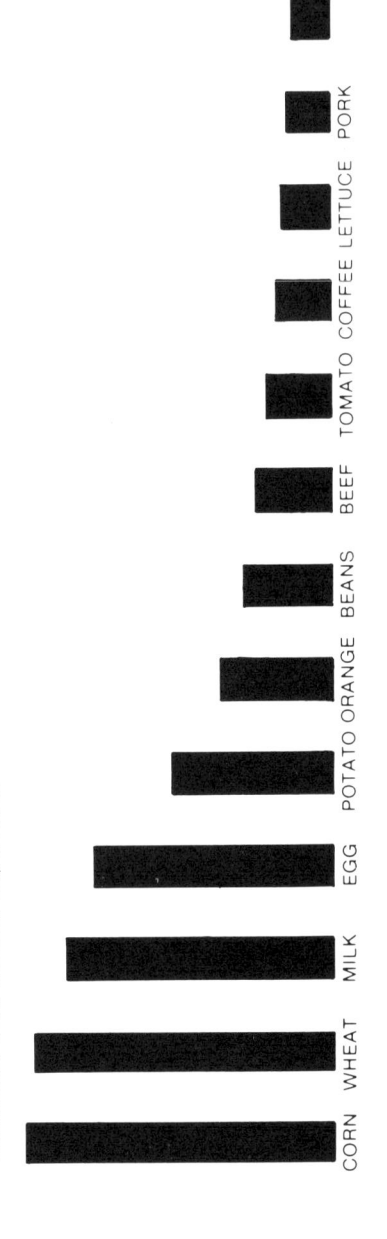

Figure 9. Types of food allergy based on the extent of involvement of the diet.

Private Practice — Chicago, Illinois

Herbert J. Rinkel visited my office for the first time in the fall of 1944 after I had been in practice but a few weeks. After an hour or so he commented, "Ted, I don't think that you are diagnosing your allergy to corn." Upon answering that I had not seen a case, he pulled from his pocket a small typewritten sheet of not more than a dozen lines of the sources of corn in the American diet.

However, all that I had needed was a mirror, for I immediately diagnosed my own allergies to corn, wheat and all other cereal grains by applying what I had recently learned from Herb — details of the individual food test and the corn sources. The avoidance of cereal grains not only relieved my frequent headaches and uncontrollable intermittent somnolence, but also provided a needed boost in energy and returned productive evenings for the first time in several years. Prior to this time, I had only known that I was sensitive to maple and peanut.

Systemic Physical Manifestations of Allergy

Inasmuch as most of the localized manifestations of allergy had been described by 1944, I became increasingly interested in systemic allergic syndromes. I was aided in this by investigating all new patients by means of the individual food test, the technique of which entailed avoidance of the food in question for at least four days prior to test feeding under observation. Several such office ingestion tests were performed daily. One by one, the following physical syndromes were appreciated and described:

HEADACHE. The first of these constitutional or general effects of food allergy was headache. Although headaches resulting from allergic reactions to foods had been described independently by Rowe and by Vaughan in 1927, (see Appendix, pages 338 and 344) and I had written on this subject in 1935,[1] there had been little general knowledge or acceptance of these views. Unfortunately, this statement still holds 60 years later!

FATIGUE. Despite both Rowe and Rinkel having described fatigue as an allergic manifestation (see Appendix, pages 334 and 338), these observations had also escaped general acceptance and application. My first detailed description of headache and associated fatigue in an extremely milk sensitive patient subjected to two previous craniotomies, was published in the Journal of the American Medical Association in 1944.[18] This prompted Albert Rowe to write me a four-page letter in which he said that this was the first confirmation of his description of allergic toxemia which had been published in this country. This letter is reproduced herewith.

October 30, 1944
My dear Dr. Randolph:
  Your article on allergic fatigue and headaches, needless to say, is of great interest to me. Realizing as I have for 18 years or more the

frequency of allergic fatigue due to chronic food allergy and discussing it in various articles and also in my various books on allergy, I am happy at last to find your confirmation in such an excellent contribution. As you probably know the only other article published on allergic toxemia and fatigue was one by Moreno in El Dia Medico, Volume 12, No. 47. It is interesting that a young associate of Moreno is now studying with me at my allergy clinic in our medical school. I agree with you that most patients with such fatigue have other allergic manifestations but I used diet trial for the study of such possible food allergy in all patients with fatigue that is not obviously due to some other cause.

I am also pleased with your emphasis of the negative skin reactions in food sensitive patients. I feel that this is so important that I have discussed the fallibility of skin tests on the very first page of my book on elimination diets as you have probably noticed. As I read your article it is evident that if my elimination diets had been continued in 1939 the role of food allergy would have been ascertained and the two operations and much unnecessary hospitalization, invalidism and suffering would have been prevented.

The fact that food allergens probably stay in the blood stream for days or even weeks after the elimination of the allergenic foods and that refractoriness follows severe allergic reactivity thus explaining relief between attacks certainly makes it necessary to use an elimination diet for a longer period than two or three days as advised by Alvarez. I have realized this for many years which accounts for my emphasis on the maintenance of nutrition with all trial diets and my emphasis of the use of elimination diets for a longer period than a few days. I am finding many patients suffering with allergy often to practically all fruits and many of these patients are victims of allergic fatigue. Gradually about five years ago I developed excessive allergic fatigue and after about one year of diet manipulation I found that I was allergic to every fruit and flavor except vanilla. This discovery has reinstated my energy and strength and ability to carry on in this interesting specialty. I have seen many other patients with such food allergy and have discussed it in regard to urticaria in my book in 1936 and in the recent edition of my elimination diets. I intend to write an article on food allergy within the next six or eight months discussing the many manifestations in fruit sensitive individuals who of course are in various degrees allergic to other foods. The gradual development of a positive delayed, erythematous reaction to milk is of great interest. I believe I have had such an experience. I have certainly seen constitutional reactions from the intradermal injections of food allergens in patients who have failed to react in the usual manner.

I am especially happy that you have pointed out the importance of the elimination of every bit of a food to which a person is allergic. Harry Alexander, for one, feels that butter can be used in trial diets and that possibly small amounts of cream and wheat can also be used. In the last edition of my book on elimination diets I have emphasized the elimination of every trace of an excluded food during the initial period of diet study, since a maximum degree of allergy must be assumed until the allergenic foods are determined, thereafter the tolerance for those foods can be ascertained. I also discuss in my chapter on the elimination diets the time necessary for diet trial, pointing out that when recurrent attacks of asthma, headaches or other allergic manifestations recur every one to four or six weeks that it is necessary to use diet trial for a longer period than the usual period for relief, and how important it is therefore to have elimination diets which can be accurately prepared that maintain nutrition and that satisfy patients as much as possible while such diet trial continues.

As you probably know I am quite certain that the fundamental underlying cause of chronic ulcerative colitis is food allergy. I have one case of seasonal ulcerative colitis due to pollen allergy which I have followed now for about four years with excellent control due to pollen desensitization. The fundamental lesion, in my opinion, is an eczematous-like reaction in the bowel complicated in varying degrees with secondary infection, anemia, hypoproteinemia, avitaminosis and digestive insufficiency. It is entirely possible that the fibril changes in the colon may be the result of chronic persistent, unrelenting allergic reactivity which as Cohen has demonstrated experimentally can result from repeated allergic reactions with the deposition of scar tissue. In fact, I believe that there is a good deal more irreversible tissue reaction in patients with chronic food allergy and other types of chronic allergy than we imagine. I have a most interesting case now of recurrent corneal ulcers due to food allergy which finally produced the complete scarring of the cornea with entire loss of sight except for slight light perception. With the elimination diet the extremely severe pain, lacrimation and photophobia gradually disappeared in three or four weeks and in the last two or three months gradual perception of motion, color and some form has occurred, but I believe that much of the scarring is irreversible and we will probably have to transplant a cornea, hoping for a good result.

The semiconsciousness in your patient could be included in the syndrome known as an epileptic equivalent. I feel that almost all idiopathic epilepsy should be studied with trial diet. I have several patients greatly relieved or nearly entirely relieved with the elimination of allergenic foods; even if only one out of ten receive such benefit such diet manipulation is entirely justifiable. In case you have not seen my recent article on allergy of the central nervous system, I am including one; I am also including a reprint of my original article on allergic toxemia and one on chronic ulcerative colitis.

For a number of years I have hoped that other men would interest themselves in the importance of food sensitization, realizing the infrequency of the skin reaction and the necessity of diet trial. It is cheering to know that you are convinced of these facts and that you are practicing in such an important medical center as Chicago. Your emphasis of the many manifestations of food allergy and the necessity of diet trial in the diagnosis and control of food sensitization will mean much, not only to the men in the field of allergy but also to its many sufferers.

As I read this letter over I find I have failed to comment on the return of your patient's symptoms when she was in the formula room. Apparently this was due to the inhalation of milk allergens. I am seeing more and more evidence of development of symptoms from the inhalation of food allergens, especially when patients are working in kitchens or eating in restaurants. I also wish to confirm your skepticism of the many diagnoses of psychoneurosis in patients suffering with allergy and with other manifestations of allergy. For some years I felt that I might be a neurotic individual only to find some 15 or 20 years ago that food allergy was entirely responsible for the difficulty. Unfortunately, too many men are specializing in the psychic treatment of disease who have no conception of the effects of allergy on the nervous system which I have discussed in the all too brief report, which I am sending you.

Again complimenting you on your recent article and on your understanding of the subject of food allergy, I am

Most sincerely yours,
Albert H. Rowe, M.D.

There were several additional presentations on fatigue during this time.[20,23,25,30,32,34,39,85] One of these, entitled: "Allergy as Causative Factor of Fatigue, Irritability and Behavior Problems in Children," published in the Journal of Pediatrics in 1947, included the first case report of hyperactivity in children as an allergic manifestation.[34]

MYALGIA, with and without associated headache and fatigue, emerged as a major allergic manifestation in 1948,[41,42] and subsequently,[48,67] although this was not published in detail until 1951.[83] This delay resulted from this manuscript, first submitted for publication in 1948, having been rejected six times before it was finally published in 1951. In the meantime, it had been necessary to retype this "dog-eared" manuscript repeatedly. One rejection slip from the editor of Medicine, who later became editor of the JAMA, was noteworthy, in that he said his editorial board takes a dim view of purely descriptive articles.

ARTHRALGIA, ARTHRITIS. Clinical observations of the significance of arthralgia and arthritis were made during this period, confirming the earlier reports of Kaufman (Kaufman W. *The Common Form of Joint Dysfunction; Its Incidence and Treatment.* Brattleboro, Vermont: E.L. Heldreth & Co., 1949)

and Zeller (see Appendix, page 347). It should be emphasized that Kaufman's observations of joint dysfunction were based on objective measurements of involved joints before, during and after dietary changes. Kaufman employed elimination diets in demonstrating these food induced changes in joint functions and associated manifestations characterized by pain, fatigue and mental symptoms, pointing out pitfalls in this approach. He did not use the method of individual food testing as advocated by Rinkel.

CARDIOVASCULAR MANIFESTATIONS. Fluid retention syndromes[34] and many other cardiovascular manifestations were described during this period, confirming and extending the clinical observations of many others. These were best summarized in our book, *Food Allergy,* written with Rinkel and Zeller in 1948 but not published until early 1951.[75]

The Book — *Food Allergy*

Mr. Thomas of Charles C. Thomas, Publisher, read my scientific exhibit describing food allergy in 1946[28] and commented spontaneously: "This is important new material. You should write a book on food allergy." I answered that I was exhibiting Rinkel's point of view, having only confirmed and extended it moderately. Although he then suggested that we should write a book together, I did nothing about it immediately.

After seeing this exhibit a second time early that fall, he asked me again when I was going to write this food allergy book. I promised to get in touch with Dr. Rinkel and did so as follows:

> Dear Herb: At the meeting this fall of the Academy of Ophthalmology and Otolaryngology, I had occasion to see Charles C. Thomas again — the medical publisher from Springfield. I had known him for a number of years since attempting to arrange for publishing my diabetic diet chart.[4,6] After reviewing my exhibit for a second time, he asked me, point-blank, when I was going to write a book on allergy, setting forth the concept of food allergy in a workable fashion. Would you be interested in a joint undertaking as senior author?

A week later, I received a reply, starting off as follows: "I am delighted at your invitation to join as senior author on the food allergy book and do think that it would be worthwhile to keep this monograph on foods, and a book on inhalants, separate."

We spent much time during 1947, 1948 and 1949 in preparing this manuscript. More details of the systemic physical manifestations, described only briefly above, were presented more extensively in this book. Herb wrote the first part, I wrote most of the middle and Zeller supervised the case reports which were contributed by all of us. Zeller's criticisms were good, as was the overall critic, George Seibold of Wichita Falls, Texas. It was finally published in early 1951.[75]

Systemic Cerebral Manifestations of Allergy

"BRAIN-FAG" AND DEPRESSION. Cerebral and behavioral syndromes on the basis of food allergy began to be apparent soon after starting practice in Chicago in the course of the routine use of the individual food test for the diagnosis of specific food allergy. Indeed, strange and previously unheard of things began to occur in the food test room. For instance, a young woman with a history of multiple localized and systemic illnesses, including periods of hyperactivity followed by depression, had had a long history of eating two eggs for breakfast, daily. Shortly after eating a test dose of eggs, for the first time in five days, I noticed that she ceased reading and became increasingly restless. Soon she got up and began pacing aimlessly about the food test room. Then, suddenly and without other advanced warning, she started crying in front of several other patients. At this point, she was taken to another room. She continued to cry for the following two hours.

Inasmuch as it was difficult to believe that this could be a food induced bipolar psychotic reaction, especially since to my knowledge this had not been described previously, I suggested the following: 1) continue to eat eggs daily, as previously, on returning home; 2) avoid all eggs and egg products for four days before returning in two weeks, at which time the egg test would be repeated. She complied, stating that she was worse for two days before returning to her previous level of chronic manifestations. When she avoided eggs again in preparation for returning, she experienced a similar withdrawal accentuation of symptoms as previously. Following the test ingestion of eggs, her initial stimulatory hyperactivity was again followed by an almost identical crying depression.

In view of the fact that I had observed several such demonstrable bipolar test responses to foods and a lesser number of unipolar reactions on an outpatient basis following specific avoidance, I was prepared for evaluating a more advanced case when a 32-year-old secretary was first seen in late 1949. She had previously been diagnosed variously as psychoneurosis, hysterical fugue state, anxiety and depression as well as early schizophrenia. Although she was too confused and amnesic to provide a history, I learned from her referring physician that her current problems were constant headaches and intermittent periods of extreme hyperactivity, anxiety and ataxia simulating alcoholic inebriation associated with amnesia and followed by severe depression.

She was hospitalized and placed on a diet avoiding probable food allergens. When it was found that she was drinking coffee almost constantly for relief of headache, this was also avoided. She remained withdrawn, seclusive and too confused to give a history for three days before reverting to normal behavior, during which time she campaigned actively for General Eisenhower's candidacy for the presidency. A history then revealed that she had been subject to these mania-anxiety-depressive episodes since the age of 14 years and that these had been superimposed on a background of perennial rhinitis, urticaria, coughing, headaches, myalgia and fatigue. An accentuation of rhinitis and hives usually preceded and followed, temporarily, the duration of these acute psychotic episodes. Although no foods were suspected, she had been drinking approximately 40 cups of coffee per day, each sweetened with two teaspoonfuls of beet sugar, for partial relief of all symptoms.

In making hospital rounds during the evening meal of the fifth day of hospitalization, I noticed that she was eating beets as a vegetable — the first material of beet origin consumed during the previous five days. Two and one-half hours later, she suddenly lapsed into a temporary stuporous sleep, from which she remained disoriented, confused and without recent memory for three days, and which was further characterized by progressive manic-ataxic behavior and anxiety, followed by residual depression. Her otherwise low-grade chronic rhinitis and urticaria were accentuated immediately before and, especially, immediately after regaining full consciousness. Recovery from the psychotic phase occurred suddenly — within the span of a few minutes, and her flat nonresponsive affect reverted to a normal expression and responsiveness.

Specific diagnosis of this initial advanced hospitalized patient required 10 weeks. Equally severe but even more immediate reactions developed at two hours after ingestion of beet sugar. Slightly less severe reactions were induced following test feedings with coffee, and with chocolate. Less advanced reactions occurred after test feedings of wheat and of potato. All other commonly eaten foods were test negative. Several single-blind feedings of beets and of beet sugar invariably produced a similar reaction pattern, whereas pomegranate juice (with a purposeful "slip" in the single-blind technique) failed to do so.

In all reactions she craved sweets, and several times in the hospital during lesser reactions she obtained beet sugar sweetened candy and promptly lapsed into a more advanced reaction.

After hospitalization, she experienced two severe reactions traced in retrospect to the accidental intake of beet sugar. Cane sugar was tolerated. She voluntarily returned to the hospital two months later for the purpose of inducing and photographing this type of reaction, after the blind intubation of beets and beet sugar. This motion picture is apparently the first experimentally induced acute psychotic episode of this degree of severity ever recorded. At least this is the consensus of those who have seen this motion picture.[74,79,130,149]

In advanced cases, such as this, lesser stimulatory and withdrawal levels were often telescoped, in comparison with less severe and advanced responses which developed more slowly and usually terminated at lesser levels. Also, in contrast with the clear-cut stimulatory-withdrawal phases as illustrated in this patient, lesser digitations between stimulatory and withdrawal levels were observed. For instance, I happened to be in her room during the dinner hour one evening as she was manifesting such a lesser reaction to cocoa (unsweetened chocolate) when she was attempting to eat her evening meal of chicken. I was fascinated by this alternating behavior occurring on a 20 minute wave length. For instance, I first noticed that instead of sitting quietly in bed and eating, as one would expect, she ate her chicken while wandering about the room for 10 minutes. Then she pushed the tray aside, laid down, and seemed to take a nap for 8-10 minutes before getting up again, wandering about her room, eating a leg of chicken in Henry the Eighth style. Intrigued by this behavior, which I had never seen nor heard of previously, I observed three or four of these 20 minute cycles before a most fascinating thing occurred. Being thirsty and drinking water frequently during these alternating hyperactive perambulations, she attempted to pour a glass of water from her opaque jug, only to find that it was empty. Shortly after she had asked her nurse to refill her jug, she lapsed into one of her recurring short stupors. Upon awakening, she criticized the nurse for not having brought the water. Whereupon, the nurse poured out a glassful!

It was only then that I realized that she had been depressed and amnesic during these quiet inactive intervals. Questioning her confirmed this fact. I then learned that both the degree and duration of the hyperactive phase varied coincidentally with the depth and duration of the depressed phase. After a few more 20 minute cycles, she remained persistently depressed and amnesic. The length of this alternating cycle in the development of reactions to specific foods seems to characterize certain individuals, at least at certain times. In subsequently observed patients, this cycle has varied from as long as two hours to as short as two minutes. In the first case, a nymphomaniac, hospitalized while her husband was recuperating, slept for an hour, then paced the hospital corridors for an hour throughout the night while in a slowly developed food reaction. The second patient exhibited epileptiform seizures every two minutes, being alternately relaxed between such hyperactive intervals.[200]

Lesser stimulatory and withdrawal levels in this initial patient, persisting for relatively longer periods of time before advancing and not developing beyond a given level, permitted an adequate general description of these constituent levels of reaction. Although this taxonomy of reacting levels of what was later designated ecologic mental illness was reported preliminarily in 1950[74] and exhibited in 1956,[110] and later, it has undergone repeated revisions as clinical experience accumulated.

From the standpoint of the acceptance and application of this exteriorization and demonstration of causality in mental and behavioral disturbances, it is of interest that this point of view was first presented at the Staff Meeting of Wesley Memorial Hospital, Chicago, Illinois, and published in abstract form in 1950,[74] the same year that psychotropic drugs developed. This point of view remained relatively noncompetitive for several years as compared with the excessive promotion and relative ease of application of such drug approaches. Only when the complications of psychotropic drugs became evident did the application of ecologic mental illness receive greater attention.

It startled me to realize, as a result of these first cases, that I might be on the threshold of an important far-reaching concept revealing undescribed relationships between allergy and psychiatry. Indeed, all of my past interests in and knowledge of mental and behavioral disturbances seemed to point in that direction. Several other clinical observations will be mentioned more briefly that not only substantiated this preliminary hypothesis, but also opened up other areas for investigation. Moreover, and it is difficult to point this out in a chronologic narrative, many of these developing concepts and observations were occurring at about the same time.

For instance, a middle-aged nurse who had a similar array of multiple symptoms, had been drinking milk excessively for most of her life. Toward the end of an individual food test with milk, she became progressively stimulated, moderately excited and slightly garrulous. Noticing that she was somewhat wobbly on getting up to leave the office, I inquired if she felt able to drive home. She assured me in a somewhat too positive a manner that she felt "fine." This person had moved to Chicago recently and was still unsure of herself in driving on the Outer Drive. This time, however, she managed this drive with utter abandon and self-confidence, only to start crying as she reached her home in Evanston! Moreover, she continued to cry most of that night. In short, this case was similar to the previous one except that the hyperactive-stimulatory and withdrawal-depressive phases occurred over a more elongated period of time.

Another younger woman in her twenties who had a similar history of multiple alternately disabling complaints, was similarly overconfident and slightly ataxic at the completion of her milk ingestion test in the office. Although her Romberg test (ataxia when standing with eyes closed and feet close together) was positive when she left the office, she also assured me that she was "fine." After retaining her in the office for a time, she was allowed to leave for home. An hour later, however, she was arrested for "drunken driving" in one of the far south suburban areas. When accused of this, she responded by telling the police, in her southern drawl, that all that she had had to drink was a couple of glasses of milk in her doctor's office. The police had difficulty in believing this, although she did not have an odor of alcohol on her breath. When the police called me I confirmed this unlikely story, as far as they were concerned. They then asked me to write them a letter on my letterhead confirming this. I simply explained in detail how a highly sensitive patient in the course of an acute test reaction to milk might have symptoms simulating alcoholic inebriation!

A middle-aged woman with a similar history and similar findings at the completion of her wheat test in my Evanston office, drove home on Ridge Avenue — a notoriously narrow street — against traffic at the height of the evening rush hour. She was startled to find that she was in her own home, when the last that she remembered she had been in my office. Obviously, she had driven home in a blackout — i.e., if her car was in her garage. She quickly looked out the windows — the garage was all right. With fear and trembling she rushed out to the garage, opened the door — the car was fine — without a scratch.

Another woman patient developed sustained pathological laughter early in the course of a food ingestion test with eggs! This in turn was followed by depression.

It should be emphasized that many of these systemic allergic physical and mental syndromes had been described earlier by Rowe, either in his books or detailed articles. Vaughan had also described some of these syndromes in his books or other manuscripts (see Appendix, pages 337 and 343.)

Kaufman pointed out that although many diverse clinical manifestations of food allergy may occur in persons with joint dysfunction, three syndromes occur frequently in response to the ingestion of an offending food or foods, namely: a) Allergic Pain Syndrome, b) Allergic Fatigue Syndrome, and c) Allergic Mental Syndrome. By the allergic pain syndrome he referred to mild, moderate or severe generalized pain in somatic muscle, tendon, periosteum and periarticular structures. Although this did not include allergic headaches, as he described it preliminarily, many of his case illustrations included headaches. His description of the allergic fatigue syndrome included the common association with cervical lymphadenopathy, lymphocytosis and hypothermia, ordinarily not relieved by rest. His account of the allergic mental syndrome was, for the most part, descriptively similar to what Albert H. Rowe and I had described as commonly being associated with headache and fatigue. Although I later had referred to these lesser cerebral and behavioral manifestations as "brain-fag," this term, not having been used in titles, had not been indexed. Consequently, it had developed little medical meaning. Kaufman had also pointed out that in addition to allergic mental syndrome, a few patients with joint dysfunction had also manifested a primary neuropsychiatric disturbance.

Clinical observations of this type in the late 1940s led to an investigation of the allergenicity of starches, sugars, alcohols and other food fractions, the addictive use of which often resulted in such sustained stimulatory responses as hyperactivity, obesity and alcoholism.

Food Fractions and Stimulatory Manifestations

In preparing the food allergy book, I was impressed with the relative absence of documentation of the allergenicity or nonallergenicity of food fractions. The basic principles of food allergy having been worked out by Duke, Rowe, Rinkel and Coca, this seemed to be a productive area for medical investigation.

GELATIN. Beef gelatin was found to precipitate acute reactions in three of four beef sensitive patients when administered orally under correct conditions of testing. The experimental intravenous injection of ossine gelatin was associated with similar reactions, but was not attempted in the third reacting patient because of the relative severity of symptoms following the ingestion of beef gelatin. None of these four patients was known to be pork sensitive, as determined clinically. All also failed to react adversely to the oral administration of pork gelatin or the intravenous administration of porcine gelatin. This was reported preliminarily in 1947.[37]

OIL. Two cases of sensitivity to cottonseed oil were found in the absence of sensitivity to cottonseed protein, as determined by negative skin tests with extracts of cottonseed. But in both instances, the experimental or inadvertant ingestion of cottonseed oil, as encountered in commercially available foods, was shown to result in the precipitation of allergic symptoms.[55]

STARCH. The starch form of a food is changed to sugar slowly by the action of saliva. Under these circumstances of relatively slow absorption, individuals rarely suspect that they may be reacting to one or more starches in their diet. Only when consumed in amounts two or threefold greater than their accustomed intake, such as might occur in a spaghetti dinner or when more rapidly absorbed than normal as in the presence of alcohol, is the starch form of a food apt to be suspected when it is consumed in the daily diet. Because of this masking phenomenon, ordinary amounts of starch in the daily diet are not only not suspected, but are usually regarded as favorite or essential parts of their daily meals. Consequently, the majority of chronically ill patients manifesting allergic symptoms and seeking medical advice are usually found to be reacting to one or more foods encountered in their starch form.

Foods packed with a liquid phase in contact with cornstarch sized inner surfaces of food containers absorbed sufficient corn to result in reactions in highly corn sensitive patients. The presence of such a starched surface was indicated by positive iodine reactions for starch. Although this "black and white" study of paper cartons was presented on the program of the Annual Meeting of the American Academy of Allergy in 1946, it was rejected for publication in the Journal of Allergy by the unanimous decision of its Editorial Board. It was published in the Journal of the American Dietetics Association in 1948.[40]

Even the exceedingly small amounts of cornstarch employed as an excipient, in persons sensitive to corn but otherwise completely avoiding corn, may result in acute test reactions. One corn sensitive patient, despite avoiding other sources of corn, continued with minor chronic symptoms while taking one grain of desiccated thyroid daily. She improved when visiting away from home for five days in the absence of her thyroid tablets. Upon returning, she washed her clothes, hung them on the line outside and took her thyroid tablet. She apparently passed into a stuporous sleep. A violent storm came up, whipping her clothes to shreds. Her neighbors assuming that something was wrong, called repeatedly, pounded her door and finally broke into her house — only to find her sleeping. An article — the first report on the allergenicity of excipients — was published in 1950.[62]

SUGAR. Corn, beet and cane sugars also induce specific reactions when these sugars are ingested by or injected intravenously into patients demonstrated to have been highly sensitive to their constituent foods.[45,50,63,64,65]

1944-1950

Especially important was the fact that dextrose derived from corn by the simple acid hydrolysis of cornstarch induced acute reactions in selected corn sensitive patients otherwise avoiding sources of corn in their edible foods. The same specifically holds for the ingestion of beet and cane sugar (both sucrose). Indeed, in selected extremely susceptible patients formerly addicted to an excessive intake of a specific sugar, who have avoided such a food preliminarily and then revert to its use under test conditions, acute reactions may be induced.

There is a practical point here in dealing with hospitalized patients extremely susceptible to one or more specific sugars requiring intravenous alimentation. Consider the case of a specifically undiagnosed patient, highly susceptible to corn, who has been without food intake for several days because of intestinal obstruction, who immediately upon emergency hospitalization for surgery is started on intravenous corn sugar (dextrose), coincidentally with surgery. Unless one is aware that an acute test reaction to corn sugar may occur under these circumstances, such a reaction is usually interpreted as a surgical complication. It was because of this type of experience that this problem was reported in 1950 in the Archives of Surgery.[65] In the event that the diagnosis of a high degree of individual susceptibility to corn, cane or beet is known prior to intravenous alimentation with a sugar solution, it is preferable to employ one to which individual susceptibility does not exist. For instance, invert sugar of cane origin or fructose of beet origin are available for intravenous alimentation in corn sensitive patients.

There are two major points to remember in respect to the application of this information. Each of the above statements was prefaced by the contingency of a high degree of specific allergy or individual susceptibility. (Those terms are used synonymously in the first part of this book.) Probably the majority of specifically less sensitive or susceptible persons, especially those receiving the sugar in question in their daily food intake, would experience minimal or no apparent reaction to its intravenous use.

If one inquires from manufacturers of intravenous solutions about their content of specific sugars, one can expect to encounter misunderstandings about the activity of specific sugars under these circumstances, as one is usually dealing with people whose basic training has been in analytical chemistry. As far as these chemically trained people are concerned, there is no difference between cane and beet sugars as, chemically, they are both sucrose. Moreover, these products have been crystallized and refined to a degree that any residual activity would have been minimized if not entirely lost! They might also claim that whether fructose is derived from splitting sucrose or directly from cornstarch is also immaterial. Biologically trained clinicians, dealing with extremely susceptible patients, know that such assertions are not valid.

Corn sugar also possesses unique physical properties in its own right, in addition to ordinarily being less expensive than sucrose of either beet or cane origin. These physical properties of corn sugar (dextrose, dextrin, glucose, fructose and corn sweeteners) are: 1) Absence of deliquescence (hydroscopic tendency) in their crystalline form, in contrast to the tendency of sucrose of either cane or beet origin to attract water and to become sticky on exposure to air. This property accounts for the fact that most hard candies are manufactured exclusively from corn sugar. The same holds for all-day suckers, most throat discs, lozenges and cough drops because of their ease of packaging; 2) Another advantage of corn sugar is that it crystallizes in thin flakes which dissolve quickly in the mouth, imparting a smooth velvety taste feature to chocolate candy. In contrast, sucrose of cane or beet origin crystallizes in relatively large polyhedrals which not only dissolve more slowly, but also impart a sandy or gritty character to chocolate candy. These relatively desirable physical features of corn sugars, in addition to their price differentials, explain why corn derived sweeteners comprise more than 40 percent of sugar currently consumed in this country. For a more detailed presentation of this subject, the reader is referred to Beatrice Trum Hunter's recent book, *The Sugar Trap and How to Avoid It* (Boston: Houghton Mifflin Company, 1982).

A physician's wife and dietician, age 30, had been subject to episodes of nausea, vomiting and diarrhea for eight years, severe fatigue, generalized myalgia, and headaches for three years, and intermittent periods of unexplained dysuria and tenesmus prior to her initial visit in 1948. A very severe episode of persistent vomiting, retching, headache, anorexia, and myalgia required continuous intravenous dextrose alimentation as a nutritional measure for a period of two weeks. As a result of a telephone consultation, having suggested the possibility of sustained allergic-type reaction to corn sugar, she was brought by ambulance to Chicago — a distance of 400 miles — during which therapy with intravenous five percent dextrose and water was continued. Upon arrival she was apathetic and unable to speak or respond except by blinking upon request either once or twice, indicating either yes or no. Corn in all forms was avoided immediately on a monofood intake per meal of less probable food allergens. By the end of the second hospital day she was improved, and by the end of a week she was ambulatory and comfortable.

A month later, 25 cc. of five percent dextrose (corn sugar) was administered intravenously. Three minutes later she complained of a dull aching pain in the right nape coincident with excessive perspiration and flushing. Initial chilling, noted at 13 minutes, was followed by nasal stuffiness and myalgia of the shoulder girdle two minutes later. A recurrence occurred at 30 minutes with all previous symptoms increasing during the following hours. She remained extremely weak, tired and nervous for 24 hours with some residual fatigue the following day. A review of the course of events during earlier hospitalizations revealed several periods of heretofore unexplained illnesses. Several days later this patient agreed to repeat the corn sugar intravenous test, but 25 cc. of isotonic sodium chloride was administered instead, in the absence of any evidence of reaction. Similar severe reactions usually constitutional in type, as was the case in this patient, had been observed following intravenous corn sugar by the time of this report and a review of the literature was published in 1950.[65]

HYPERACTIVITY. I am credited with having reported the first case of hyperactivity in childhood on the basis of food allergy,[34] although at the time I was neither aware of the seriousness nor the frequency of this problem. However, from what has been learned later, this problem was apparently less common in 1947 than since this time.

OBESITY. Ever since food allergy had first been described by Rinkel, its relationship to the development and persistence of obesity has been noted. Although I first reported obesity as a manifestation of food allergy in 1947,[35] this subject was not reported more extensively by me until 1956.[111] Foods most commonly involved are cereal grains (especially wheat and corn), potato, milk and sugars. Crystalline sugars of corn, cane and/or beet are particularly apt to be incriminated because of their tendency to be associated with their addictive intake. Recognizing that their cravings for these foods high in calories is similar to the behavior of alcoholics, such addicted persons often refer to themselves as "wheat-o-holics," "choc-o-holics" or "sweet-o-holics" or "sweet buffs." Despite some victims of masked food allergy recognizing the foods to which they are reacting addictively, the majority of specifically addicted weight gaining patients do not associate any particular food or foods as being responsible for their gradually increasing obesity.

One striking example of obesity comes to mind when I was visiting a patient in one of this area's largest psychiatric hospitals. I noticed an enormous hulk of a person approaching at a distance in one of these extremely long halls. Getting closer, I could make out that she was shuffling something on her enormous bosom, it being held here by envelopes under each arm. Still closer, I could make out that the material being shuffled on her anterior "shelf" was about a dozen slices of bread. From time to time she would stop long enough to take a bite of one of them. Upon inquiring about her, I was told that she was one of the "trustees" of the place, employed in making deliveries of large envelopes held tightly under each abundant upper arm. These envelopes were helpful in confining the bread to her "shelf." Typically, she would start off in the morning or at noon with a loaded "shelf," gradually consuming the bread by the end of a half-day. In this sense, this person was an excellent example of food addiction, hyperactivity and obesity.

The general inability of obese patients to follow reduction diets in which the intake of specific foods is reduced but not avoided is notorious. Indeed some patients choose physicians who do not weigh them on each return visit.

ALCOHOL AND ALCOHOLISM. Due to the fact that these two items are so inextricably tied together, they will be considered conjointly. Because of similarities in the way the human body handles sugars and alcohols, an investigation of the allergenicity of alcoholic beverages was a natural sequel to a study of the specificity of sugars in relation to allergic manifestations.

Since it seemed strange that this had not already been done, I spent considerable time in reviewing the studies of alcohol and alcoholism between 1947 and 1950 before starting this study.

The first indication of an unexpected find in this endeavor was a book written by Francis Washington Everard Hare in 1912 entitled, simply, *On Alcoholism* (see Appendix, page 331). In this outstanding book he referred to his earlier, even more remarkable book, *The Food Factor in Disease,* published in 1905. After looking unsuccessfully for available copies of this two-volume 1,000 page book in England, Australia (where it was written), Canada and this country, I advertised for two years before finding copies of these books. I will go into other details of this outstanding early medical investigator and his contributions to alcohol, alcoholism, nutrition, and what later came to be called clinical ecology, as my wife and I are preparing a biography of Dr. Hare, some major points of which will be described later.

This preliminary survey of what had been written in English on alcohol and alcoholism revealed virtually nothing on the subjects of the allergenicity of alcoholic extracts and distillates as well as the role played by specific foods in alcoholism. However, interest in these related subjects was galvanized by the following case report of a patient who was first seen in mid-1948. The timed pattern and characteristic features of his symptoms — typical of both food allergy and alcoholism — influenced me to pursue these topics. Indeed, this case served to stimulate interest in studying relationships between these two conditions.

The earlier hay fever of this 37-year-old man had been characterized by a profuse watery nasal discharge, which continued after his rhinitis became perennial. This not only included the late summer occurrence of his watery rhinorrhea, but also accentuated rhinitis upon arising each morning, before lunch and late in the afternoon. In recent years, this patient had increased his intake of blended bourbon whisky to a quart daily by the time that he joined Alcoholics Anonymous in 1941. Within a short period after abstaining from alcoholic beverages his perennial nasal symptoms improved remarkably, along with relief from his morning headaches, fatigue and fuzzy thinking. This enabled him to return to his work with an advertising agency.

But following a slip in alcohol avoidance in 1944, he reverted to his former intake of bourbon whisky as well as the timed manifestations of combined food allergy and alcoholism. After a few months of progressive daily drinking, associated with a recurrence of all of his previous disabling symptoms, especially chronic fatigue, all of his activities were monopolized by maintenance of his drinking routines.

But by early 1946, a sudden change in his tolerance for bourbon whisky developed, in that a single drink was now followed by generalized itching and nausea. Two drinks were followed by vomiting. Not being able to drink as formerly, he again joined Alcoholics Anonymous in October 1946. But despite abstaining, he remained extremely weak and generally debilitated with profuse constant rhinorrhea and fatigue. He had been seen by several different physicians, with neither understanding of his problem nor relief of his symptoms. He was first seen in my office in early June 1948, having been abstaining for nearly two years.

Recognizing that he was probably reacting to cereal grains and other ingredients entering the manufacture of blended bourbon whisky, the intake of wheat was avoided preliminarily to performing an individual food ingestion test on an outpatient basis. From the very start of avoiding wheat he developed a severe withdrawal characterized by such extreme fatigue that he was unable to walk from his bed to the bathroom. When he called to report this development, I urged him to crawl to the bathroom if necessary for the next day or so. His fatigue lifted significantly the third day and had disappeared by the fifth day. Immediately following an individual food ingestion test with wheat, he developed a progressive accentuation of fatigue with muscle tautness and aching.

With the preliminary avoidance of corn, his fatigue was again accentuated temporarily. A feeding test under observation with corn was associated with progressive fatigue, dizziness, confusion and ataxia. His Romberg test being positive at the end of this test, and fearing that he might break his abstinence from alcoholic beverages under these circumstances, I placed him in a taxi and admonished the driver to take him directly to his home without any stops. I called his wife who met him at the door and nursed him through another hangover. He called in with this comment the following morning: "It's funny to have a hangover 21 months after stopping drinking." He went on to say that there was absolutely no difference in the fatigue, lassitude, dizziness and uneasiness associated with gagging. He also said: "Coming off binges toward the end, I remember feeling exactly like this. One drink would relieve this just like that (snap of the fingers)."

In summary, this patient was found to be sensitive to all cereal grains, yeast, cane, grape (the food ingredients of blended bourbon whisky) as well as to potato, milk, egg, lettuce, orange, and coffee. Allergy to coffee is exceedingly common amongst abstaining alcoholics. He also reacted to various pollens, fungi, house dust and animal danders.

This patient, an active organizing member of local Alcoholics Anonymous chapters, facilitated my visits to several meetings of neighborhood groups of abstaining alcoholics. During these meetings, which were centered about the coffee pot in the living room of homes, one person at a time came out to the kitchen where I recorded the histories of 44 persons in respect to their drinking habits and allergic manifestations. The following points relating food allergy and alcoholism were revealed by this experience: 1) Former problem drinkers tended to have chosen and used specific types of alcoholic beverages; 2) A positive present or past history of allergy — omitting the drinking problem itself from consideration — was common; 3) There was a definite trend for those with the most strongly positive histories in this respect to have developed the characteristic features of addictive drinking at younger ages and after relatively shorter periods of drinking experience than occurred in those with less positive histories of clinical allergy; 4) There was a tendency for both localized manifestations (rhinitis, asthma, pruritus, etc.) and constitutional effects (fatigue, headache, myalgia, arthralgia, arthritis, and other symptoms of cerebral dysfunction) to recur as part of the hangover response; 5) One or more of these chronic symptom patterns persisted in the majority of the abstaining alcoholics interviewed; 6) All continued to eat the foods represented in the manufacture of their favorite alcoholic beverage and continued to manifest the typical craving for the immediate beneficial effects formerly secured from drinking; 7) Candy was the most commonly used food substitute following abstinence from drinking. These observations were reported preliminarily in 1950,[72] and extensively in 1956.[111] Not only was candy containing corn sugar usually present, along with the inevitable coffee pot at the Alcoholics Anonymous sessions, but the majority of abstainers lined their pockets with it each morning and ate along on it through the day as needed; 8) Most of the addictive drinkers interviewed had long since ceased to drink and eat simultaneously, having tried that a few times in their earlier days. This increased effect of an alcoholic beverage in the presence of edible foods is best explained on the basis of a greatly enhanced speed of absorption of foods in the presence of alcohol, as reported by Dees. (Dees SC. An Experimental Study of the Effect of Alcohol and Alcoholic Beverages on Allergic Reactions. Ann Allergy 7:185, 1949).

A few other case reports, illustrative of the relationship between allergy to given foods, their alcoholic extracts and distillates and incipient alcoholism, will be cited briefly starting with my personal experience. As an intern, I was subject to chronic fatigue, occasional periods of overwhelming somnolence and intermittent unexplained headache. When attending beer drinking parties with my intern associates, I invariably vomited after drinking two beers. I might add that beer was the only alcoholic beverage we could afford when earning $25 per month. The only way I was able to carry on at these parties was to go to the bathroom intermittently and vomit as quietly as possible. I would then be able to continue drinking for about an hour, but the following day I felt generally shaky and tired, always having a ravenous appetite and usually eating two breakfasts and two lunches. Slightly later, when finances permitted drinking bourbon whisky, I tried drinking three days in a row on the occasion of a weekend holiday wedding of one of my best friends. I discovered that my morning hangovers, characterized by headache, shakiness and fatigue, could be alleviated immediately by another drink of bourbon whisky.

As far as alcoholic beverages were concerned, I was able to drink rum occasionally without apparent ill effects, but later my cereal grain intolerance spread to include cane. From the beginning, however, I was able to drink Cognac (brandy from the Cognac region of France), containing grape, yeast and probably beet sugar, without adverse effects. Wines of less than 13 percent alcohol content, whether imported or domestic, were also tolerated. But fortified wines (greater than 13 percent alcohol) and domestic brandy (known to have contained added corn sugar) were usually associated with corn reactions.

Realizing for the first time that I was an incipient alcoholic, I never tried drinking on successive days again. And because of the general shakiness following the use of alcoholic beverages (at that time this meant bourbon whisky) I drank more moderately and rarely thereafter. One other observation is worth noting. While still single and living in a club, dinner had been delayed one evening while waiting for a late guest. In the meantime, the members of our bachelors' table waited in the bar, where I had two drinks of bourbon whisky. Being hungry and sitting in front of a large bowl of popcorn, I also ate much of that. The following morning I had a greater than usual headache after drinking, but additionally a vague sense of upper abdominal distress. By noon, rebound lower right abdominal tenderness had developed. Recognizing this clinical picture as probable acute appendicitis, I mentioned to the nurse in charge of the allergy clinic where I had worked that morning, that I was having an appendectomy that afternoon and would not be back for a few days. An acutely inflamed appendix was removed surgically three hours later. In view of findings to be described shortly, this excessive intake and rapid absorption of corn, both in terms of the amount of popcorn eaten and in the presence of bourbon whisky, and profuse eosinophilia of the removed appendix, this episode of acute appendicitis and allergy to corn were probably related.

As previously stated, until a few weeks after starting practice in Chicago in 1944, when I had been posing as an expert in food allergy, I did not know that I had a major food allergy problem or that allergy to corn even existed. Indeed, I had been using corn in an elimination diet for the detection of food allergy before Herb Rinkel visited my office in the summer of 1944 and told me of his own corn allergy. He also left a copy of his initial instructions for avoiding corn. I immediately diagnosed my own allergy to corn, wheat and all other cereal grains. As a result of avoiding cereal grains in all forms — both edible and potable — I have remained in excellent health, in contrast to my spotty medical history prior to this change in my diet.

Dr. Rinkel and I, in conjunction with our numerous corn sensitive patients, worked out the sources of corn in the American diet during the following five years, with the active noncooperation at the time and later coercion at the hands of the corn industry.

I happened to be on hand when a younger member of my family had her first drink of bourbon whisky. She became exceedingly sick immediately and has not tried repeating that experience since. She was later found also to have been extremely sensitive to wheat and corn. After using rice only once in three days as a cereal substitute for several months and not suspecting it, rice was avoided for five days prior to attending a family reunion. She developed an acute immediate incapacitating reaction after eating rice for the first time in five days, complicated by acute torticollis lasting for several additional days.[67]

Private Practice — Chicago, Illinois

Although Rinkel did not publish his observations on the prophylactic use of the rotary diversified diet until 1947, I had been aware of his use of this principle earlier. I was still experimenting with different schedules when I had advised my sister, naively, to minimize the development of allergy to rice by eating it only once in three days. This clinical experience in her case brought my routines in line with Rinkel's experience.

Another 20-year-old woman, with a history of eczema, hay fever, and more recent fatigue and myalgia, began going steadily with a beer drinking alcoholic. Initially, she also became sick and vomited after a few drinks each evening. But as she persisted in nightly beer drinking, she required more and more beer before becoming ill. In short her "tolerance" for beer improved to such a degree that she no longer became sick. Indeed, progressive amounts of beer were necessary before she even felt a "buzz." Although she had the impression that she was doing very well with her beer drinking, in view of her shaky start, she began to wonder if she might be hooked on it. She found to her amazement that now she became ill only when she did *not* drink. But drinking nightly in order to remain well failed in her case as she soon became morose and depressed, especially premenstrually. At this juncture she became a patient and was found to be highly sensitive to corn, wheat, coffee and a few other foods. Only by the avoidance of these dietary items in both their edible and potable forms was she able to again remain symptom-free and productive.

A social drinking business man observed that he was able to drink a given brand of straight bourbon whisky enjoyably and without apparent ill effects, whereas certain cheaper blended whiskys invariably made him sick. He felt sufficiently sure of this observation that he supplied a bottle of his tolerated booze to some of his friends for his use when visiting them. In the event that he became ill at their homes, he knew that they had filled his bottle with something else. This was explained when he was found to be extremely sensitive to potato but not sensitive to cereal grains. For a time after World War II cheaper blends of whisky were often adulterated with potato spirits.

Observations of these effects of specifically derived alcoholic beverages in known food sensitive social drinkers, incipient alcoholics and advanced alcoholics, led to the hypothesis that alcoholic extracts and distillates not only carried the allergenicity of the foods entering their manufacture, but because of the effect of alcohol as a drug (accentuating its speed of absorption) this food fraction was the most allergenic of all others. Moreover, this point along with a knowledge of the specific foods entering the production of different alcoholic beverages in social drinkers proved to be helpful in suspecting the existence of specific food allergy in discerning social drinkers. The diagnostic significance of this point of view in taking medical histories was reported in 1951.[78]

For instance, repeated clinical observations in myself in respect to the activity of various fractions of corn in inducing reactions shortly after the diagnosis of sensitivity to corn, while otherwise avoiding all sources of corn in the diet, are of interest. Not being sensitive to yeast, grape or beet, I was able to drink Cognac brandy. But if such a drink had been poured in an unrinsed jigger which had previously been used in measuring a bourbon drink, I experienced an intense headache in three minutes. Corn sugar obtained accidentally or deliberately precipitated a headache between 10 and 12 minutes. Cornstarch was followed by the onset of a similar reaction between 20 and 25 minutes. Corn oil produced mild scratchiness of the throat between two and three hours after ingestion.

At about this time, the food-related etiology of alcoholism became increasingly evident. Experimental confirmation of this concept, as well as the role of specific alcoholic beverages in known food sensitive patients, depended on the availability of pure samples of alcoholic distillates of fermentable foods for test purposes. I explained this concept of alcoholism to the research and technical directors of a large Midwestern manufacturer of alcoholic beverages in the hope of obtaining pure samples of alcoholic distillates. They listened attentively but with little apparent interest to my description of craving in food allergy and how satisfying this craving often led to obesity, as well as similar cravings for alcoholic beverages derived from specific foods and how satisfying that craving often manifested in alcoholism. But when I mentioned that there were several clinical observations that did not seem to fit this food-related concept of alcoholism, their interest increased as they inquired about these apparent exceptions. As I related these exceptions, they provided the technical answers which supported this interpretation of alcoholism.

For instance, I could not understand why patients sensitive to corn reacted to Scotch whisky, when corn does not grow in Scotland. Answer: This country and the Argentine ship boat loads of corn to Scotland; this is fermented into grain whisky which is blended with pot-stilled malt whisky to make blended Scotch whisky. Also why did my grape sensitive patients react to Puerto Rican rum but not to Jamaican rum? Answer: Puerto Rican rum is imported to the continental United States in hogshead lots and blended up to two and one-half percent with grape sherry before it is bottled, whereas the laws of Jamaica stipulate that only bottled rum may be exported. Also, why did corn sensitive patients react to applejack brandy? I had written this firm, the leading manufacturer of applejack brandy in the Eastern United States, about that and how this firm had denied that there was any corn entering the manufacture of this product. Answer: Although they were not certain, they suggested that I write again and inquire about what sugars were employed in the manufacture of the caramel used for maintenance of uniformity of color of the final product. Before writing this letter, I tested several additional corn sensitive patients with applejack. Confirming this view, I suggested that the writer of the previous letter probably perjured the company's letterhead, as I was certain that their product contained some form of corn. I then inquired as to the food sources of the caramel added for color uniformity. In due time I received a long letter of apology, as the president of the company needed to write the manufacturer of the caramel to learn that it was made from 50 percent corn sugar and 50 percent cane sugar.

To determine why the corn sensitive patient also reacted to a widely sold brand of manufactured and bottled in California brandy was even more intricate. This will be described in the following chapter.

Having the liquor manufacturers in a favorable position to enlist their cooperation, I asked if they would prepare distillates at 100 and 190 proof of all fermentable foods. They graciously agreed to do so, fermenting the mash from a single inoculum which had been grown on the same media. These pure distillates from a dozen foods reacted in parallel with the same foods from which they were derived when they were tested by ingestion under correct conditions in known specifically food sensitive individuals. These findings were reported in abstract form in 1950.[72]

In the course of these clinical observations, it occurred to me that testing with a nonfood derived alcohol would be desirable as a control. Only two such sources were available: synthetically derived ethyl alcohol manufactured from ethylene gas and methyl alcohol derived from the sulfite wastes of Swedish paper mills, which was said to have been used to some extent for edible purposes in that country, but banned by law for drinking in the United States. After checking the pharmacologic and toxicologic safety of synthetic ethyl alcohol, 15 cc. of a 100 proof sample diluted with 100 cc. of compatible water was ingested experimentally by 10 persons, including myself, who had reacted acutely to one or more of the special food derived distillates. When I had shown no evidence of reaction to this synthetic ethyl alcohol, I assumed that it would be safe to be used in others, but not so! Three persons developed unmistakable evidence of reaction and one person promptly went into a state of shock — for no apparent reason. This experience only became evident several years later. This will be reported subsequently in the course of this chronologic narrative.

Results of testing known food sensitive patients with various food fractions, including alcoholic extracts and distillates, were reported for the first time in 1951.[78] Suffice it to say here that most food fractions carry the allergenicity of the food from which they are derived, when checked under correct conditions of testing, as reported preliminarily in our book, *Food Allergy*.[75] Most of these observations in respect to the allergenicity of given sugars and alcohols had not been completed by the time that this manuscript had been submitted for publication in late 1948.

Inhalant Allergy: Its Testing and Therapy

It should be emphasized that I had employed traditional approaches to the diagnosis and treatment of inhalant allergy since first becoming interested in this field while working in the University of Michigan Allergy Clinic (1934-1937), the Massachusetts General Allergy Clinic (1937-1939), while practicing in Milwaukee (1939-1942), and again in the Allergy Clinic at the University of Michigan (1942-1944), as well as when first starting practice in Chicago in 1944. Both scratch and intradermal skin tests were performed with 1-10 serial dilutions of allergenic extracts, employing multiple tests simultaneously and reading the results in terms of negative to ++++ reactions. Treatment was started with relatively small therapeutic doses of extracts of pollens, molds, dust, animal danders and other biological materials, as determined by details of the allergy history, and building up progressively to maximal tolerated doses per patient.

I visited the office of Herbert J. Rinkel in Kansas City, Missouri for the first time in the fall of 1944 at the end of the fall ragweed pollinating season. At that time, Dr. Rinkel had just completed a comparative study of treating ragweed hay fever, using 1-10 serial dilutions of pollen extracts as compared with using 1-5 serial dilutions, in order to find which was the more effective in determining the end point of reaction between successive intradermal tests. He had concluded that the serial 1-5 dilutions were by far the more accurate and he had just completed making over his test solutions with 1-5 dilutions and writing up how to carry through with this approach for the next pollen season. Although he gave me mimeographed copies of these instructions, I am unable to find them currently. He defined the end point of reaction in performing the intradermal injection of 0.01 cc. each of serial 1-5 dilutions which resulted in the first wheal of 7.0 mm. in diameter and erythema of 20 mm. diameter, and which was followed by progressive whealing with the application of successively stronger concentrations. For details the reader should refer to the reports listed in the Appendix, page 334.

I was impressed with his results and changed my technique of inhalant allergy management to the 1-5 system of dilutions in keeping with his other recommendations. Having employed this system since that time, I have found this more accurate and predictable in determining where to start inhalant allergy treatment than the previous system entailing an excessive number of injections in reaching an effective therapeutic level. It was particularly effective in the management of coseasonal pollinosis, as recommended by Rinkel.

Although I have never reported the therapeutic results of treatment of pollinosis, mold and animal dander sensitivity, I made a special study of house dust sensitivity, titrating for the end point of dust sensitivity at weekly intervals for several months in several cases treated on a perennial basis. Differences in the end points of reactions at different seasons of the year sometimes varied by as much as 125 times. I concluded that the house dust season, as far as variations in symptoms and the end points of reaction were concerned, was the opposite of the baseball season (mid-April to late October).[61] Also, selected cases of ragweed pollinosis varied sometimes in respect to their end points of reaction by as much as 3,000 times at different times of the year, selected patients reacting much more strongly during the season than outside of the specific season. These observations were never reported.

Alternative Mechanisms of Allergy

As previously stated, Dr. Arthur F. Coca had objected to changing the definition of allergy from its original descriptive and clinically oriented sense of altered reactivity occurring with time to a definition based on immunologic mechanisms. Although Coca was one of the deans of American immunologists at this time (1924-1925) he felt that this limitation, virtually excluding other interpretations, was too restrictive. Despite his vigorous objections, however, the immunologic interpretation of allergy became generally accepted.

Faced with the responsibility of preparing lectures for medical students on the subject of allergy in 1942, I reviewed the status of this field as it had existed in 1925 and, since concurring with Coca's views, taught the immunologic theory of allergy to the medical students, encouraging them to look for other interpretations. Since that time, I had followed the same advice seeking other possible mechanisms of the allergy phenomenon. Several fragments of information had emerged by 1950 when Dr. Harry G. Clark, a Detroit surgeon who was interested in electrolytes and other applications of physiology, consulted me as a patient. Since his contribution helped to integrate these fragments of information, these will be described as they existed prior to Dr. Clark's help in bridging these disconnected pieces of information.

Perhaps, most significantly, the immunologic interpretation of allergy failed to explain the extreme speed of the acute allergic reaction, as antigens, antibodies and their mediators were measurable by 1950. Also, basic features of food and drug allergy remained unexplained immunologically by the midcentury point. Although Rowe had paid lip service to a probable immunologic explanation of food allergy, neither he nor others had demonstrated the existence of these relationships in the majority of food sensitive patients. Rinkel simply described his clinical observations in food allergy, without attempting to interpret their underlying mechanisms. Coca denied the immunologic basis of the common form of food allergy by the title of his presentation: *Familial Nonreaginic Food Allergy,* published in its first edition in 1943.

More particularly, there had been no satisfactory interpretation of the masking phenomenon of food allergy — namely, the apparent suppression of symptoms associated with the cumulative intake of a given food by a highly sensitive person, in contrast to the acute immediate reaction demonstrable in the same person when exposed to the same dose of the same food following a temporary period of its avoidance. Not only had the fallibility of skin testing in food allergy been recognized independently by Rowe, Coca and Rinkel by 1950, this had also been my experience which had prompted me to dispense with skin testing with foods upon starting practice in Chicago. Neither had drug allergy, as represented by aspirin sensitivity, been explained as to its mechanism.

The function of eosinophils in allergic reactions to foods and drugs also remained obscure. Indeed, changes in eosinophils occurring in acute reactions to test doses of drugs and foods, as determined by accurate techniques, had just been described by 1950.[31,47,49,51] For example, eosinophils in the peripheral blood of a patient with a diagnosis of bronchial asthma and known to be highly sensitive to sulfathiazole, showed the following changes after the oral challenge of 2.0 grams of this sulfonamide. From a pretest level of 8 percent eosinophils, these cells decreased to a low of 4 percent at the height of a severe attack of asthma, only to rise to over 30 percent as the acute attack subsided.[21] I had observed similar but less marked changes in eosinophil levels of the peripheral blood of specifically sensitive patients following the test ingestion of a food allergen.[31]

The function of lymphocytes in allergic reactions was an even greater mystery by 1950, despite some preliminary clues. For instance, leukopenia had been observed as a common finding in chronically ill patients, especially those manifesting generalized or systemic effects[29]. Another clue that the lymphocyte might be involved in some way was the observation that atypical lymphocytes, as observed morphologically in the stained blood smear, often characterized the peripheral blood in instances of chronic food allergy.[10,15]

In a given unreported case, followed by sequential blood counts before and following the induction of an acute test response to an allergenic food, atypical lymphocytes increased during and following such reactions, sometimes reaching 20 percent of the total leukocyte count. Fitting into the idea that there might be some malfunctions of lymphocytes in allergy was also the occurrence of lymphoid nodules in the posterior pharynx of allergic individuals, especially in children, who also often presented markedly enlarged and edematous tonsils and adenoids.

It had been known for several years that measures resulting in increased oxygenation of tissues of the body were helpful in the treatment of allergic conditions. For instance, oxygen administration for bronchial asthma, headache and fatigue had been useful. There also had been recognized benefits derived from regular vigorous physical exercise especially for those subject to low-grade headaches, chronic fatigue, mild muscle and joint aches and pains, as well as lesser and greater degrees of depression. These measures were most often observed by chronically ill persons themselves. For instance, a syndrome in which those suffering from such lesser allergies became staunch devotees of the athletic clubs on the excuse that they felt better and were more productive when exercising regularly, came out in the histories of some of these cases. Probably the same explanation accounted for regular swimmers, runners and gymnasts, as long as they were able to maintain such schedules. With advancement of their illnesses, however, to a degree that such routines could not be maintained, many deteriorated rapidly.

In an effort to find some measurement for this apparent beneficial effect of increased oxygenation of tissues in clinical allergy, I obtained the necessary equipment and examined the scleral blood vessels of patients before, during and following food induced clinical reactions. Although increased sludging of red blood cells (intravascular agglutination of erythrocytes) often occurred in such reactions, these changes were not specific for allergic responses, as they also occurred in infections, as had been reported by others. This occurrence of clumps of red blood cells in the circulating blood decreased the ability of red blood cells to carry oxygen on their surfaces as well as the removal of carbon dioxide from such reacting tissues. Although I had referred to these observations a few times they remained unpublished, as it was impossible by means of techniques available at the time to photograph these changes or to measure the degrees of these variations in the circulation of scleral blood vessels except by means of one plus to four plus estimates.

From these various observations it was reasonable to assume that oxygenation of reacting tissues in allergy might be reduced to a degree to account for some of the associated symptomatology of such reactions. I was at this juncture in my thinking of other possible mechanisms of allergy when Dr. Clark arrived as a patient in 1950. I will give a brief account of his background and medical history before describing his contribution to this effort.

Dr. Harry G. Clark had been an energetic surgeon, having built up a large surgically oriented practice in suburban Detroit, Michigan. He had graduated from the University of Toronto Medical School, where some of the early work in electrolytes had been done. He had also developed the application of this knowledge in his surgical practice. Academically he had carried this point of view into an avocation, frequently visiting oceanographic laboratories where work was being done on unicellular forms of life. He argued that if one knew what was happening in one-cell organisms, the same general principles might be applicable in the care of his surgery patients.

With time, however, Dr. Clark became increasingly tired and was unable to keep up with his surgery schedule. He was referred to me because of his fatigue and related systemic manifestations. He had also been subject to unexplained episodes of diarrhea for 20 years, bouts of bronchitis for 7-8 years, and rhinitis for 3-4 years. His fatigue was sometimes accompanied by discouragement and depression. Pork, milk and wheat had been suspected. The fact that his nonseasonal fatigue and related symptoms were accentuated on certain days, especially in the mornings, suggested the likelihood of chronic food allergy. Consequently, he was hospitalized on a diet avoiding his most frequently eaten foods with the rotation of all other foods so that no given dietary item was eaten more often than once in four days. He not only improved rapidly on this program, but also reacted acutely when each of several of his favorite foods was returned to his diet.

He was fascinated by this experience and by observations of other patients being investigated similarly. Indeed, he likened this experience to the observations of other patients being treated with ACTH and cortisone at the time. He not only read widely on the subjects of allergy and immunology but also quizzed me in detail as to what was known about underlying mechanisms. After I had given him the background, as related above, he began to speculate about mechanisms. He concluded, first, that in view of the speed of these acute reactions, changes in electrolytes must be involved. Based on the fact that the pathology of allergy is characterized by edema, and observations from the marine biology laboratory that unicellular forms of life swell when acidified, he concluded that intracellular edema developed in acute allergic reactions more rapidly than it could be neutralized by the available alkali existing at the sites of such reactions.

On the basis of this reasoning, we began giving the bicarbonate salts of sodium and potassium in the proportions existing in the body (2/3 $NaHCO_3$ and 1/3 $KHCO_3$) orally in the treatment of acute reactions. It worked. We then tried various combinations, but the original proportions worked as well or better than alternatives. Between us, we then constructed the remainder of what we first called the acid-anoxia-endocrine theory of allergy. This was first published in abstract form in 1950.[68,69]

The remainder of this theory is as follows: Since cellular functions in unicellular and multicellular forms of life are basically similar, and since the body seems to try to break down, neutralize or eliminate foreign materials to which it is susceptible as rapidly as possible, and since the end products of the break down of foods and other foreign materials are largely acid, acid end products of catabolism accumulate intracellularly more rapidly than they may be neutralized by the relatively more alkaline extracellular fluid. Indeed, a low-grade reaction may not develop beyond this point.

In a more advanced reaction, other factors are apparently brought into play. Exhaustion of the available extracellular alkali seems to serve chemotactically to attract eosinophils from the peripheral blood. Presumably, these are lysed and their highly alkaline granules (otherwise they would not take an eosin stain with such avidity), augment the local alkali reserve. In an expanding and/or an increasingly severe reaction, the size of the edematous response apparently spreads by direct extension. Intravascular sludging of erythrocytes may further reduce the transport of oxygen to such an acidic, edematous and inflamed area, from which carbon dioxide is removed less effectively for the same reason. The result, presumably, is an expanding and/or increasingly severe allergic inflammation which may become generalized.

If the combination of acidity and hypoxia is registered in a lowering of the pH (hydrogen ion concentration) of the blood, this presumably is a powerful stimulus for the outpouring of the relatively alkaline secretions (endogenous ACTH) from the pituitary-adrenal axis, as the third line of defense. Anywhere in this process, the body's buffering system is apparently brought into play. The first buffer to be involved is the bicarbonate buffer. The effectiveness of the alkali salts of sodium and potassium bicarbonate, in alleviating allergic inflammation is apparently explained by this overall mechanism. Perhaps that is the principal reason that alkali therapy is more helpful during the first 24 hours of an acute allergic reaction than later.

It is important to remember that although alkali therapy is effective in the treatment of most acute allergic reactions within the first several hours, it may accentuate the manifestations of such chronic reactions as protracted bronchial asthma, depression and others. If alkalinization of such chronic reactions is to be considered, this must be monitored by a knowledge of the pH. An integral part of alkalinization for acute allergic responses is providing an abundance of water. For instance, a dose of 10-12 grams of $NaHCO_3$ 2 parts and $KHCO_3$ 1 part should be taken in 600 to 1000 cc. of water. Renal or cardiac failure or hypertension contraindicates this therapy. In selected instances, such as in those who retain fluid readily, the use of equal parts sodium and potassium bicarbonate is desirable.

Adding to the effectiveness of orally administered alkali salts in the treatment of food allergy is the saline laxative effect commonly produced which tends to empty the gastrointestinal tract. Indeed, if such a laxative action is not induced within approximately three hours after ingestion of a reacting food, symptoms may recur; this sometimes indicates an additional oral dose.

In instances of persistent vomiting or unconsciousness preventing the oral administration of alkali salts, this mixture may be given either rectally or sodium bicarbonate may be injected intravenously. The dose most commonly used intravenously is 100-150 cc. containing between 7.5-12 grams. This dose may be administered relatively slowly or rapidly. As one approaches the desired end point, unconscious, stuporous patients, or those in shock, tend to regain consciousness. Administration is continued until the patient becomes very thirsty. The injection is then discontinued. Administration of several glasses of water orally (600-800 cc.) is an integral part of intravenous therapy. Maximal improvement does not ordinarily occur until this water is received. The intravenous route of administration was first reported in abstract in 1954.[105]

It is of interest that approximately a century ago sodium bicarbonate was a popular home remedy for many lesser ills. This resulted in statements from medical authorities of its hazards. As an apparent result this practice became relatively less common. I remember that my grandparents were advocates of the practice of dipping into the soda jar for the alleviation of minor ills.

The Midcentury Controversial Status of Food Allergy

Since coming to Chicago in 1944, I had been an instructor in the Department of Internal Medicine and the subdivision of Allergy at Northwestern University Medical School. There was little interest in the subject of food allergy at this institution or elsewhere in Chicago at this time. For the most part, skin tests with food extracts were performed mechanically and interpreted literally. As I became aware of the existence of allergy to corn, I began diagnosing corn sensitive patients in the Allergy Clinic of the Medical School. Although I attempted to demonstrate such cases to my immediate superior, Dr. Samuel M. Feinberg, he was always too busy to see such cases, apparently preferring to continue to be able to say that he had never seen a case of allergy to corn.

A similar situation existed in respect to mental and behavioral responses demonstrated to be on the basis of reactions to specific foods. Although medical students were fascinated by these demonstrations and developments, not so for most members of the medical staff. For instance, when I showed the motion picture of an acute psychotic episode following the blind intubation of beets and beet sugar at a Medical Staff Meeting at Wesley Memorial Hospital where the pictures were taken, the chairman of the Medical Department commented that this person was obviously an hysterical woman. I replied to the effect that I would not disagree with the designation of hysteria if by that he meant an entity of unknown etiology, but if his use of that term carried the connotation of suggestion in this case, I disagreed strongly.

Under such circumstances, it was not surprising that the Medical School failed to approve any of my manuscripts for publication. Neither was I permitted to list my affiliations with Northwestern University Medical School nor Wesley Memorial Hospital where these observations were made. Thereafter, I simply published from my office address in the absence of academic and hospital affiliations.

Critics of food allergy from both within and from outside of the medical profession were becoming increasingly vocal by the midcentury point. Since I had been active in publicizing the roll of allergy to corn as the leading cause of food reactions, as well as extending the medical applications of food allergy, much of this criticism was directed toward me and my medical activities.

Because I felt that this criticism stemmed largely from the paradoxical addictive nature of food allergy not understood by these critics, the best defense would be to refer to this subject as food addiction rather than food allergy, presenting detailed case histories in support of this position. Consequently, a monograph on food addiction and its manifestations was prepared between 1949 and 1952.

The controversial status of food allergy existing in 1949 is also indicated by what happened in the hearings for reaching a standard of identity of bread and related products by the Food and Drug Administration in Washington, D.C. Since the deadline for applying to submit testimony at this hearing expired two weeks prior to the Annual Meeting of the American College of Allergists, I applied for and received an appointment to testify. When I offered to turn this appointment over to anyone the Directors of the College might wish to name, they refused to appoint anyone. Upon appearing at this hearing, I was surprised to find five other allergists from various cities surrounding Chicago scheduled to testify immediately following my testimony.

Only then did I understand why I had seen the Director of the Corn Products Research Foundation talking with other physicians at the College of Allergists Meeting. Various junkets to Washington had been set up by the corn industry for the obvious purpose of negating my testimony. I submitted that *all* sources of given foods, including specifically designated sugars, entering the manufacture of processed foods be declared on the labels of these products. Immediately following my testimony the other allergists, introduced by the Corn Products Research Foundation, stated that they had not seen a case of allergy to corn and that this was not a major problem. This entire testimony is a matter of documented public record.[43]

When the effectiveness of adrenocorticotropic hormone (ACTH) and cortisone in the treatment of rheumatoid arthritis was announced in Life Magazine in early 1949, I applied to Armour and Company for ACTH and to Merck, Sharp and Dome for cortisone, knowing that from Zeller's and Kaufman's observations of the roles of specific foods in arthritis, as well as my own experience, that these agents would be effective in the management of other allergies.

Upon appearing at the Armour Laboratories for my appointment, the Medical Director, Dr. John R. Mote, was looking for an accurate method of counting eosinophils in the prospect that this might be useful in determining dosage. At this point, I pulled out from my briefcase reprints describing a counting chamber technique for this purpose, including a comparison of the counting chamber and stained film techniques.[9,14,16,19,22,24,47,49,51] I walked out with my pockets bulging with ACTH bottles for the purpose of counting eosinophils. Availability of supplies permitted use by physicians in only four locations. I was given a monopoly in the central U.S.A. for a period of six months. Supplies of cortisone and adrenal cortex extract (ACE) were also obtained for comparative observations. After participating in the First and Second ACTH Conferences,[53,69] clinical observations of the effectiveness of these materials in various allergic syndromes were also reported.[53,54,56,57,58,60,77]

Obtaining supplies of adrenocorticotropic hormone (ACTH), cortisone and adrenal cortex extract (ACE) for treatment of allergic manifestations was largely a matter of luck in being at the right place at the right time with the right method (counting chamber technique of enumerating eosinophils). A greater, but unpublicized, distinction is the fact that since 1953 I have not started patients on steroids except rarely for therapy of lupus or collagen disease. In general, far more ecologically sound and clinically effective long-term therapeutic results may be obtained from the application of specifically focused diagnostic and therapeutic measures.

Approximately six months later, a Research Grant for the Study of Food Allergy from Swift and Company for the previous five years came up for renewal. Despite the fact that the person in charge of grants at that institution had recommended its extension, this decision was reversed at the Executive Committee level, so I was told, as a result of pressures from the corn industry.

Although representatives of Northwestern University Medical School were much more interested in my activities with ACTH and related steroids than previously, I continued to work from an office basis where patients could be diagnosed and followed more accurately than at a charity clinic. Shortly thereafter, my tenure at the Medical School was terminated on the charge that I was "a pernicious influence for medical students." This action also terminated my teaching of allergy in the Department of Ophthalmology and the Department of Otolaryngology at the Medical School which I had been doing for several years. Neither was I permitted to accept an appointment in the Otolaryngology Department offered by Dr. George E. Shambaugh, Jr., head of that department. In commenting on this expulsion later, Dr. Shambaugh likened this experience to that of Semmelweis approximately a century earlier (Ahead of Their Time. Arch Otolaryngol 79:118, 1964, Editorial ©1964, American Medical Association). Since there might be interest in this spontaneous comment, it is being reproduced herewith:

Ahead of Their Time

The life of Semmelweis, so beautifully depicted in the novel by Morton Thompson, is the story of the discovery of the infectious nature of puerperal sepsis and the possibility of controlling this ancient scourge by aseptic technique. The discovery was made in 1847, thirty-two years before Pasteur described cocci in chains as the cause of puerperal sepsis. It occurred in the wards of the Vienna General Hospital where the maternal mortality often exceeded 30 percent among women delivered by doctors who passed freely back and forth, without washing their hands, between the labor ward and the autopsy room stacked with fresh cadavers dead of puerperal fever. Semmelweis connected the lower mortality among women delivered by midwives with the fact that the latter did not perform autopsies. He then observed identical postmortem findings in a former instructor who died of cadaveric poisoning after dissecting a victim of puerperal sepsis, and he deduced the infectious nature of both diseases, carried by contaminated hands and instruments. When he introduced a meticulous sterile technique with rigorously enforced thorough washing of the hands in a chloride of lime solution before vaginal examination, the mortality among women delivered by doctors fell, for the first time, below the rate among women delivered by midwives.

Because the ideas of Semmelweis were completely unorthodox, he found himself bitterly opposed by the conservative traditionalists, including the head of his department, and he was, literally, "kicked off" the staff of the hospital. He returned to his native Hungary where he succeeded in reducing deaths from puerperal sepsis to an unprecedented fraction of one percent. Applying the same methods to gynecological surgery he originated surgical aseptic technique. The opposition by the leading surgeons and obstetricians, and even by the great pathologist Virchow, continued until Semmelweis finally died in an insane asylum, his ideas still not accepted by the leaders of European medicine.

A fate not always so tragic and extreme, but nevertheless similar, has befallen scientists throughout the ages who have dared to deviate from the traditional patterns of thought. We like to think that in our enlightened era such attitudes cannot exist. Have we forgotten the concerted opposition to Julius Lempert by the senior and most highly respected members of our specialty in the early years of endaural and fenestration surgery? When was he invited to present before the American Otological Society his work on the greatest single contribution to otology of this century?

Like Semmelweis, whose pupils carried his ideas to distant lands, Lempert taught the endaural approach and the fenestration operation to young physicians from every corner of the globe. It appears to be characteristic that acceptance of the new ideas of a great man begins away from home, and last of all, in his own city. Thus with Semmelweis and Lempert.

Can one discern the same pattern with two or three other brilliant and original individuals who have contributed mightily to otolaryngology in our time? French Hansel comes immediately to mind; the late Herbert Rinkel, a student of Hansel was another, while Theron Randolph, coauthor of Rinkel's monumental work on food allergy, lost his academic and hospital appointments because his teachings "were a pernicious influence on medical students."

The amazing effectiveness of the minute optimum dosage therapy of Hansel and Rinkel for inhalant allergies becomes apparent to any who gives this method a fair trial, yet the majority of allergists have not tried it, any more than the contemporaries of Semmelweis would try his simple methods of aseptic technique.

Freedom of inquiry in medicine, so necessary to permit the germination and flowering of new ideas, can flourish only when doctors are not completely subservient to established authority, one of the eventual and probably unavoidable outcomes of government dominant medicine, and medical research. One can imagine how quickly an end would have been put to Lempert's earliest endaural and fenestration operations had he worked under one of the older, most highly respected professors! For that matter, could Samuel Rosen have made his great contribution of reviving stapes mobilization had he not been engaged in private practice? It is fortunate indeed for the progress of medical science that complete socialization of medical practice has not occurred in America, and that the Lemperts, the Rosens, the Hansels, Rinkels, and Randolphs can still pursue their unorthodox and pioneering investigations in their private practices. Could there be a relationship between the abundant flowering of medical science in America in recent years and the prevalence in America of free enterprise private practice?

Wondering if I might have any legal recourse in this decision, I consulted one of my patients, a lawyer, who advised me not to do anything about it. Only after his death did I learn that his firm had long been retained as legal counsel by the Medical School!

Shortly following this Bread Hearing, Dr. Mary Loveless, who had also submitted written testimony at the time of this hearing, organized a Symposium on Food Allergy to be published in the Journal of Allergy. Herbert J. Rinkel had been invited to present the positive side of this subject but, fearing a trap, refused. I accepted, thinking it better to present the subject accurately than otherwise. Immediately after my presentation, "Concepts of Food Allergy Important in Specific Diagnosis,"[70] Dr. Loveless and several others who had submitted contrary views at the Bread Hearings, presented their reports. This apparently purposefully one-sided "Symposium" was published in the Journal of Allergy in 1950. In contrast to my positive side of this subject, consisting of seven pages,[70] this presentation was followed by seven articles totalling 41 pages of negative findings in respect to the clinical significance of food allergy in general, and allergy to corn and corn sugar in particular.

After this, and on the apparent supposition that I was "anti-corn," a representative of the Sugar Research Foundation approached me in the hope that I might be helpful in aiding the cane and beet sugar industries to regain some of the markets they had lost to corn sugar because of price differentials in the sugar market during and following World War II. Of course, there was nothing I could do. However, this interview provided an opening which I had considered for some time, namely, to determine if alpha d'glucose prepared from splitting sucrose might or might not induce reactive symptoms in corn sensitive patients known to react adversely to alpha d'glucose manufactured from cornstarch. The representative of the Sugar Research Foundation agreed to prepare this.

A few months later, a half-pound of crystalline sugar marked only as alpha d'glucose arrived in the mail from Cambridge, Massachusetts in the absence of correspondence indicating its specific origin. I assumed that this had been manufactured from cane sugar, because at this time there were no beet sugar mills in the United States within 800 miles of eastern Massachusetts. Thinking that these circumstances provided an excellent opportunity to determine the identity of this sample double-blindly, I set up the following experiment. Nine private patients were asked to donate two hours of their time for an experiment. All had either been diagnosed specifically within the previous few weeks or were known to be highly sensitive to corn, cane or beet sugar in the absence of sensitivity to the two alternate sugars. Of the three patients in each group, care was exercised in performing the tests in the food test room one at a time. Neither the corn sensitive nor the cane sensitive patients reacted to this unknown sugar. But all three of the beet sensitive patients reacted acutely; one developed a sharp attack of bronchial asthma, another a severe headache and the third generalized urticaria. These results were also reported in 1950.[63]

The history in the third patient had been especially baffling. For several years she had been subject to urticaria yearly starting July 5th and persisting through to about January 15th, but only very rarely and for no apparent reason otherwise. According to this patient, and I had to agree, I had been a complete failure in helping her solve her hive problem, whereas others who she had referred to my office had worked out well. Knowing that this family moved to their summer home in southern Michigan on the shore of Lake Michigan July 4th yearly, I assumed that she must be reacting to some fungus in her summer home or area, there being no pollen exposure exactly corresponding to these dates. But on repeated occasions she had failed to react to extracts of pollens, dusts, molds, insects, animal danders or other extracts of inhaled allergens.

In desperation at being ribbed repeatedly as a failure in her case, I recorded a new detailed medical history. This solved the problem. During the War this patient had been a sugar hoarder, having stashed 300 pounds of beet sugar in the attic of her summer home. On July 5th of each year she would bring down 25 pounds of beet sugar for the summer. In the fall she would take the remainder to Chicago, using it to make candy and cookies for a yearly holidays "extravaganza." By January 15th supplies would be exhausted and she would return to using cane sugar until the following itchy and puffy summer. But I nearly lost this patient for good when I muttered something about retribution and absence of patriotism. Nevertheless, she was so grateful to have comfortable summers, falls and holiday seasons that she volunteered for this double-blind experiment.

Chapter III — 1950

Further Anecdotes — Coca, Rowe and Rinkel

Further Anecdotes — Coca, Rowe and Rinkel

One of the aims of this book is to report personal anecdotes involving Coca, Rowe and Rinkel. Several of these occurred in the years 1950 and 1951 — a common high point of activity for all of us in this new field.

This period was between the second and third editions of Coca's book, *Familial Nonreaginic Food Allergy*. During the first part of this interval, Rinkel and I were correcting, proofing and indexing our common book, *Food Allergy*.[75] Rowe was at the midpoint of a 60 year span of his productive medical career.

Thus far I have refrained from relating anecdotes which cast others in unfavorable lights. But since everyone is prone to have certain hangups or blind spots, perhaps these should be combined in one place. These fall into several categories: those having to do with delayed acceptance and application of findings reported by others. Due to the stress of accustomed routines, one often simply fails to take the necessary time to try out something new and different. Even though one may be familiar with the addictive features of allergy and clinical ecology, a person may be unwilling or unable to discipline himself to test out and control his own addictions. Lastly, supersalesmanship often rebounds negatively.

A few of these points will be taken up in respect to the pioneers in this field with whom I had the opportunity of working closely. A good place to start is with some of my own hangups.

I was slow to pick up on the significance of allergy to yeast until becoming immersed in studying alcoholism, despite this subject having been described earlier. The same holds for candidiasis which I did not start diagnosing until 1977. Although I had been diagnosing and treating for sensitivity to TOE (Trichophyton, Oidiomycin and Monilia), I never pursued it as thoroughly as Herb Rinkel had done. Being over impressed with the significance of reactions to foods and chemicals, I minimized a comparable emphasis on pollens, dusts and other particulate biological inhalants for a few years, especially in the absence of a positive history.

Although I had avoided cereal grains carefully for 40 years as well as certain other foods, I had used honey sweetened chocolate intermittently for its desirable stimulatory effect, despite the fact that I usually was slightly more tired the following day. But when faced with an impossible deadline, I found that I could work about two hours additionally under these circumstances — if not used too frequently. If used too often or in too large amounts, intense sharply localized head pain would result.

Early on, after learning of the clinical significance of food allergy, I found that overenthusiasm in telling other physicians commonly backfired. Merely describing my clinical experience and answering questions were far more helpful and effective than overstating the merits of food allergy. The same has held for the chemical susceptibility problem.

Dr. Coca thoroughly antagonized his contemporaries in immunology when he became interested in food allergy as determined on the basis of the technique of pulse acceleration, described in detail later. He was so consumed in this interest, he could not understand why his former friends and associates were so disinterested.

When I visited Albert Rowe's office and hospital for the first time in 1950, I arrived at his hospital a few minutes before he came in. While waiting, I met another member of this hospital staff who had been my junior intern at the University Hospital in Ann Arbor, Michigan 15 years earlier. Despite the fact that he knew that I was to visit Dr. Rowe, he launched into a highly critical account of his fellow staff member to the effect that Albert Rowe was driving the others crazy in the medical department by his persistence in discussing the food allergy aspects of virtually every medical problem which was presented before the medical department. He went on at some length to say that he always attended and that he always brought in food allergy in some way — in short that they simply could not cope with this!

As he continued his tirade, I pondered as to how I should respond to it. Finally, when I had the chance, I said that I had made a point of reading everything Dr. Rowe had written as listed in the Quarterly Cumulative Index as well as his three books. Moreover, I had been able to confirm all of his clinical observations with the exception of his report that, in general, allergic manifestations tended to be accentuated at sea level as compared with improvement inland or in the mountains. I had not been able to study this. I commented farther that, in my opinion, Albert Rowe was an amazingly astute physician. My friend was so obviously dismayed that he had some good reason to leave shortly.

As we made rounds all was in good order except for one patient who was receiving an intravenous infusion, to whom Dr. Rowe referred to as a failure. Although she had been there over two weeks, he had been unable to clear her symptoms sufficiently to discharge her. He discussed in detail her medical history and her failure to respond. It was not until later in the day that he brought up the case of the recalcitrant hospital patient and asked if I had any suggestions. I commented that I had had similar experiences with patients not responding until intravenous corn sugar (dextrose) had been discontinued and all other corn derivatives removed from such a person's diet. I also mentioned that I had submitted a manuscript on this problem but which was still in press.[65] He welcomed this suggestion and said he was stopping this intravenous administration immediately. He also thanked me for having previously pointed out a few minor sources of corn in his elimination diets, such as corn sugar in most commercially processed ham and bacon as well as in excipients of pharmaceutical preparations.[62] Although Albert did not tell me at this time that he was corn sensitive, I had received a letter from Herb Rinkel dated December 11, 1949 bearing on this point from which I will quote: "It is indeed interesting to learn that Albert Rowe is corn sensitive. I think that he has gradually been seeing the light in respect to corn with the recent emphasis that has been placed on corn. It stimulates lots of these fellows to think about it, you know. They go home and begin to check it and the first thing you know they find it is true." Albert told me later that this patient had improved within a few days following the avoidance of all sources of corn so that she was able to leave the hospital. Figure 10 shows Albert Rowe at a College of Allergists meeting about this time.

Figure 10. Albert Rowe is at the extreme right. Seated to his left around the table are Helen C. Hayden, William G. Crook, Hilda M. Hensel, Theron G. Randolph, Mrs. Schafer, Walter L. Schafer, Janet M. (Tudy) Randolph, George S. Frauenberger and Helen Frauenberger.

The above experience indicates that although Coca, Rowe and Rinkel all knew and seemed to appreciate each other's contributions to their common interest, there was little personal contact or correspondence between them. This fact highlights the first joint meeting of Coca, Rowe and Rinkel prior to a meeting of the American Academy of Allergy in New York City in early February 1951, which Joseph Interlandi and I also attended.

As far as can be determined from anything any of them said, neither Coca, Rowe nor Rinkel had ever visited the offices of the others. Living, respectively, in the states of New York, Missouri and California, the only personal contact that they apparently had had was at medical meetings. These were relatively few and far between as both Coca and Rinkel had largely given up attending medical meetings except if they happened to be on the programs.

This meeting was set up by Dr. Coca in conjunction with Leonard Green, a friend of his and the manufacturer of Dust-Seal, a petroleum hydrolysate used for the treatment of fabrics and other dust retaining and generating materials. Coca had set up the meeting but none of us had been informed of its purpose or agenda. However, this soon became evident as immediately after lunch Dr. Coca called the meeting to order and launched into a tirade of the medical profession to the effect that medical academia in the United States, and most physicians here, were helpless and hopeless in accepting medical facts or techniques by means of which clinical observations may be demonstrated. Since what he said is similar to the preface to the second edition of his book, *Familial Nonreaginic Food Allergy,* published in 1945 (Springfield: Charles C. Thomas), I will quote from this source:

> In the preface of a publication an author is permitted, by custom, to intrude his more personal thoughts.
>
> For taking advantage of that custom in this second edition I plead a considerable provocation at the hands of many personal acquaintances among experienced allergists and other medical specialists from whom I could reasonably expect at least an unprejudiced hearing, if not a generous cooperation.
>
> The attitude of most of these towards the first edition of this monograph has been that of a skepticism so uncompromising that I have not even been invited to demonstrate the new method of examination described therein.
>
> It is quite out of the question to attribute this attitude to any personal prejudice; no, the reason for it is that the medical profession is again faced with scientific findings and their consequences that are so far out of line with settled concepts as apparently to represent the impossible.
>
> The following preliminary conclusions drawn from my own study and the reports of Locke and of Price will illustrate the wide divergence of these findings from accepted medical dogma:
>
> 1. The level and range of the normal pulse rate is a physiological constant in each individual, varying widely in different individuals.

2. The most common cause of variations in the individual from this normal constant is familial nonreaginic food allergy (idioblapsis).

3. Upwards of 80 percent of the white population are hereditarily affected with idioblaptic allergy.

4. Idioblapsis is probably a lethal character, the most important primary cause of noninfectious disease, and a predisposing cause of some infectious disease.

These four major conclusions are of a sufficiently revolutionary nature to explain the hesitation of experienced medical specialists to waste much time with the described method of diagnosis and treatment. A few tell me that they have "tried it" unsuccessfully but only two have seriously asked me to help them with their difficulties in its use, much as I should like to do so. (One of these was the late Warren T. Vaughan, who sent me a beautifully complete pulse-diet record, from which it could be seen that the patient was intermittently exposed to an environmental pulse accelerating allergen.)

However, I know that an interest in this matter is stirring here and there and confirmation of the fundamental facts is already beginning to appear.

Coca went on to say that in view of these uncompromising medical attitudes he would continue to write for the public in the hope that patients would call this work to the attention of their doctors, inasmuch as there seemed to be no other way to reach physicians through alternative channels.

Dr. Coca then asked for our opinions on this approach. Before describing our replies, I should point out our relative ages at the time: Coca was 75, nine years prior to his death in 1959. Rowe was 61, Rinkel 54, I was 44 and Interlandi was 31. This question came out of the blue, as none of us was given any advance indication of its nature. This question was posed to us in the descending order of our relative ages.

Dr. Rowe hedged, saying that he had not yet given up hope that this work would be accepted by the medical profession, citing evidence of increasing interest in the subject by selected representative physicians, but admitting that there still was much apparent opposition to this view.

Dr. Rinkel also equivocated, without very much pro or con discussion.

Then it was my turn. Although I had been rebuffed repeatedly by 1951, both individually and academically, apparently I had not had enough by then, so I also waffled, as did Joe Interlandi. But to show that there is a saturation point in all of this, by the time that I had reached Coca's age of 70, when he began to write for the public, so did I, starting 10 years ago.

Unfortunately, this meeting occurred shortly before I began photographing events such as this. Although Leonard Green arranged for a photograph and pictures were taken, I never received a copy.

Realizing that there was apparently no photographic record of this historic meeting in the 1960s or early 1970s, by then both Dr. Coca and Mr. Green had died, I wrote to Mrs. Coca and Mrs. Green about the possibility of a

1950

picture which I had not received. Neither of them nor the photographer who Leonard Green had been accustomed to using at that time knew anything about such a picture.

To the best of anyone's knowledge, this was the first and only time these three pioneers of food allergy ever met to discuss the subject of food allergy. It is also significant that although each of them was sensitive to given foods, each determined this by means of different techniques. Rowe had worked out his medical problems by means of elimination diets, Rinkel by the use of individual food ingestion tests and Coca by means of the pulse technique. Each had also been stimulated by working out their own problem, to develop the field of food allergy.

There is one other anecdote involving Dr. Coca which should be mentioned. Since Dr. Coca had expressed interest in seeing my motion picture entitled, An Acute Psychotic Episode Following the Intubation of an Allergenic Food — Beets, I arranged to show this in his home in 1956. Leonard Green met Tudy and me in New York City and drove us out to Oradell, New Jersey. Although I had objected to the application of Dust-Seal in homes of those susceptible to environmental chemicals, Dr. Coca had given no credence to this objection and assured me that his Dust-Sealed home (all floor coverings, upholstered furniture and curtains) would be well tolerated by Tudy. Having our doubts, we were prepared in advance for any medical emergency.

Tudy watched the 40-minute motion picture sitting on the carpet and leaning up against an upholstered chair. I noticed that she was nodding and dopey when lights were turned on at the end of the film. Shortly thereafter she lapsed into an unconscious stupor from which she could not be aroused, to the consternation of Dr. Coca, whereupon I simply carried her into a bedroom and administered sodium bicarbonate intravenously. Upon reviving, she drank the usual three or four glasses of water, following which she was again normal. Under the circumstances, neither Dr. Coca nor I discussed Tudy's episode to any extent.

Tudy and I were never able to accept invitations in the respective homes of either Dr. Rinkel or Dr. Zeller, as they both lived in completely gas equipped residences. This not only cut down on our contacts with both of them, but neither did either give any credence to the existence of the chemical susceptibility problem. Nor did Dr. Rinkel nor Dr. Zeller concur with my findings of the frequency of the allergenicity of specific sugars and alcoholic extracts or distillates or with their manifestations or with my emphasis on mental and behavioral problems as allergic reactions. Dr. Rinkel never picked up on the interpretation of masked food allergy as an addictive manifestation and was critical of my not emphasizing the term masked food allergy.

Herb Rinkel, Mike Zeller and I wrote most of our *Food Allergy* book at Blanchard Lodge in Boulder Canyon, Colorado in 1948. During a rest period with our feet on the table Herb commented: "Isn't it strange that although Albert Rowe had had the insight and knowledge to put food allergy together, that he went so far and stopped, as if he had confronted an invisible wall." We likened this to the territoriality of birds flying to the end of their self-imposed range and going no farther.

I was reminded of this discussion on several subsequent occasions after my interests had spread to include alcoholism and certain other addictive responses, "brain-fag," hyperactivity, depression and, especially, the chemical susceptibility problem, with which he never concurred. Indeed, he was so annoyed with me for substituting the term, food addiction, for masked food allergy that our relationship was never the same thereafter.

Herb had many visitors in his office. It became known amongst them that if they were inclined to leave a gift that he would be glad to receive either motion picture color film or his favorite brand of brandy. It was well-known among his contemporaries that Herb was sensitive to corn. I knew that this favorite brandy contained corn sugar, on the basis of my own adverse reaction to it as well as that of many of my patients. Although I told Herb of this, he acted as if I had not said anything.

Meanwhile I had also learned that both Federal regulations and California law permitted the addition of sugars to brandy manufactured in that state. Having had a prolonged correspondence with this manufacturer in which the presence of corn sugar in this product was neither admitted nor denied, I drove to this area and posed as a tourist. Also, having become more knowledgeable about alcoholic beverages by then, I knew what questions to ask. I merely inquired upon reaching the research laboratory what sugars had been added to this product. The answer was Numoline, one of the trade names for corn sugar. I told Herb this story on another occasion, and again he acted as if I had said nothing.

1950

Chapter IV — 1951-1953

Other Environmental Exposures

The Chemical Susceptibility Problem

Smoke, fumes and London fog were mentioned by Henry Hyde Salter as causes of asthma in 1882 (*The Pathology and Treatment of Asthma.* New York: William Wood & Co., First American Edition). Coca also commented briefly on fumes from motor exhausts and evaporating paints as inducing migraine in his book published in 1943 (see Appendix, page 315). E. A. Brown published an article entitled, "Persistent Cough and Bronchospasm Due to Exposure to Fumes from Range Oil," (Annals of Allergy, 1949) in which he also pointed out that incomplete combustion of kerosene emitted formaldehyde and other aldehydes. David Morris recently reviewed the subject of formaldehyde extending back to 1909 (Clinical Ecology 1:27, 1982). S. D. Lockey, Sr. described "Allergic Reactions to FD & C Dyes in Coloring and Identifying Agents in Medications," (Bull. Lancaster General Hospital, September, 1948). F. W. Wittich reported reactions to pesticides in an article entitled, "Respiratory Tract Allergic Effects from Chemical Pollution," (Arch. Industrial Hygiene 2:329, 1950).

Although the existence of the chemical susceptibility problem had been emerging for several years, as mentioned, full realization of its existence, scope and clinical significance, as far as I was concerned, did not occur until April 1951. The first patient, a physician's wife and cosmetics saleswomen, was seen in 1947 presenting the manifestations of rhinitis, asthma, headache, fatigue, irritability, depression and marked variations in weight and intermittent episodes of loss of consciousness. Many of the environmental exposures which she had suspected as being related to her multiple complaints had not been described previously, in view of my knowledge and experience at the time. Indeed, her suspects were mostly meaningless at the time that they were recorded as data, independently of any offhand interpretation. As she turned out to be an accurate observer and excellent record keeper, I added her to a small group of research patients who were followed closely in the absence of economic considerations. Fifty single spaced typewritten pages of clinical observations had accumulated by April 11, 1951 when I had the opportunity of reading this record with continuity in the patient's presence, having suspected that there might be some common denominator in this mass of strange information. Only then did the clinical significance of these observations become apparent.

During this four-year period, the medical significance of ambient air pollution had been accumulating in her case. For instance, each time that she drove by automobile from southern Michigan to Chicago, she became acutely ill in passing through the industrially polluted areas of northwestern Indiana and the south side of Metropolitan Chicago. She had learned to stay on the top floor of the tallest hotel in Chicago where she recovered more rapidly than elsewhere. But even so, one and one-half to two days were necessary before she improved sufficiently to be able to come to my office. A few times between 1948 and 1950 I saw her on the 26th floor of her hotel when it was still possible to look down on the ceiling of air pollution existing in Chicago.

It should be pointed out that the opportunity of studying the medical aspects of ambient air pollution in Chicago was unusual for several reasons: 1) its location on the southwest corner of Lake Michigan, 2) the fact that the major sources of industrial air pollution were south and southeast of the city, and 3) the major focus of automotive and diesel exhausts was in the center of the area where traffic and rail lines converged. These advantages of Chicago, in contrast to other major American cities, in studying the medical aspects of air pollution will be evident as I describe, briefly, unique meteorological conditions existing here.

Chicago is located on one of the major storm tracks which move from west to east or east-northeast across the continent at variable rates of several hundred miles per day. A storm is characterized by a low pressure center and counterclockwise winds. This means that with a low pressure area located in eastern Iowa, south and southeast relatively warm winds bring industrial pollution from northwestern Indiana and the industrial south side of Metropolitan Chicago into the center of the city and its northern suburbs. This also means that patients having the chemical susceptibility problem become progressively ill. These relationships are shown diagrammatically in Figure 11. Moreover, the single best index of air pollution here is visibility. In general, if the horizon on Lake Michigan is fuzzy or obscured, ambient air pollution tends to be relatively high.

But as soon as the center of this storm passes through Chicago, all of this changes. When the center of this low pressure area is located in southern Michigan, counterclockwise north and northeast winds sweep 400 miles of Lake Michigan bringing the Chicago area relatively cool, chemically less contaminated air. This also means that patients who had been reacting to ambient air pollutants soon became relatively well and productive. These relationships are shown diagrammatically in Figure 12. Coincident with these

Other Environmental Exposures

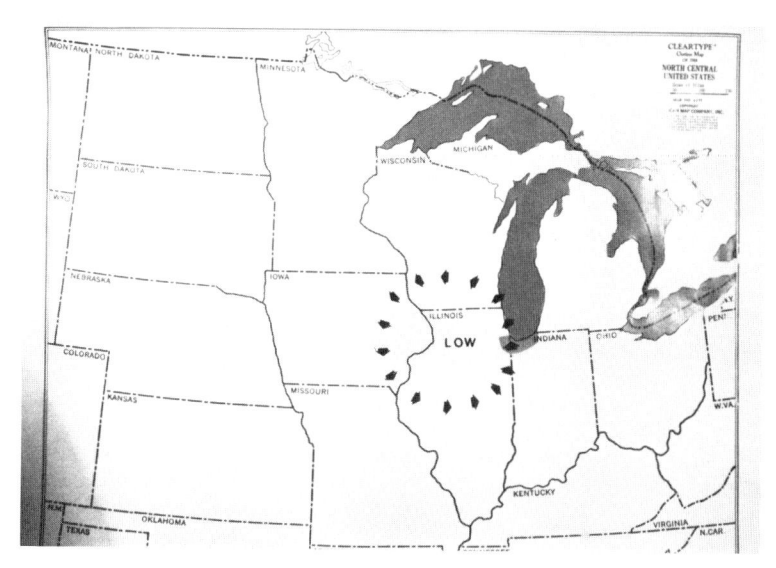

Figure 11. Low pressure center to the west, bringing industrial air pollution into Chicago area.

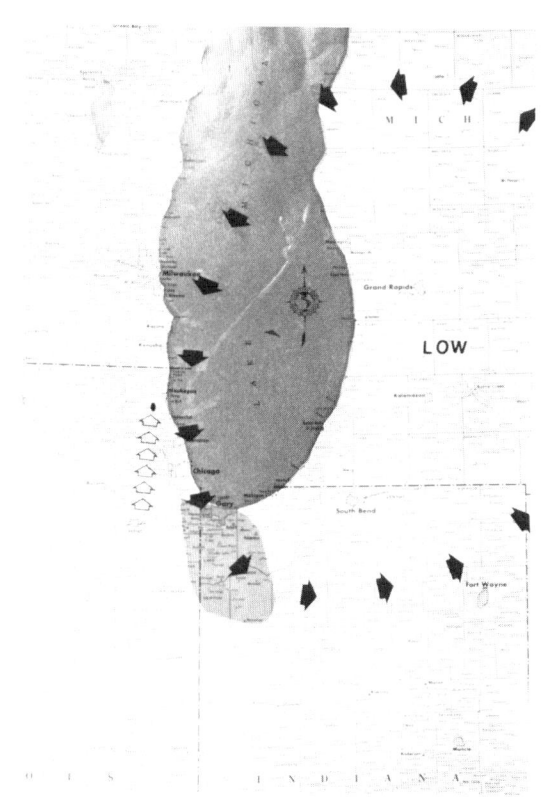

Figure 12. Low pressure center to the east, bringing cool, less contaminated air to Chicago area.

meteorological changes and clinical improvements in many patients, the horizon on Lake Michigan also became sharp and clear.

By the time I had finished rereading one-third of her history and progress notes I realized that there was, indeed, a common denominator extending through this report. It had to do with man-made combustion products and derivatives of gas, oil and coal, in contrast to environmental chemicals as identified by their structural chemical formulae. Once this realization dawned on me, I recorded the first chemical susceptibility history. Because of the detail involved, this led to the development of a questionnaire for taking this type of history. During the following few years, this questionnaire was revised approximately 25 times before reaching its present form. These observations were published in the form of preliminary abstracts in 1952[88] and 1954.[96] The major subdivisions appreciated initially are indicated by a series of five abstracts as follows: Allergic Type Reactions to Industrial Solvents and Liquid Fuels,[99] Mosquito Abatement Fogs and Mists,[100] Motor Exhausts,[101] Indoor Utility Gas and Oil Fumes,[102] and Chemical Additives of Foods and Drugs.[103]

It should be emphasized that various aspects of this chemical susceptibility problem did not develop simultaneously. Each time that I attempted to present the subject in its apparent totality, some new previously unrecognized aspect of the chemical susceptibility problem became apparent which required its integration into the overall presentation. What was considered as the final form at the time, was published in a series of four connected manuscripts in 1961 under the title of Human Ecology and Susceptibility to the Chemical Environment.[129,132,133,134] These articles were indexed and republished in book form in 1962.[137] The major subdivisions at this time and since have consisted of ambient or outdoor air pollution, domiciliary or indoor air pollution, food additives and contaminants, water additives and contaminants and synthetically derived drugs.

The first case report in this book is the physician's wife and cosmetics saleswoman. The second case report is my present wife, Tudy. My first wife, Janet Sibley, and I were married in 1941, separated in 1949 and divorced in 1952, following which she and our three sons moved to another part of the country when the boys were five, six and eight years of age. Consequently, I had little influence on their early lives. My second wife, Tudy, came to me as a patient in early 1953. She had been a hyperactive child with a ruddy complexion and subject to rhinitis, sick headaches, and car sickness with superimposed episodes of loss of consciousness. She became extremely uncoordinated, irritable and asthmatic when exposed to utility gas, even in small amounts, with delayed arthritis, depression and sometimes cardiac irregularity. Greater exposures were followed immediately by marked redness of her face, crossing of eyes and sudden unconsciousness with residual arthritis and depression persisting for two to three days. After her medical problems had been worked out, I hired her in the office because of her understanding of the chemical susceptibility problem and ability to talk with patients about it. We were married in late 1954.

These two case reports are cited only briefly here in order to introduce this subject, as they are reported in detail in the book, *Human Ecology and Susceptibility to the Chemical Environment*.[137] I might add that it has been difficult to introduce the subject of individual susceptibility to common environmental chemical exposures for several reasons. Most physicians have been trained to appreciate science in relation to medicine in terms of analytical approaches both to the patient and to the patient's surroundings. In contrast, this approach involves synthesis and demonstrable dynamic interrelationships between selected susceptible persons and their intake and surroundings. This concept is not easy to grasp the first time it is encountered.

For instance, the first time that I attempted to present this chemical susceptibility problem to the Annual Meeting of the American College of Allergists in 1954, it was received with considerable good-natured but critically focused smiling and snickering. Sensing this smoldering derision on the part of the audience, I reckoned that I might as well give them something more tangible to laugh at. I then mentioned that in my experience, the gas kitchen range was the most hazardous device in the American home and that I had already removed several hundred of them. As one might expect, this brought down the house.

Another reason mitigating against acceptance of the chemical susceptibility problem by allergists is the fact that it is not, apparently, mediated by immunologic mechanisms. In keeping with the widely accepted immunologic interpretation of allergy at this time, obvious nonimmunologic clinical observations were not acceptable. Indeed, this was the gist of the criticism that followed the above presentation. Dr. George Waldbott of Detroit was the first to criticize this view, saying that this was not allergy and should not have been accepted for the program. My reply went something like this. These patients are relatively common. They usually present other aspects of allergy such as sensitivity to pollens, dusts, danders, foods and other materials. Patients with this type of chemical problem will be coming to your offices with their complaints. Simply to state that a part of their problem is allergic and that the same or similar manifestations brought on by environmental chemical exposures are not allergic makes no sense. You should remember that I did not talk about sensitivity to environmental chemicals — I was careful to speak of individual susceptibility to chemicals — in the hope that I might avoid this type of controversy, as the concept of allergy is gradually extended to include exposures which ordinarily do not involve immunologic mechanisms. Within a year or so after this, Dr. Waldbott became interested in reactions to fluorides.

It should also be emphasized that this concept of adverse clinical responses occurring in specifically susceptible persons and involving multiple combustion products and derivatives of gas, oil and coal, did not occur simultaneously. For instance, the relative importance of indoor air pollution emanating from the combustion of gas fired and oil burning utilities within the home was not appreciated initially. Having recognized intolerance to such exposures earlier, the first patient had moved to an all-electric home in the country some time before becoming a patient and did not register these utility related exposures as major complaints. However, as other patients were studied within the first year, it soon became apparent that fumes from gas and fuel oil utilities, as well as pesticide exposures, were major problems within the home. Neither were formaldehyde exposures in the home recognized initially, as will be described later in this presentation.

It might also be stated that at this stage of the development of clinical allergy, the three most commonly encountered materials responsible for individual susceptibility and chronic illnesses occurring in my practice were biological inhaled particles (pollens, spores, etc.), foods, and environmental chemicals, in that order. The subsequent relative significance of the chemical susceptibility problem is indicated by the fact that this order has since been reversed. Moreover, of the three areas in question, environmental chemical exposures are also presently associated with higher degrees of individual susceptibility and relatively greater persistence of susceptibility as well as more advanced clinical syndromes. Also, once individual susceptibility to one or a few environmental chemical exposures has developed, it almost invariably tends to spread to involve other combustion products and derivatives of gas, oil and coal, to which one happens to be exposed in significant amounts — despite the common notion that these materials are ordinarily regarded as being GRAS (generally regarded as safe). Finally, of the three areas under consideration (biological inhalants, foods and chemicals), the chemical susceptibility problem has come to be far more often associated with impaired productivity and disability than the others.

Other Pros and Cons

My status as far as other physicians were concerned remained divided, most local allergists being opposed. This included a few of my most loyal friends and supporters. For instance, Dr. George A. Zindler of Battle Creek, Michigan had been a long standing friend since our medical training. He and his wife had become patients and he had developed this type of practice in southwestern Michigan, taking over many of my patients in that area. But despite my long standing and close association with him, he advised me to "pipe down," as I was getting "too hot." This advice startled me as he had followed through and confirmed many of my clinical observations. In looking into the possible reasons for this criticism, two factors seemed to emerge. It apparently originated with the wives of other physicians in that locality, who had been critical of Mrs. Zindler because of her husband's activities. She, in turn, prevailed upon her husband to become more subdued. And he prevailed upon me! The second reason was his immediate objection to acceptance of the chemical susceptibility problem, with the initial comment: "What are you trying to do? All these things that you say cannot be tolerated — gas heating and cooking, automotive exhausts, perfumes, traffic fumes, and all — one cannot live avoiding all of these things." However, he began to change when he noticed that Tudy and I could eat out and have a good time. But conviction came when he realized that he had this chemical susceptibility problem himself. Indeed, he found, to his dismay, that he was ill every time that he attempted to make rounds in one of his hospitals.

I did hold back on publishing for a time, but continued with investigative activities. For example, I postponed submitting the manuscript on food addiction for publication, despite it having been completed. Indeed, it never was published. In an overall sense, this was unfortunate as it contained detailed medical histories of the addictive aspects of food allergy with examples of its application to obesity, social drinking and compulsive drinking, although the food ingredients entering the production of alcoholic beverages and generalizations regarding obesity and alcoholism were published later.

I was elected president of the Chicago Allergy Society, an office held for one year. But my application for advancement to fellowship in the American College of Physicians was blackballed locally, despite being sponsored by a friend in the same office building and president of the Chicago Medical Society at the time, Dr. Willard O. Thompson.

I presented the subject of food allergy in relation to mental illness at the staff meeting of the Milwaukee Sanitarium in Wauwatosa, Wisconsin where I was well acquainted. Dr. Josef Kindwall, Chief of Staff, introduced me with the paraphrased admonition: "Be not the last to take on the new untried nor yet the first to lay the old aside." The discussion was particularly helpful. Dr. Kindwall suggested that preliminary fasting might be easier in preparing patients for subsequent food testing than avoiding single foods. After reading on this subject and agreeing with its safety, I later admitted six therapeutic failures to the St. Francis Hospital and fasted them simultaneously. This had the disadvantage that they knew each other. They also were aware of the fact that I had never fasted anyone prior to this time. Not surprisingly, they magnified their withdrawal complaints to a degree that I chickened out on the third day and fed all of them.

The following three times that this subject was presented to audiences where I was not known, the psychiatrists and neurologists of Columbus, Ohio, the general practitioners of Illinois and a national group of college health physicians, the response was the same — none. Although on each occasion I had refrained from criticisms of psychiatrists and other physicians and had stepped back from the podium to reply to questions, none was forthcoming.

There were several other events during this period of 1951-1953 which had a bearing on my personal and professional activities. In the absence of my family, I simply devoted full time to my practice and research activities but

I was handicapped by the unavailability of research funds, decreased income, increased expense of maintaining two households plus alimony and child support, and the absence of a hospital appointment. Fortunately, some of these deficiencies were soon to improve.

Lee Winfield Alberts, a lawyer and loyal patient, offered to establish a foundation for medical research and to serve as its secretary, gratis, for a period of five years. This generous offer was accepted and the Rockwell M. Kempton Medical Research Fund was chartered in the State of Illinois. This was named after my late uncle, a Saginaw pediatrician, from whom I had borrowed funds for my medical education. This was a milestone in my medical investigative activities as, during the 35 years of its existence, nearly a half million dollars have been contributed to it.

During 1953, I obtained a hospital staff appointment at the St. Francis Hospital in Evanston, Illinois and moved my home nearby. I was remarried in 1954 to Janet Mitchell Walker, hereafter to be referred to as Tudy.

Francis Hare Discovered

Although I had known of Francis Hare earlier, I really became acquainted with his life and medical accomplishments after reading his books in the early 1950s.

Francis Hare, whose pictures are shown in Figure 13, described many of the clinical features of food allergy a quarter century before the publications of Coca, Duke, Rowe and Rinkel who are usually credited with having developed this field. Indeed, Hare died in 1928, and Rowe's first book, *Food Allergy,* was published in 1931.

Figure 13. Francis Hare, M.D. as a young and older man.

Other Environmental Exposures

Francis Washington Everard Hare — hereafter simply to be referred to by his chosen appellation, Francis Hare — described many of the manifestations of food addiction, including obesity and alcoholism. These original observations were made in Brisbane, Queensland, Australia — a frontier outpost at this time — the turn of the century, and were reported in a two-volume 1000 page book, *The Food Factor in Disease* (see Appendix, page 330).

My wife, Tudy, and I are in the process of writing a biography of this remarkable physician. But since we will not be able to complete this immediately, a preliminary synopsis of this material will be presented here. This forthcoming book will probably be entitled: "The Life, Times and Medical Contributions of Francis Hare." Tudy will be preparing the personal and social parts. I will describe the medical aspects.

How we heard about this and the circumstances that prompted the beginning of this effort in behalf of Francis Hare will be described briefly. As previously mentioned, I became interested in alcoholism as probably the ultimate in food addiction in the late 1940s, as a result of attempting to study the ability of alcoholic extracts and distillates of specific foods to carry the allergenicity of those foods. Wishing to be certain that this general subject of alcohol and food allergy had not already been described, I made a search of the world's medical and popular literature on this subject. In the course of this effort, I came across the book, *On Alcoholism,* written by Francis Hare in 1912. More will be said about this later in describing Francis Hare's life's accomplishments chronologically.

In this book, *On Alcoholism,* were references to his earlier publication, *The Food Factor in Disease.* I was fascinated by scanning these books in medical libraries, reaching the preliminary opinion that this man's remarkable early accomplishments in the fields of nutrition, alcoholism, allergy and chronic illness had been completely overlooked and apparently forgotten by those writing documentaries on these subjects. Realizing that I needed possession of these books to become more familiar with them, but having no idea of their rarity, the next time I was in New York I dropped by the Argosy Book Store and inquired about them. Volumes I and II of *Food Factor in Disease,* were brought out promptly, for which I paid the nominal sum of $10 and left on a train trip between two or three talks in the Eastern United States. Being fascinated by the early remarkable insights in these books, I underlined in pencil certain points of special interest. All this occurred in 1952 when I was 46 years of age. But one look at this "trembling hand" gives the general impression that the underliner must have been about 90, and what he underlined may not have been very important.

The relative rarity of these books is emphasized by the fact that other available copies have not been found in any old book store in England, Australia, New Zealand, Canada or the United States. Indeed, I returned to the Argosy Book Store looking for other copies, only to be told that they had had this in only once and these copies were sold to a Chicago physician in

1952. Neither has repeated advertising turned up additional copies. But after advertising for the book, *On Alcoholism,* for two years, three copies arrived within a week.

We didn't think much more about this until traveling in Australia in 1970 took us to Brisbane, accidentally, because of Tudy's chemical susceptibility problem. This visit to Brisbane and the Barrier Reef had been substituted for a scheduled trip to New Zealand, which we had cancelled because of the mandatory pesticide spraying on all planes entering that country due to the fear of bringing hoof-and-mouth disease into their agricultural economy. Having a few days to spare in Brisbane we decided to look up Francis Hare, briefly, before going on to the Reef. Needless to say, we were so intrigued by what we found, we never got to the Reef. We decided then and there to obtain the necessary documentation to write up Francis Hare. During our remaining period in Australia, we were assisted graciously by the medical librarians in several Australian libraries. Putting information together from many sources, including on site visits to Australia, Ireland, Scotland, England and Canada, the life of this remarkable physician unfolded something like this.

Francis Hare had been born in Dublin, the son of a lawyer who died when he was a small boy. His primary education had been received at Fettes College near Edinburgh, Scotland. Tracing his subsequent activities has been greatly facilitated by the fact that this institution had maintained a record of all places where Francis Hare had lived and worked. Following this chronologic account, we have since visited all of these places, taking pictures of many of them. As this chronologic course of events will be described in detail later, only his medical contributions will be outlined here as we knew of them in late 1970, following our Australian trip in February of the same year. Other than for a presentation of this subject on the program of the Society for Clinical Ecology on one occasion and later published,[387] this material has not been presented previously.

After Francis Hare completed his medical education in Newcastle and training stints at St. Thomas and Guy's Hospitals in London, he embarked on a one-way passage to Brisbane, Australia as the ship's surgeon in 1892, arriving there in the midst of a raging epidemic of typhoid fever. In his position as house officer of the Brisbane General Hospital in charge of the fever ward, he was astonished by the morbidity and mortality of this disease.

Remembering what he had read earlier about the treatment of typhoid fever in Germany which described how extremely febrile patients had improved by being submerged intermittently in a tank of cool water, he set up and applied this program. Despite outcries about the inhumane nature of the treatment, he carried on and significantly reduced the staggering death rate of this illness. After this commotion had subsided Francis Hare, single-handedly, was credited for this performance. He published his findings in book form in 1898 (see Appendix, page 330).

Later, as a member of the staff of the Diamantina Hospital for Chronic Illness in Brisbane, he had the opportunity to make basic observations of the role of foods in disease which he reported in his book, *The Food Factor in Disease,* (see Appendix, page 330). He reports how his interest in foods had started in 1889 when he prescribed a diet largely excluding fats, carbohydrates and saccharine alcoholic drinks and consisting largely of protein foods, for the treatment of obesity. In the meantime, he lost sight of this patient for two years. He then learned that this person had adhered to the prescribed diet, not on account of his obesity, which had come under control, but because he had quite ceased to suffer from periodic headaches. Thinking of headache as a disorder of the brain, he regarded the cessation of headaches in this case as a coincidence. It was not until an almost identical result occurred in a second case that he began to suspect a direct causal relationship between the food supply, obesity and headaches. He concluded that migraine was a food disease, related to the accumulation of carbonaceous material in the blood and eventuating in hyperpyraemia (fuel) in excess of the capacity of the organism for physiological disposal.

Although this theory of hyperpyraemia was based initially on a few therapeutic observations and was purely provisional, he soon found that it correlated with and explained a large number of isolated and heretofore inexplicable paroxysmal affections. Hare also confirmed Hyde Salter's observation that such paroxysmal afflictions as headaches, dyspepsia and asthma, frequently alternated. Closely related to dyspepsia were headaches, catarrhal conditions of the mucous membrane, epilepsy, angina, and many others attributable to hyperpyraemia. He went on to describe how a modified degree of abstention from carbohydrates completely removed the recurring tendency for symptoms of dysentery. He commented that a small quantity of such a stimulus as alcohol may incite diarrhea and vomiting. He also stated that temporary abstention from food (fasting) may be the shortest route to relief of all these afflictions.

In respect to obesity, Hare stated that the majority of such patients have hearty appetites for the more highly carbonaceous food stuffs, the sugars and starches. He refers to sugar as perhaps the worst article of the diet and likened sugar gluttony to alcoholism. Although emphasizing beneficial effects from the dietary reduction of carbohydrates, he pointed out that reduction of proteins sometimes relieves migraine, but leaves others unaffected and may exacerbate some cases.

In respect to alcoholic beverages, Hare pointed out in his book, *The Food Factor in Disease,* that substances undergoing rapid absorption, such as wine or alcohol of any kind may produce asthma within a minute or two. He also said that the frequency of morning headaches was proverbial and he agreed that between two and three in the morning was the most common time for recurrence of coughing and asthma.

Hare was also impressed by the observation of how a small daily addition to the carbonaceous intake was sufficient to reinduce attacks. Two or three lumps of cane sugar, an ounce of butter, a pint of extra milk, a small plate of porridge, an extra ounce of bread or toast, etc., may suffice to bring on typical migraine paroxysms.

One of his patients reported typically of food reactions as follows: "At first my attacks would come on suddenly and unexpectedly, several hours after a meal, or I would awake with one in the morning. The attack would last three days, and on the third day I would be very weak, faint from want of food yet unable to eat, and the slightest exertion would induce weariness. Rather quickly, within a few hours, the anorexia would disappear and, after one good meal, I would feel quite well again."

Another patient commented: "As to the influence of diet on the attacks, there is no doubt. By missing a meal or by reducing my diet to a very small amount, I would often succeed in avoiding an attack, or in making it much less severe. I also found that excess in carbohydrates was especially harmful and, amongst these, sugar in any form appeared to produce the most rapid bad effects. I am naturally fond of porridge, potatoes, puddings, jams, cakes, etc. I repeatedly transgressed and was convinced of their harmfulness against my will by innumerable experiences...."

Another said: "Bush holidays practically always gave me relief. Sometimes the relief would be immediate; at other times I would have to be very careful of my diet the first day. After that, I had a good appetite, ate freely, and put on weight. I often developed a craving for the articles of food which most disagreed with me at home — porridge, puddings, potatoes, jam and even treacle — and ate them freely with impunity. I was much puzzled to account for the effect of these excursions."

In discussing neurasthenia, Hare said that most of the cases of neurasthenia which have come under his notice have been amongst women and in the great majority of these there has been a marked deficiency of protein in the diet. Hare commented favorably on the Weir-Mitchell treatment of neurasthenia, which consisted of a short fast, after which nothing but milk is allowed for a week or so, and this in moderate quantities.

In respect to fasting and insanity, Hare cites the following case reported by George Keith: "A man had been in an asylum for the greater part of his life. From time to time he refused food, and it was put into his stomach by a tube. At last his health quite failed, so much so that it was considered useless to force food any longer upon him, and he was left to die in peace. He did not die and, more than that, he recovered his sanity and was able to leave the asylum." Hare went on to say that he had heard of a lunatic who escaped from his asylum and wandered about for several days without food, recovering his sanity before he was found.

Hare commented on postprandial depression by citing the case of an elderly gentleman complaining of intense mental depression, recurring daily between the hours of 10:00 A.M. and noon. His breakfast consisted of porridge and milk with sugar, followed by eggs and bacon, or a chop or steak, also bread and butter and marmalade. The more depressed he became, the bigger the breakfast he ate, in order to "keep his strength up." As a result, his depression intensified further. He was ordered to have a lighter breakfast, avoiding porridge, sugar and bread, eating a small chop or steak or one egg with a little bacon. The depression ceased and did not return.

Hare agreed with George Savage, who first described the alternation of psychoses with what were later referred to as allergies (rhinitis, asthma, eczema and headache), that in many cases, "The depression is most marked in the very early morning, and persists until midday, when it slowly becomes less, and by six or seven o'clock has all but passed off, and during the evening no one would suppose that the patient had been intensely melancholic in the early morning. These cycles of depression recur and recur with pretty regular precision." Hare also reported similar cycles in the sane, citing a friend who awakens at three or four a.m. worrying about insuperable monetary troubles which improve greatly after breakfast.

In respect to arthritis, Hare also said that it is quite common to see cases rapidly improve concurrently with decreasing weight when placed on mainly a protein diet. In addition to arthritis, he also pointed out that the patient often retires to bed feeling in his usual health, but on awakening in the morning he discovers some recurrence of his malady; it may be muscular pains or stiffness, the beginning of a headache, or more or less severe pain in some joint or adjacent area, such as a stiff neck, lumbago, or a burning phalangeal joint. These troubles or some of them may have come on in the night but have not been sufficient to distort sleep. Cramps in the calves of the legs are especially prone to vex gouty patients at night, and sometimes for several nights precede a severe attack. Epilepsy, neuralgia, spasmodic asthma, gastralgia, angina pectoris, laryngeal stridor and hemicrania are all prone to disturb sufferers during the early hours of sleep, or immediately on awakening.

Francis Hare's Contributions to Alcoholism

Francis Hare married and returned to England with his wife and his manuscript, *The Food Factor in Disease,* in 1903. He became Medical Director of the Norwood Sanitarium in Beckenham, England, located directly south of London. He reported his experience there in another book, *On Alcoholism,* published in 1912 (see Appendix, page 331). Having reviewed most books on the subjects of alcohol and alcoholism published in English, as well as the world's literature on this subject which has been collected, abstracted and published by the staff of the Quarterly Journal of Studies of Alcohol, I believe this is one of the most outstanding books ever published on this subject.

Hare's presentation of acquired tolerance to alcoholic beverages was especially good. He stated that tolerance arose through a long course of steady uninterrupted drinking. The amounts required to produce mild or marked euphoria gradually increased with time until eventually the alcoholic is endangered by severe nervous complications. Hare apparently was one of the first to point out that the establishment of tolerance also carried as its corollary, intolerance of sudden abstinence. Because of the danger of delirium tremens associated with the sudden withdrawal of alcoholic beverages in long-standing steady drinkers, Hare disagreed with most other medical authorities of the time who recommended abrupt withdrawal. Although he employed sudden complete abstinence in dipsomania in which appreciable tolerance had not developed, he agreed with most of his patients that it was better to taper off in those instances in which tolerance had been acquired. The general "uproar" that this recommendation precipitated in the morality dominated views of the time is indicated by correspondence on this topic in the medical press.

Hare observed that there were certain periods in the day when the craving for alcoholic beverages was most marked. Early in the morning, an hour or so before lunch and, especially, before dinner were the times that alcoholics were accustomed to drink most freely. These were also the times after withdrawal of spirituous beverages that abstaining alcoholics were most apt to relapse. It was for this reason that afternoon tea, "together with some solid food," was insisted upon as a sanitarium routine.

Hare also commented that sugar and sweet things generally held a special inverse relation to alcoholism, pointing out the well-known observation that alcoholism usually killed the desire for sweets — the taste for which nearly always returned within two or three weeks after the final withdrawal of alcoholic beverages. He apparently regarded this recurrence as an evidence of "improvement," stating that obvious enjoyment of the sweet course was accepted in the sanitarium as a "satisfactory index of recovery." He concluded that at least in some cases, the ingestion of sweets was capable of controlling for a time the craving for alcohol.

Hare stated that the majority of chronic alcoholics suffered eventually from anorexia, gastric catarrh or some other form of dyspepsia and that not infrequently the development of these symptoms had led them to seek treatment. Persistent gastric catarrh related to what Hare referred to as chronic alcoholism consisting only of loss of appetite and discomfort after eating. However, nausea and vomiting might arise either from heavier drinking than usual or, especially, from the sudden or too rapid withdrawal of alcohol in those who had *developed a tolerance to it*. Hare distinguished this chronic course of events from a more acute gastric catarrh characterized by nausea, retching and raising of huge quantities of mucous, which often develops in intermittent drinkers toward the end of a paroxysm and which often terminated it. He recommended treating this total inability to retain either solids or liquids developing in intermittent drinkers in the *absence of acquired tolerance* by sudden and complete abstinence, rather than making any attempt to taper off the intake of alcoholic beverages.

Although chronic alcoholics may have suffered temporarily in the course of a gradual withdrawal from alcohol, Hare was astonished in how quickly they recovered. At the same time, the recent abstainer was especially prone to indulge to excess in those articles of the diet, the taste for which alcoholism was well-known to have destroyed, namely, sweets, puddings and rich carbohydrates generally. Suffering no distress after meals, Hare describes how they continued to indulge in sweets.

Hare went on to say that a rapid increase in weight might be all that occurred. But, especially in those who had not gained weight, there was a tendency, often at regular intervals, for the sudden recurrence of anorexia, more marked "bilious" attacks characterized additionally by nausea, vomiting, diarrhea and/or sick headache. He apparently found, empirically, that the best treatment for this complication consisted of reducing the intake of sugars and starches combined with regular muscular exercise between meals. The recurrent morning headache, with or without anorexia, was best handled by eating a small evening meal from which sugars and most starches had been eliminated and taking no other nourishment after this 7:00 P.M. feeding.

Hare also stressed that such delayed effects of abstinence frequently arose *for the first time* during convalescence from alcoholism. But, *in addition,* there was apt to be a recurrence, under identical circumstances, of asthma, angina pectoris, acne, eczema and other skin manifestations to which patients had been liable in the past, but which had remained in abeyance during the period of alcoholism.

This writer also emphasized the role of infections, especially influenza, which were followed by fatigue and depression which, in turn, often led to alcoholism originally or to recurrences in abstaining alcoholics. He also described alternations between alcoholism, depression and other forms of insanity, as well as with the various manifestations later referred to as allergies, as previously noted, confirming Savage's earlier observations in this respect.

In my experience, influenza is often followed by the onset of various allergic syndromes and alternation between the manifestations of allergy, alcoholism and mental illnesses is a common occurrence. Indeed, the tandem sequences of these relationships is so clear in selected instances that at times one is interpreted as the apparent cause of the other — a matter discussed by both Hare and Savage.

Hare apparently failed to recognize clinical differences between different alcoholic beverages and specific foods from which they were derived in either the development or treatment of various chronic illnesses and alcoholism. Nevertheless, he did relate, and was apparently the first to do so, that sugars and starches in general bore a unique relationship to such chronic syndromes and alcoholism. He also pointed out that many of the clinical features of alcoholism might be reproduced in abstaining alcoholics by the regular intake of sugars and starches.

In short, Hare recognized, but did not state in so many words, that alcoholism was simply an advanced manifestation of a food related chronic illness which was capable of being perpetuated in a lesser key after abstinence from alcoholic beverages by the regular inclusion of sugars and starches in the diet. By reference to sugar in Australia at the turn of the century, he undoubtedly referred only to cane sugar, certainly not to corn sugar and probably not to beet sugar.

In this respect it is significant that, in my experience, individual susceptibility to such sugars — corn or maize (dextrose and glucose), beet or cane (sucrose) — and such starches — corn or maize, wheat, barley, malt, rye, oats, rice and potato — top the list of foods to which individual susceptibility exists. These very foodstuffs are most capable of inciting and perpetuating specifically adapted (addicted) advanced clinical syndromes of which alcoholism is a striking example.

In this sense, alcohol may be considered as a jet-propelled vehicle for the transmission of specific foods across absorbing membranes. If this interpretation is correct, the clinical phenomenon of alcoholism may be considered as the acme of the food addiction process. Therefore, it becomes important for experts in alcoholism to become familar with the dynamic nature of food addiction.

Summary of the Medical Contributions of Francis Hare

The contributions of Duke, Rowe and Rinkel to an understanding of the concept of food allergy/food addiction, as will be described, were so outstanding and essentially revolutionary that one is left with the impression that these brilliant clinical observers were entirely responsible for it. Indeed, this interpretation is apparently true in the sense that these pioneers were apparently unaware of the earlier observations of Francis Hare.

Although Duke, Rowe and Rinkel — all working independently — recognized the specificity of food allergy, as far as given items of the diet were concerned, they apparently failed to associate this phenomenon with obesity and alcoholism which Francis Hare had done much earlier. But despite Hare having connected responses to classes of foods such as cereal grains and sugars with obesity, alcoholism and other syndromes, he only mentioned briefly the roles played by specific foods.

There are several reasons for describing the medical contributions of Francis Hare in detail:

1) To my knowledge, his contributions to what was later called food allergy/food addiction have been totally neglected by the modern pioneers of this subject.

2) His holistic view of nutrition has also been lost in competition with the current vogue for analytic dietetics which had started in Germany at about the same time.

3) Hare is to be credited in having related the intake of foods with the stimulatory manifestations of ecologic responses to dietary factors manifesting in obesity and alcoholism and epileptiform seizures. Although he associated such lesser stimulatory responses as auras, insomnia, night-terrors and a wide range of muscle spasms with seizures related to food intake, he did not describe the associated hyperactivity so often characterizing this group of manifestations.

4) His observations of virtually the full range of the localized and systemic manifestations of allergic/ecologic disturbances at this time is simply an amazing accomplishment.

Although the phenomenon of sensitization to specific foods did not seem to occur to him, sensitization in the sense of an individual's increased proneness for carbohydrate foods as a group occurring with the passage of time is described both generally and is illustrated in several case reports. For instance, he commented that once certain reactive patterns had been laid down, such as the manifestation of seizures, with the elapse of time symptom patterns developed increasingly readily. He mentioned, moreover, that the same observations were applicable to various other recurrent paroxysmal neuroses.

It may be said in summary that Hare's observations of the clinical effects of food groups, especially cereal grains, sugar, alcohol and potato, as well as mixtures containing such carbohydrates, represent an important interim phase in the development of the more specific present concepts of food allergy/food addiction.

Arthur F. Coca, M.D. — 1875-1959

The medical contributions of Doctors Coca, Rowe and Rinkel have recently been described in a separate report.[393] Although many of their more important references are included in that report, their complete bibliographies, as obtained from the Quarterly Cumulative Index Medicus, are listed in the Appendix of this book, Dr. Coca's on page 311. Dr. Coca's picture is shown in Figure 14.

Figure 14. Arthur F. Coca, M.D.

1951-1953

As stated earlier, my interest in environmental medicine was stimulated from having heard Dr. Arthur Fernandez Coca speak in defense of nonimmunologic factors in allergy in 1933 while I was still in medical school. The next time that I saw Dr. Coca was in 1938 while he was the invited speaker for one of the meetings of the Sneeze, Wheeze and Itch Club, later renamed the Boston Allergy Society. Although I don't remember the topic of his presentation, the one thing that I recall about him was that he arrived carrying a box under his arm. I happened to be on hand to help take care of his hat and coat. When I inquired if he would like to leave his box also, he said, "No." At the dinner before the meeting I noticed that he refused the plate when it was served asking, instead, for an empty plate. At that point, he simply removed his food from his shoebox and began eating it without comment.

Planning on being in New York City for a few days in 1943, I inquired if I might visit Dr. Coca. Upon learning at that time that I was smoking a cigar a day, he agreed — if I would cease smoking for 10 days and have all clothes brought with me recently cleaned. Planning to meet me at the New Jersey side of the George Washington bridge, I was reading the Sunday New York Times as he drove up, looking askance at my newspaper as I brought it into his car. Upon arriving at his home he asked me to leave the newspaper in the car. Shortly after arriving, Mrs. Coca and I went out to dinner. She said that Arthur never ate out, regarding eating as a private venture, rarely if ever to be shared with others. In the course of my long subsequent conversations with Mrs. Coca I learned of Dr. Coca's background as follows:

Dr. Coca had been on the faculty of Cornell University Medical Center since 1910 (age 35), founding the Journal of Immunology in 1915 and serving as its editor for many years. During this long period, he had remained the leading theoretician in the field of allergy, carrying on a prolonged debate with Dr. Robert Doerr of Switzerland in regard to the definition of the term, allergy. I will not go into the details of this difference of opinion, except to say that this has recently been described in detail by Dr. Alsoph Corwin who had direct access to the German language correspondence involved. (Corwin AH. The Allergy Fallacy: Clinical Consequences. Clinical Ecology 3:177, 1986). Since Dr. Coca's thesis had been published in Tice's Practice of Medicine — a loose-leaf publication — in 1920, it had been virtually unavailable for the past half-century because earlier sections were destroyed as they were replaced by new pages.

Under these circumstances, there is very little documentation in the readily available medical literature about this controversy over the definition of allergy. About the only account that I have found has been a summary which appeared on page five of *Allergy,* by E. Urbach and P. M. Gottlieb in 1943 (New York: Grune & Stratton) as follows: "Under the leadership of Doerr, immunologists and allergists abroad united (1925) in recognizing as truly allergic only such reactions as are based on an antigen-antibody reaction."

Other Environmental Exposures

But Coca's definition of allergy, as a condition of hypersensitiveness* in which an antigen-antibody response has not been shown to be the underlying cause of the symptoms that characterize it, was the opposite of Doerr's and the view finally accepted by academicians in this country.

Dr. Coca married his laboratory assistant, Ella Grove, in 1930. This second marriage of Dr. Coca came about in an interesting "scientific" fashion. It seems that Dr. Coca had been running a series of animal experiments requiring operations on the animals when he left for several days on a medical trip, having instructed a surgeon to operate his animals, assisted by his diener, Ella Grove. Each operated animal dying, his laboratory assistant, thinking that she could do no worse, operated the remainder of the animals, saving them all! When Dr. Coca heard of this upon returning, he was impressed and assisted his assistant in the next scheduled operation. Dr. Coca being even more greatly impressed, proposed over the operating table, then and there! (I can't recall whether this account was told me by Dr. Coca, Mrs. Coca or both of them.)

Two years after his marriage to Ella Grove, he retired from Cornell and became medical director of Lederle Laboratories at the age of 57 in 1932. In 1935, at the age of 60, Dr. Coca started the Allergy Discussion Group consisting of physicians and scientists interested in immunology, but by 1940 his major interests had changed from immunology to migraine, hypertension and food allergy — to the utter consternation and dismay of many of his scientist friends, including several of those in his discussion group who were nonplussed by this change from a highly esteemed scientist to a clinician fascinated by clinical observations in a not too savory a branch of medicine. Indeed, four of them have recorded their impressions of this metamorphosis. (Chase MW. Irreverent Recollections from Cooke and Coca 1928-78. J Allerg Clin Immunol 64:306, 1979; Tuft L. The Allergy Round Table

---

*Unfortunately, Coca's long article on Hypersensitiveness (91 pages) — an important landmark in the history of allergy — has not been republished as yet. Those interested in the controversial history of allergy should read this presentation — in many ways the antithesis of what has come to be known as the presently accepted meaning of this term.

Briefly, Coca considered the phenomenon of hypersensitiveness to consist of two categories: the first to include only those responses that have been shown to be due to the interaction of antigen and antibody; the second to embrace all other forms of hypersensitiveness — that is, those in which the cooperation of true antibodies has not been demonstrated. For the latter category is reserved the term allergy. Coca's proposed classification consisted simply of: Anaphylaxis, on the one hand, and allergy on the other.

Coca emphasized that allergy was elicited by a wide variety of substances having many more features in common than merely the negative one of an unproved antigen-antibody mechanism. Manifestations are both general and local, the latter being more frequently observed and more widely studied than the former. The responsible allergens were also of widely varied nature, some possessing antigenic properties, many lacking these completely; some being directly poisonous when administered in sufficient amount, such as some drugs and chemical substances, others being entirely harmless to most individuals even when administered in large quantities, for example, the horse serum of diphtheria antitoxin, food substances or the various pollens.

Discussion Group. New England and Regional Allergy Proceedings 6:279, 1985; Cohen SG. Firsts in Allergy, IV. The Contributions of Arthur F. Coca, M.D. New England and Regional Allergy Proceedings 6:285, 1985; Criep LH. Arthur F. Coca, Distinguished Mentor, Colleague and Friend. New England and Regional Allergy Proceedings 6:301, 1985)

    Dr. Coca's new found interest in foods seems to have stemmed in considerable part from Mrs. Coca's observations. Noticing during his Lederle tenure that Mrs. Coca was being preoccupied with taking her own pulse and wondering if she was degenerating into a profound hypochondriac (her behavior not being explained otherwise) he finally asked her for an explanation. She then told him that she had observed that following several different foods her pulse rate often became greatly accelerated, sometimes as high as 160 beats per minute. She told him that she had purposely not said anything to him about this as she wished to obtain more information about it before stating what she believed to be an important finding. Dr. Coca then began taking his own pulse in the prospect that he might find some answer for his hypertension and migraine headaches. Indeed, he found that he had only four foods which did not accelerate his pulse rate.

    Dr. Coca's interest in allergic reactions to foods and other environmental exposures not associated with immunologically measurable parameters led to the publication of his book in 1943, *Familial Nonreaginic Food Allergy*. (See Appendix, page 315) This book was bitterly criticized by his contemporaries. His resentment of this criticism is indicated by the foreword to the second edition which was published in 1945. A third and enlarged edition was published in 1953. In these books, Dr. Coca made major original contributions to both localized and systemic manifestations of allergy, especially the latter. Although headache, fatigue, rheumatism and arthritis had been described as systemic syndromes of allergy previously, Coca confirmed these findings and reported additionally on food related manifestations of hypertension, psychoses and seizures. In addition to his interest in foods, Dr. Coca was the first to note that such environmental chemicals as fumes from natural gas, gasoline, automotive exhausts, perfumes and other hydrocarbon exposures were also responsible for precipitating acute reactions in selected highly susceptible allergy patients. Despite criticisms directed toward him, it should be noted that all of these early clinical observations have since been confirmed and extended.

    In response to my request for more information, I received the following letter from Dr. Coca:

Arthur F. Coca, M.D.
425 Grant Avenue
Oradell, New Jersey
March 2, 1953

Dear Dr. Randolph:
    Your concern about priorities in the observation of "fume-allergens"

is comprehensible but, I think, exaggerated. You can forget *me,* at any rate, in that connection. Long before I was even aware of food-allergy I read a report of migraine and epileptic seizures induced by the inhalation of fumes from some product of petroleum or coal-tar. It seems to me that the effect was not recognized as allergic. I regret (not very deeply) that I cannot send you the reference.

So, I am not at all interested in any priority even if I could wangle one by some hook or crook, but I can back you, by personal experience, in your useful thesis.

I have occasional four to 68 hour attacks of paroxysmal tachycardia (with chest oppression, constant *extra systoles* and initial polyuria — one liter (plus) in the first hour). These *always* follow brief exposure — once only 20 seconds — to *fumes* of floor-wax, automobile paint, fresh newsprint and earth odor. I had fifteen attacks in 1952.

All good wishes from us both.

Sincerely,
Arthur F. Coca, M.D.

In summary, both Dr. Coca's interpretation of the mechanisms of allergy and his clinical findings constitute a continued medical interest. Indeed, Dr. Coca is to be commended in having the courage of his convictions despite intense criticism from his contemporaries and former admirers, some of which carried the implication of his mental infirmity. But having known Dr. Coca since 1942 and having carried on an active correspondence with him from that date to the end of his life, I can attest to his mental acuity. I must admit, however, that some of the profuse criticism directed toward him and his views rubbed off on me at the time. Whereas Dr. Coca usually wrote his longhand letters in green ink, shortly after our book, *Food Allergy,* written with Rinkel and Zeller[75] was published in 1951, I received a red ink letter from him. He strongly criticized our failure to mention his term, idioblapsis — a word coined by him and first used in the second edition of his book published in 1945. I recall discussing the use of this term. We could not bring ourselves to use it, in part because of the intensity of the criticism of other allergists of it and in part because it did not seem necessary to us to have an additional term for the common form of food allergy.

Dr. Coca retired from Lederle Laboratories in 1948 at the age of 73. He and Mrs. Coca then entered the private practice of allergy in his home in Oradell, New Jersey. The two of them were named jointly in this endeavor, inasmuch as from several visits to this office it was Mrs. Coca who was the clinician in charge of patients. She not only recorded the histories but seemed to be answering most of their inquiries, while Dr. Coca made many of the overall decisions.

Albert H. Rowe, M.D. — 1889-1970

Albert Holmes Rowe of Oakland, California functioned as lecturer in allergy at the University of California Medical School (San Francisco) for a period of 30 years and also served as chief of the Medical Center Allergy Society.

Albert Rowe was cofounder with Grant L. Selfridge and George Piness of the Western Society for the Study of Hay Fever, Asthma and Allergic Diseases in 1923. This original national allergy society later merged with the Society for the Study of Asthma and Allied Conditions to form what later became known as the American Academy of Allergy and Immunology.

In 1930, Dr. Rowe, in association with his wife, Mildred, translated Alimentary Anaphylaxis from the French. Dr. Rowe's picture as a younger man is shown in Figure 15.

Figure 15.
Albert H. Rowe, M.D.

Dr. Rowe's first major book, *Food Allergy: Its Manifestations, Diagnosis and Treatment*, was published in 1931 (Philadelphia: Lea & Febiger). The opening paragraph of the preface of this remarkable book is as appropriate today as when it was written — a prophesy still far from being fulfilled:

> Food allergy as a common cause of human symptomatology is gaining increasing recognition, as emphasized in the literature of the last decade and in my experience as an internist as well as an allergist. Such allergy not only enters into the etiology of many conditions necessarily treated by the general practitioner, but also into the causes of many symptoms demanding relief by all specialists. Food allergy may produce localized symptoms in any tissue or parts of the body in which it differs from inhalant types of allergy whose main manifestations occur in tissues with which those allergens have more or less definite contact. Thus food sensitization is one of the most important etiological agents known.

The reader is referred to the Appendix, page 337, for the titles of six books. The largest, published in 1937, entitled *Clinical Allergy Due to Foods, Inhalants, Contaminants, Fungi, Bacteria and other Causes — Manifestations, Diagnosis and Treatment*, is one of the most impressive in the field in respect to bibliographic references. It remains a book which is needed if one is to write on this subject and be reasonably certain not to omit many early references.

Albert Rowe is best known for his elimination diets, probably the most widely used technique for diagnosing food allergy and one of the most effective, provided adequate instructions are given and adequate compliance is obtained. The fact that these two provisions are not always followed probably accounts for failures in the use of elimination diets. Other reasons for failure are inadequate labeling of the constituents of processed foods. For instance, at one time bacon was permitted on an elimination program which he recommended, without adequate notification that commercially available bacon usually contains corn sugar.

Perhaps one of Albert Rowe's most outstanding characteristics was his ability to absorb criticism yet to remain steadfast in his endeavors. He told me one time that in this field one must learn how to take criticism. "Criticism," he said "rolls off me like water off a duck." Indeed, for about a half-century he received the bulk of the criticism that the medical establishment vented on the concept and techniques related to food allergy.

Despite earlier observations of Francis Hare (with which Rowe was apparently not familiar), Oscar M. Schloss, who reported the first case of allergy to a commonly eaten food, and William W. Duke, whose remarkable book published in 1925 really opened up this field of allergy, Albert H. Rowe deserves the accolade as the father of food allergy in this country, as well as the world. Indeed, he was recognized for this accomplishment at the First

Figure 16. Albert Rowe wearing his medal.

1951-1953

International Congress of Food and Digestive Allergy in Vichy, France in 1963 when he received an award for this contribution. Figure 16 shows Albert talking with me at that time when he was wearing his medal proudly and with good reason — he justly deserved it.

Herbert J. Rinkel, M.D. — 1896-1963

Herbert J. Rinkel enlisted in the United States Infantry at the age of 15 years, obviously falsifying his age. Despite his age, he soon became regimental photographer — a position that he continued to hold throughout his tenure of service. Apparently, his technological genius began to show at an early age.

Dr. Rinkel was married and had a small child when he entered Northwestern University Medical School after the war, graduating in 1925. Having little money, Herb and his family subsisted mainly on eggs as their principal source of protein. His father, a Kansas farmer, sent them a gross of eggs, 144 per week. Under these circumstances, it is not surprising that Herb developed a profuse rhinitis and rhinorrhea and was later found to be allergic to egg. Before this diagnosis, he had consulted several physicians about his rhinorrhea with no help. Having read Duke and Rowe about allergies, he wondered if he might be egg sensitive. Consequently, he placed six eggs in a blender and drank them without any evidence of reaction. Only several years later, after he had avoided eggs completely along with several other foods, he had a piece of angel food cake at a birthday party. Within minutes he lapsed into a state of profound physical collapse. Other physicians present were at a complete loss to explain this. Pulse, blood pressure, respiratory rate, neurological and other findings were within normal limits; unconsciousness was his only manifestation. Rinkel regained consciousness within a few minutes. The other physicians, as well as Rinkel, were astounded by this sequence of events.

In thinking about this, Rinkel wondered if this experience might indicate something of importance in the basic nature of food allergy. Perhaps if one had been eating a given food every day, or frequently and regularly, then omitted it for a period of several days, re-exposure might induce an acute, violent type of reaction. To put this concept to test, he began eating eggs again as formerly. He then omitted eggs again for five days, repeated the egg ingestion and confirmed this hyperacute test response.

At this particular time he was working in a well-known allergy clinic and found himself under very peculiar circumstances. It seems that a short time earlier the director of the clinic had sent Rinkel to an allergy meeting with instructions to photograph an exhibit on insect allergy. Rinkel had turned over the photographs to his chief who, in turn, used them in writing an article on insect allergy, submitting it to the Journal of Allergy. The director was convicted of plagiarism on the complaint of the editor of the Journal of Allergy, who had two articles on the same subject with identical photographs. As a result, the director was stripped of most of his medical society memberships. Rinkel was told that he had three months to sever all connections with this organization or he would suffer a similar fate.

It was during these three months that the birthday cake/egg reaction had been observed. Rinkel began experimenting with several unsatisfactorily treated, chronically ill patients from the clinic, keeping secret records of his observations. These patients would wait all day to see him, refusing to be seen by anyone else. Secretaries there were astounded by the obvious fact that these patients who had been in the clinic for years without responding, were getting well. Yet there was nothing in the record to indicate that their treatment had changed.

Since Rinkel had been under contract not to practice in any adjacent state, the nearest major city which attracted him was Kansas City, Missouri, where he went into practice, temporarily, with Dr. Orval Withers. At about this time, Dr. Michael Zeller of Chicago was studying with Dr. Withers. All three had known each other in medical school. Rinkel, being anxious to start his own practice, made a deal with Zeller in which Rinkel would teach Zeller all he knew about allergy while waiting for his private practice to develop.

It seems that Rinkel had been interested in reporting his findings in clinical allergy, having written on the subjects of eczema and the pathology and symptomatology of headaches by 1932. By 1936, Rinkel had confirmed and extended his observations of masked food allergy. Although these findings were reported in the Journal of the Kansas Medical Society and the Journal of the Missouri Medical Society by 1937, the major article on masked food allergy was not accepted for publication by the editor of the Journal of Allergy. (See Appendix, page 334) Rinkel was very upset by this rejection and made no further publication on this subject for the following eight years, during which time he worked out the basic nature of masked and unmasked food allergy.[29]

I first heard of this point of view after Rinkel had presented it at an instructional course in Omaha, Nebraska, sponsored by the American College of Physicians in early 1944. Up to this time, I had had the experience of feeding patients single foods which they were continuing to eat regularly. I then had patients avoid specific foods four to six days prior to test feeding — the technique of the individual food ingestion test, which Rinkel reported in Annals of Allergy in 1944. After confirming his observations of masked and unmasked food allergy, this work was also reported in Annals of Allergy. (See Appendix, page 334).

In 1948, Rinkel published his observations on the use of the rotary diversified diet in the Journal of Pediatrics. I once heard Dr. Albert Staesser, a prominent pediatric allergist in Minneapolis, say that this was one of the most important observations in the field of allergy, and that this alone was a major contribution for any person in this field (See Appendix, page 334.)

Rinkel, at the invitation of Hansel, began to appear regularly on the programs of the American Society of Ophthalmologic and Otolaryngologic

Allergy, as well as independent instructional courses. The first of the latter was arranged by Dr. Richard Stahl in Akron, Ohio. Later, there were two similar courses in Cleveland, one in Cincinnati and another in Houston, Texas. By 1957, the demand for instruction in the field of allergy was increasing rapidly. Rinkel was glad to accept these invitations, for most conventional allergists were not interested and he had been rather systematically denied access to instructional courses of the other national organizations.

The demand for teaching the so-called Rinkel approach in clinical allergy increased and led to the suggestion that he should start a yearly course. He agreed with the understanding that this course would be held in Wyoming. What later came to be called the Wyoming Postgraduate Course in Allergy and Immunology, was started in 1956 under the direction of Dr. Russell I. Williams. This important yearly event continued under Dr. Williams' direction until Dr. Rinkel's death in 1963.

Chapter V — 1954-1956

Changes in Medical Points of View and Terminology

Allergy — An Expanding Field

It had become increasingly apparent by the 1950s that etiologic observations of individually studied cases of chronic illness by means of the techniques described were yielding clinical findings at considerable variance with many accepted views in the field of allergy. In an effort to state this difference and its implications for the future of medicine, I tried to collect my thoughts on this subject by writing an article entitled, "Allergy: Whence, Whether and Whither," which was published in 1949.[44] This pointed out that newer etiologic concepts and techniques in the field of allergy not only differed from allergy as usually considered but were also spilling over and involving several other medical specialties, especially the fields of otolaryngology, psychiatry and rheumatology.

Relationships with otolaryngology will be described briefly. French K. Hansel founded the Ophthalmologic and Otolaryngologic Society of Allergy in 1941, but which lapsed during World War II. Hansel started this up again in 1947 with his friend, Herbert J. Rinkel. Herb brought me into these annual programs in 1948. Both of us, meeting increasing resistance to the presentation of our views and clinical observations in the meetings of the national allergy societies, welcomed this opportunity. Otolaryngologists also welcomed these presentations as their field had been changing since the advent and widespread application of antibiotic therapy had decreased the indications for surgical interference in the management of infectious processes involving the nose, throat, ears and sinuses. It is of interest that these trends toward less surgery and more medical management in otolaryngology were fostered by French K. Hansel and George E. Shambaugh, Jr. — the two physicians most responsible for the development and application of allergy in this field. In short, otolaryngologists were taught from the start of their interest in allergy, concepts and techniques which later were to be described as ecologically oriented.

1954-1956

Hansel's interest in the cytology and pathology of the nose and paranasal sinuses led to the allergic etiology of abnormalities of these areas. According to his bibliographic references (see Appendix, page 325) this interest started with publications in 1929 and continued for the next decade. The key cell in this inflammatory pathology was the eosinophil. This led to his development of the technique for staining nasal secretions for eosinophils and other cells, and the use of a picture of a red stained eosinophil as the logo for the American Society of Ophthalmologic and Otolaryngologic Allergy, which he founded.

Hansel started practice in St. Louis, Missouri at the depth of the depression. Under these circumstances, he spent his spare time reading about the subject of clinical allergy. He was eminently qualified for such a review from the background of his interest in pathology and clinical allergy, on the one hand, and his knowledge of French and German, on the other. I had been told that his mother was French and his father German and that he had a speaking and reading knowledge of both. His book, *Clinical Allergy*, published in 1953, remains to this day one of the four most comprehensive early bibliographic references in this field. His interest in the pathology of clinical allergy led him into the study of immunology. I recall talking with him several times about his interest in Reuben Kahn's interpretations of the immunology of allergic inflammations. In this respect, French Hansel was far more interested in immunologic mechanisms of allergy than other pioneers who contributed to the beginnings of clinical ecology whose interests were primarily based on clinical observations. Hansel not only combined these interests in bodily mechanisms and environmental exposures but also employed new technical approaches. In addition to the development and interpretation of staining nasal secretions, he experimented with low dosage treatment schedules of inhalant allergy and sublingual therapy. His contributions spanned the period between 1929 and 1966. He continued to attend medical meetings in this field until his death in 1980. Interested persons should read: French Keller Hansel and the History of Allergy by George E. Shambaugh, Jr., M.D., Arch Otolaryngol 109:126, 1983.

Second only to French K. Hansel in pioneering the subject of allergy for Otolaryngologists is George E. Shambaugh, Jr. A student of both Hansel and Coca, his interest in allergy was also initiated by the preponderance of eosinophils in nasal and sinus tissues. Certainly, as far as the Chicago Metropolitan area is concerned, Shambaugh is to be credited in having pointed out to otolaryngologists the relative desirability of the allergic, rather than the surgical, management of nasal and paranasal allergy.

For several previous years, surgically oriented otolaryngologists had been impressed by the apparent immediate symptomatic benefits from having performed submucous resections and other intranasal procedures in patients presenting allergic manifestations. But when similar or accentuated symptoms recurred after two or three months, operated patients often failed to return to their surgeons who had not been helpful in the long run. Consequently, as pointed out by Shambaugh, surgeons were not fully aware of the extent of their failures. Postsurgical scarring of the nasal ciliated mucous membranes also interfered with normal nasal function.

As an increasing number of cases of mental illness and behavioral disturbances were moved from their former descriptive designations to this relatively new, and as yet unnamed, etiologic classification, based on the demonstrable impingement of such nonpersonal factors as specific foods and environmental chemicals, the need for new terminology in this field became increasingly apparent. Nowhere was this need more marked than in those cases presenting lesser degrees of mental and cerebral dysfunction. New terms were especially needed in this field since many formerly descriptive designations were not available for use by others in view of the way they had been redefined in psychiatric dictionaries.

By the midcentury point, it had become increasingly evident that chronic fatigue and muscle aching were often associated with sufficient cerebral dysfunction to cut back materially on production and performance. Yet, there was no name for this problem greater than fatigue and myalgia but less than advanced depression. While grasping for a name for this cerebral dysfunction between fatigue and myalgia, on the one hand, and depression and other psychiatric manifestations, on the other hand, an executive solely responsible for decision making in a major government agency in Washington, D.C. consulted me in 1952.

How he happened to come in is of particular interest. This 49-year-old professional man complaining of "brain-fag," was also subject to numerous other localized and systemic allergic manifestations. Knowing two foods which would induce this phase of cerebral inefficiency, he had assumed correctly that probably other unknown foods were similarly related. For over a year he had consulted allergists from Richmond to Boston for help in solving his "brain-fag" problem. Being the recipient of more derision than assistance, he concluded that the allergists consulted were uninformed on this subject. Thrown back to his own devices, he started reading in the Congressional Library. Coming across our book, *Food Allergy,* in which I had written the chapter on fatigue and related conditions, he went out to the nearest telephone and called for an appointment.

I was amused by this obviously erudite gentleman coming up with the apparent slang expression, "brain-fag," every few sentences. But to my amazement, I found from checking dictionaries that evening that this term had been in the English language for nearly a century and fortunately had not been defined in psychiatric and medical dictionaries.

Although he had been subject to various other localized and systemic manifestations of allergy, his fatigue and associated "brain-fag" had been present progressively for 20 years. Indeed, this disability had come to dominate his life. For instance, he was barely able to get home to eat and rest at the end of each day in the hope that by resting in the evening he would be able to work constructively the following day. Although this fatigue and associated difficulty in evaluating evidence and making decisions based on correlating evidence occurred every day, it was much more pronounced on certain days. It occurred despite obtaining eight hours of sleep nightly; although accentuated by loss of sleep, it was not helped by obtaining an excess of sleep. During this time he always felt dragged out in the morning, feeling like a "flat tire" for a few hours each early forenoon. At the worst of his fatigue-"brain-fag"-syndrome, he also experienced difficulties in reading comprehension, slurring of speech, transposition of syllables and letters in speaking and writing, many slips of the tongue and mumbling of words. Aphasia at such times was associated with a marked reduction in his working vocabulary. Impairment of memory was prominent, as well as being particularly slow in mathematics. His lessened efficiency in general at such times resulted in his working in a relatively slow and plodding manner. However, he did have good days occasionally when "normally" effective in which he was fully capable of correlating, associating, and integrating bits of information — before the days of computers which were later designed to help with these functions. On such days, he would tear into this pile of unfinished folders stacked upon the difficult corner of his desk.

His first insight into causative factors which might be involved occurred when he observed a marked improvement in his headaches, fatigue and associated "brain-fag" following avoidance of milk three months prior to his initial office visit. It was after this observation that he had doubled his efforts in seeking help from allergists, but to no avail. The only major clue in respect to the roles of other commonly eaten foods which might have been acting similarly, was his observation that he only had sufficient energy and activity to perform the sexual act after two drinks of bourbon whisky. Being a native Texan and having been brought up on corn pone and grits also increased his chances of being sensitive to corn.

Sharp reactions to corn, coffee, paprika and lesser reactions to certain other foods and their avoidance while rotating test compatible foods resulted in significant control of his "brain-fag" and related systemic manifestations. Needless to say, "brain-fag" has been a part of the terminology of this field since that time, at least as far as my own vocabulary was concerned.

In reading this chronologic story of medical events, the reader should bear in mind that subtle changes in nomenclature and vocabulary were occurring continually. For instance, a persistent problem was encountered in dealing with new patients, many of whom seemed to resent the suggestion or implication that they might be allergic. I pondered for some months on how to get around this word "allergy" more acceptably. The term "food addiction" instead of "food allergy" became increasingly attractive, yet I could not bring myself to make this break for some time.

However, I did go so far as to look up the 24 best descriptive accounts of opiate addiction written from about 1860 to 1930. I was impressed by the fact that I had seen all the examples of the withdrawal effects from opiate addiction in food allergy with one exception — death of the patient. I gave serious thought to submitting this review and a monograph on food addiction for publication, but never did. This simply remains one of a considerable collection of unpublished manuscripts which, for one reason or another, seemed to be inappropriate at the time.

This experience prepared me for an emergency confrontation on this subject with a prominent business man and cultural leader in Chicago. Indeed, I was surprised to find his name on my appointment schedule and even more startled by his response to my suggestion that he probably was allergic to eggs. He came in as a result of his afternoon secretary's complaint that starting about 3:15 to 3:30 P.M. each day his dictation deteriorated in respect to its diction, thought content, clarity, relevance and continuity. This posed two types of problems for her. If she transcribed his dictation literally, he objected to it the next morning. If she typed what she thought he had meant to say, he also objected to that. In short, each afternoon he seemed "to run out of steam," becoming tired, listless, dopey and moderately confused about 3:30 P.M.

In taking his food intake history, I learned that he always had eggs for breakfast, eggs for lunch and an egg-containing dessert for dinner. His secretary then observed, astutely, that when he relaxed with the rest of the office staff and had a snack in the middle of the afternoon, his dictation remained fine. Feeling better when he did this, he raised no objection to this changed routine. But an appropriate snack for this man meant having some more eggs. This degenerated to his secretary "egging him on" each afternoon. Despite these regular pick-me-ups, he continued with a gradually increasing level of chronic fatigue and "brain-fag."

When I suggested that he might be reacting to eggs on the basis of being allergic to them he stood up, as if ready to leave the room, and said: "Doctor, I may be allergic to certain foods, but I know that I am not allergic to eggs. Instead of feeling worse, I always feel better after eating eggs. Even my secretary noticed this, as I just finished telling you." Hearing a noise in the food test room, I excused myself on this pretext, but actually for the purpose of collecting my own thoughts. Upon returning to this confrontation, I began talking to him in the lingo of drug addiction. He seemed to understand the fact that he developed a yen for eggs starting between three and four hours after the last egg-containing meal, and that if he ate some more eggs, all went well. If he did not satisfy this craving, all was not well. He then agreed to avoid eggs, preparatory to testing them by ingestion in the office.

Withdrawal effects precluded work for three days before he improved. He reacted acutely and convincingly to a test feeding of eggs. In preparation for this, he was so withdrawn that he was unable even to get to his office. By the fourth day he felt much better, and on the fifth day he had his egg test which precipitated a sharp recurrence of symptoms. Follow-up on this case has been nil, except for a chance interview on the street a few months later and all was well. In summary, this case is unusual in view of the severity of reaction to virtually a single food.

1954-1956

Referring to food allergy as food addiction has been a distinct improvement in vocabulary, especially when speaking to new patients relatively uninformed on this subject. I tried this out on medical audiences in presentations and in several scientific exhibits without serious objections.[90,91,92,95] But when speaking to the staff of a major pharmaceutical concern on this subject, several scientists objected to this extension of the term "addiction" when most experts in this field were attempting to define and limit its more general use. The fact remains, however, that many new patients, who have often referred to themselves as "food-o-holics" or "sweet-o-holics," understand and accept this designation far more readily than any other alternative expression which has been tried. In contrast, "allergy" or "allergic to" in reference to addictive behavior toward either an identified or unidentified substance strikes the uninformed new patient as backhanded and confusing.

Because allergy had become closely linked with immunologic mechanisms by the 1950s, I had been looking for a less restrictive alternative term for several years. The descriptive term, food addiction, seemed more appropriate, useful and better accepted than the designation of food allergy which it was replacing, at least in my own vocabulary. This led to the preparation of a manuscript entitled, "The Descriptive Features of Food Addiction — Compulsive Eating and Drinking," which was finally published in the Quarterly Journal of Studies on Alcohol in 1956.[111] Because of interest in this subject I will retrace steps, briefly, which led up to this publication.

Shortly after the interest in alcohol and alcoholism developed I visited the Department of Applied Physiology at Yale University Medical School where an active interest in alcoholism, especially its sociologic aspects, had existed for several years. The Quarterly Journal for Studies on Alcohol was published from here. The center for the world's literature on alcoholism was also located here. I was given the opportunity to address this staff on the subjects of food addiction and alcoholism at least twice and perhaps three times. Despite there being no special interest in the relationship between allergy and alcoholism, the editor of the Journal agreed to look at a manuscript on this subject which was submitted in 1954. Not having heard of its acceptance or rejection two years later, I called the editor to inquire about it. Apparently, he feared that this zany view of alcoholism might be correct and, if rejected, he might be criticized. But he also feared that it might be frivolous and if he published it he would be criticized. Consequently, he simply remained ambivalent. Having given him two weeks either to accept or reject it, unknowingly he called Fred Kessler of New Haven, an allergist and longstanding friend of mine. Fred apparently spoke a good word for it and for me. It was published immediately in the June issue of the Quarterly Journal for Studies on Alcohol — the leading publication in this field. But this was not the end of this ambivalence!

In the course of my visits to New Haven, Vera Efron, the archivist for the world literature on alcohol and alcoholism, had been helpful as I reviewed the publications in this field before writing on these subjects. But I was astounded to find this article classified under psychiatry in the next installment of the classified literature of alcoholism. I called her immediately in protest, only to be told that she did not know what else to do with it! Apparently, unclassified is dumped into this nondescript category. But this was still not the end of this ambivalence.

A few months later, Dr. Jellinek, dean of the experts on the etiology of alcoholism, dismissed this idea of the "masked food allergy" interpretation of craving on the basis that he was unable to understand it. (Jellinek EM. *The Disease Concept of Alcoholism.* New Haven, Connecticut and New Brunswick, New Jersey College and University Press in association with Hill House Press, 1960.) This is but another example of the negative connotation of the term, food allergy. It was because of this type of response that I prepared a monograph on food addiction containing numerous case reports on the subjects of addictive eating and addictive drinking in the hope of circumventing this type of rejection. In view of the fact that relationships between the addictive use of specific foods and alcoholic beverages derived from specific foods has not been applied to a significant degree in alcoholism, despite this subject having been reported in the alcoholism literature,[111] I hope to publish these case reports in the near future.

Finally, after seeking an alternative designation for allergy for several years, the substitute of clinical ecology was suggested and quickly accepted by me. This came about as follows. From time to time, when it became necessary to seek a new hospital appointment because of overcrowding, I had learned to go to the center where hospital bed occupancy statistics for metropolitan Chicago are kept on file. I did this in the late 1950s for the purpose of finding a hospital with a bed occupancy rate of 70 percent or less. In the course of explaining my purpose and the reasons for seeking these data to a Mr. Sibley who was in charge, he commented: "What you are really doing is clinical ecology." After reading that ecology had first been used by Ernst Haeckel, a German biologist, to explain Charles Darwin's contribution in 1868, I decided that this, indeed, was an excellent term to describe what I and others were trying to do. I was also interested that ecology had been widely used in biology for nearly a century, and was rapidly coming into popular use by 1960. Moreover, this term had virtually no use nor application in the practice of medicine. In moving the location of my suburban office in November 1960, I used the designation, clinical ecology, to describe my activities. Several medical colleagues stopped me to inquire what I was doing. After seeing the direction that this work was taking, the name of the Rockwell M. Kempton Medical Research Fund was also changed to the Human Ecology Research Foundation.

Another desirable change in differentiating clinical ecology from allergy was abandoning the words sensitivity, hypersensitivity, sensitize and sensitization — essentially synonyms for allergy in a limited immunologic sense — in favor of susceptibility, individual susceptibility and susceptible, because of their wider connotations.

1954-1956

Another change in terminology had to do with abandoning the term "stress" — used variously as a noun, verb, adjective and adverb and which no one had been able to define. Indeed, in attending the First International Congress on the Management of Stress in Monaco in 1980 no one, not even the originator Hans Selye, had been able to define the key word in the title of this meeting.

With at least a part of my medical and scientific vocabulary gradually straightening out by 1954, I was better prepared to integrate the concepts of addiction and ecology into the interpretation of the phenomenon of adaptation than if this had occurred earlier. This realization of how adaptation fitted into this scheme of environmental medicine will now be described.

Specific Adaptation in the Presence of Individual Susceptibility

As stated earlier, both my preceptor, Francis M. Rackemann, and I were impressed with the early observations of Hans Selye in describing the successive stages of adaptation to various environmental exposures occurring in laboratory animals. Dr. Rackemann was a member of the editorial board of the Journal of Allergy at the time. Whatever may have been the connection, through him or otherwise, Selye's first effort to present his concepts of adaptation to clinicians was published in a series of articles in the Journal of Allergy in 1946. Although I also read these several times, I was still unable to see how this applied to allergy or how these concepts could be employed gainfully by clinicians. Nevertheless, I was still sufficiently impressed with his observations that I followed his work closely and bought all of his books. Moreover, he and I frequently talked together during the early ACTH sessions and at other medical meetings.

It was not until my wife and I were driving back from California on a lonely straight stretch of New Mexico highway in the winter of 1954 that I suddenly came to the realization of how the stages of adaptation applied to my daily handling of cases highly susceptible to common foods and environmental chemicals! The answer was amazingly simple in view of its exceedingly long incubation period — 17 years to be exact.

In short, physicians did not ordinarily have the opportunity of exposing well patients to cumulative doses of foods or chemicals and watch them sicken initially, then become resistant, and eventually become sick again, as described by physiologists. Most patients were already sick when they were first seen by their doctors. To what they were reacting and for how long this had been occurring remained unknown. The trick in revealing this long-term and subtle impingement was simply to avoid such exogenous exposures on the basis of probability for a sufficiently long time to permit the inured patient to recover from their cumulative effects, then re-exposure of such a person to this avoided item(s) and observation for evidence of an acute reaction(s). In fact, this was precisely what we had been doing empirically in diagnosing instances of chronic food allergy, since Rinkel had described masked food sensitivity and how to recognize the specific offending food(s) by means of the deliberate food (ingestion) test, which he had finally described in detail in 1944.

In looking about for other possible examples, apparently similar mechanisms were operating in occupational sensitization dermatitis as described by Louis Schwartz in 1940. (Occupational Dermatitis: Treatment and Prevention. J Mich Med Soc 39:179, 1940) He went on to relate that when workers first began to handle solvents and other chemicals capable of inducing individual susceptibility, they often developed a dermatitis of their hands. Occupational physicians had learned by experience to keep such workers on the job, perhaps protecting them slightly by means of protective ointments or gloves, and soon they would usually become "hardened" to such a regularly repeated exposure. Although this tolerance might persist for several days in the absence of repeated exposures, occupational physicians learned empirically not to permit these workers to take a prolonged vacation. In such an event the worker had to go through the "hardening" process all over again.

This "hardening," inurement, or resistance in occupational sensitization dermatitis might persist for several years, as long as such a person remained on the same job and the character of the job had not changed. But if at any future time this dermatitis recurred while maintaining the same job exposures, this development was an absolute indication for a change of occupation. Similar 1-2-3 sequential stages of adaptation occurring in the presence of specific susceptibility may be observed in painters, who on long vacations sometimes seek out other painters and "sniff around" them at work. This also may happen to chefs who had been accustomed to cooking over a gas range daily. But, upon returning from hiking in the mountains, they might seek out gas exposures for its apparent immediate beneficial effects. Although such hardening or resistance may be maintained by intermittent exposures, too long a period of abstinence might result in the precipitation of acute test-type nonadapted reactions immediately after specific re-exposure.

Similar clinical findings may be observed even more frequently in addiction to tobacco smoke, but here the addiction cycle is much shorter in the sense that the desired temporary improvements may best be maintained by smoking hourly or even more frequently. Such a person finds it "convenient" to stretch his or her legs during the middle of a two-hour theater performance — having a cigarette "coincidentally." They often get up to smoke during the night or smoke before breakfast. Such addictive smoking patterns readily become ingrained behavioral responses. Moreover, this conditioning may persist for long periods after the true addictive aspects have worn off. For instance, the wife of a company underling who had been addicted to both potato and tobacco broke the stranglehold of these addictants during the hospital program of comprehensive environmental control.[166] She had been home about a week, avoiding both and doing well, when she saw the boss's wife coming up the walk for a visit. Hating this woman and knowing how difficult this interview would be, she grabbed and lighted a cigarette to help her through it. However, she remained acutely ill from this test reaction for the following three days.

The extreme degree of individual susceptibility in some of these specifically addicted states leads to interesting variations in dosages of the addictant at times. Instances of addiction to coffee may be cited as examples. One person addicted to coffee and to beet sugar, drinking between 40 and 45 cups of coffee per day has been described previously. In contrast, a nurse extremely addicted to coffee found it so difficult to avoid overdosage in maintaining optimal beneficial effects that she drank coffee only by the teaspoonful. Others similarly addicted to coffee simply sip their favorite beverage through their working hours. Some alcoholics also find it beneficial to sip through the day on a similar schedule.

There are also wide variations in the duration of withdrawal effects after the avoidance of specific addictants. One food addicted person voluntarily fasted a month without improvement in his chronic manifestations. An enema performed at the end of this period revealed the identity of addictive foods in the enema return. Even in the course of the hospital program of comprehensive environmental control preceded by emptying the alimentary canal, patients who had been "saturated" by occupational chemicals may maintain their chronic level of symptoms for as long as 10 days before clearing. The majority of food addicts under these circumstances will clear their chronic manifestations between three and five days of fasting under controlled environmental conditions.

The decision to incorporate the concept of adaptation as accentuated by individual susceptibility in the interpretation of clinical reactions to foods, exogenous chemical exposures and other human-environmental interrelationships was made by 1955 when a scientific exhibit was prepared for the Annual Meeting of the American Medical Association that year. After rejection of this initial application, this exhibit, prepared with eight other physicians who had been working closely with me, was presented at the Annual Meetings of the American Medical Association and the American Psychiatric Association, both in Chicago in 1956 in the months of May and June, respectively.[108,109]

The two sides of a brochure describing this exhibit, Specific Adaptive Illness — An Effective Nonpsychiatric Approach to Mental and Related Ills (Without Drugs), are shown in Figures 17 and 18. Figure 17 includes a summary and the names of the exhibitors. A brief resume of each physician follows:

C. Richard Ahroon, M.D. of Bloomington, Illinois, an internist and allergist, became interested in the application of this point of view from having seen an earlier scientific exhibit in 1953. Thereafter, until his untimely death in 1958, he had visited my Chicago office every Wednesday. In the course of these visits his clinical reactions to foods (especially cereal grains) and environmental chemical exposures (especially automotive exhausts) were worked

Medical Points of View and Terminology

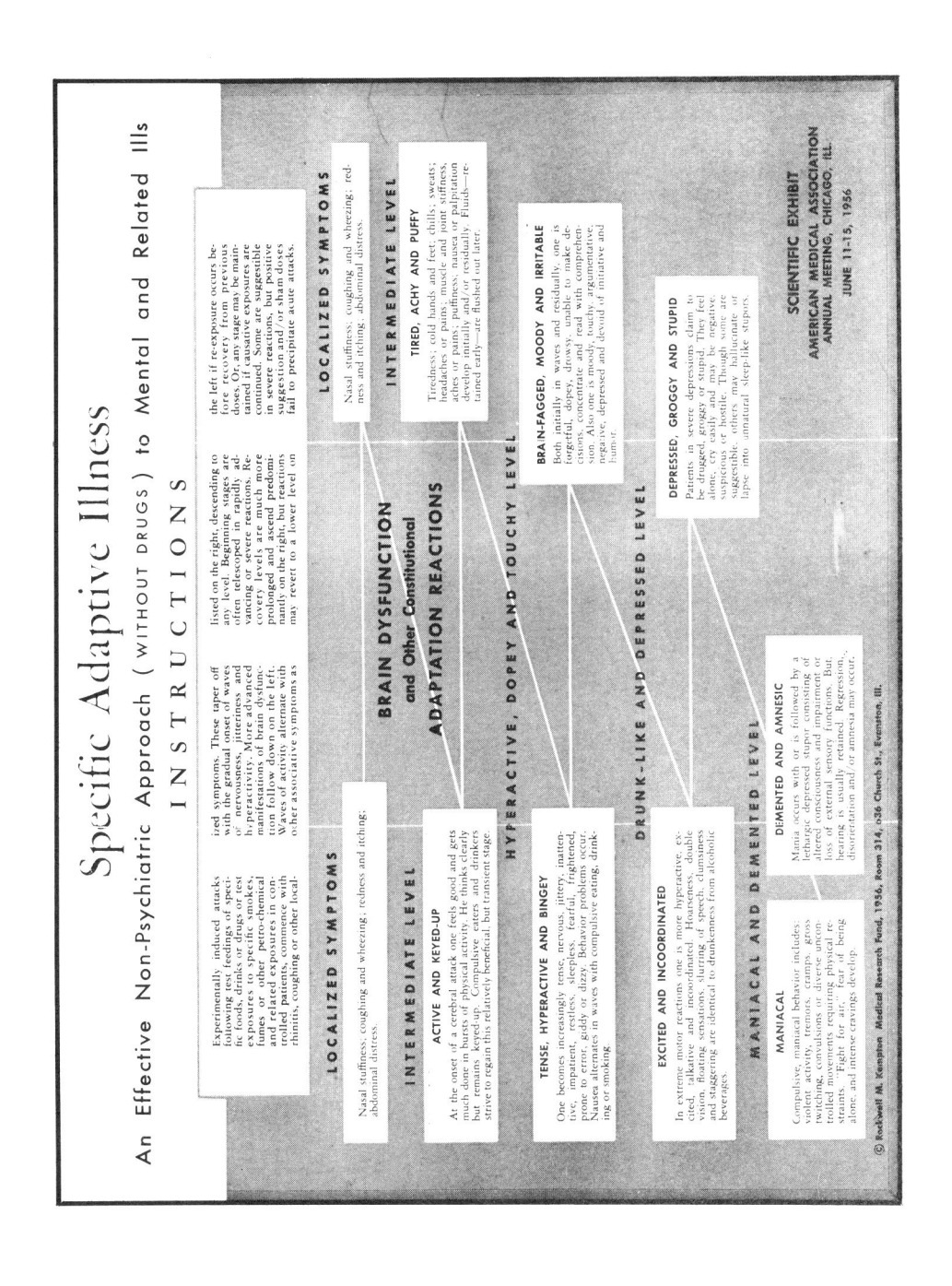

Figure 17. Side 1 of brochure, *Specific Adaptive Illness*.

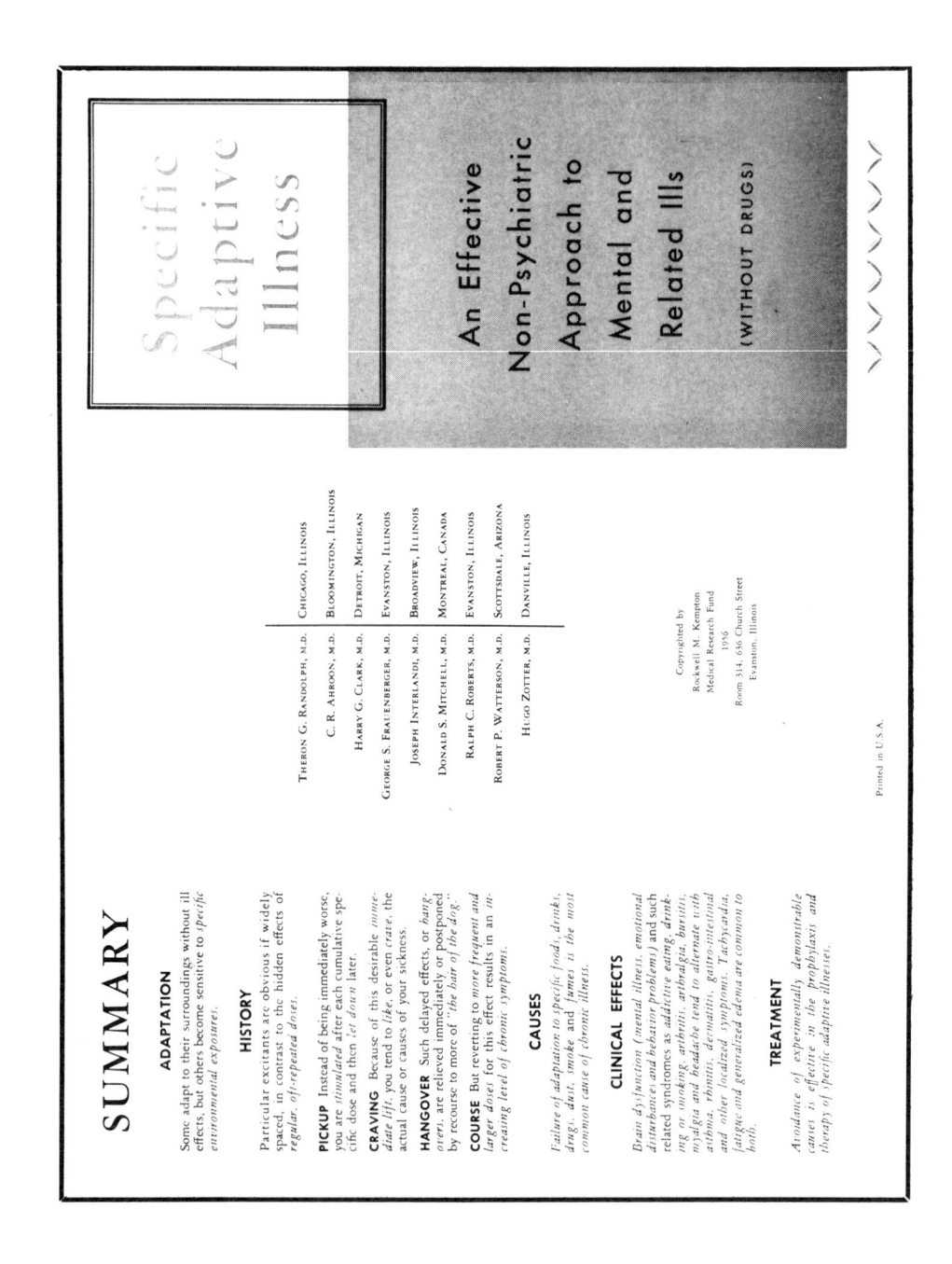

Figure 18. Side 2 of brochure, *Specific Adaptive Illness*.

out. A major medical problem all of his life had been stuttering, acute recurrences of which followed intermittent exposures to wheat or corn or after driving into Chicago on Ogden Avenue — at that time a major diesel truck route. Indeed, allergic-type reactions to these materials were apparently responsible for his death.

A few days before his death, while his wife had been hospitalized, he had been less able to follow his restricted diet than ordinarily. The night before an acute coronary occlusion, neighbors had brought in a dinner for him containing both wheat and corn. Although he reacted more acutely than ordinarily, this was not regarded as unusual until the onset of acute chest pain during the night. He was hospitalized immediately with a diagnosis of myocardial infarction. Apparently, being unconscious upon hospitalization, or shortly following, an intravenous infusion of 10 percent dextrose (corn sugar) was started. He was still unconscious when I arrived at noon the following day. Although I stopped the intravenous immediately, he never regained consciousness.

Dr. Harry G. Clark's history had been given briefly in connection with the development of the acid-anoxia interpretation of allergic inflammation. He was also highly susceptible to cereal grains. Dr. Clark died in 1962.

Dr. George S. Frauenberger, classmate from Medical school, for many years had been my closest personal friend and medical associate in the Chicago area. He was also highly intolerant to multiple foods and was subject to arrythmia. He died in 1969 of generalized carcinomatosis apparently resulting from the metastasis of a localized skin lesion of his leg having been cauterized in the absence of a biopsy identifying its nature.

Dr. Joseph Interlandi, a Chicago allergist, had become interested in this work as a result of viewing an early scientific exhibit. He had also visited my office once weekly for a period of three years during the late 1940s. He practiced clinical ecology in Chicago for many years and later was elected president of the Society for Clinical Ecology. He died in 1982 of an acute coronary occlusion at the age of 60 years.

Dr. Donald S. Mitchell, a dermatologist from Montreal, Quebec, Canada, developed a workable concept of food allergy when in charge of dermatology in the Royal Canadian Navy during World War II. He helped work out the technique of comprehensive environmental control and joined Dr. Richard Mackarness and me in a scientific exhibit in Edinburgh, Scotland at the combined British and Canadian Medical Associations Meetings in 1959. He had spent several months in my office in 1957 and again in 1963. He died in 1984 at the age of 82 of cancer of the prostate with metastasis.

Dr. Ralph C. Roberts, an internist, and my personal physician and closest medical friend and greatest source of referred patients for many years, served as Secretary of the Human Ecology Research Foundation from its inception to the present. Dr. Roberts remained loyal to me and this point of view throughout the controversy with Northwestern Medical school and thereafter. For the past three years Dr. Roberts has been retired to southern

New Mexico.

Dr. Robert P. Watterson, a general practitioner of Scottsdale, Arizona, had applied this medical point of view for several years. He died in 1965 as a result of an accidental fire occurring in a gas fired sauna.

Dr. Hugo Zotter, a psychiatrist, died in 1971.

Clinical Levels of Reaction

Some of my contemporaries will remember this (See Figure 18) as the exhibit with strings attached, the strings representing developmental interrelationships between induction and recovery levels of experimentally induced acute reactions. It should be noted that the layout of this exhibit, containing five levels and reading from left to right, is quite different from the current layout consisting of four levels and reading from stimulatory levels listed at the top and withdrawal or recovery levels presented on the lower half of the page, as will be illustrated later.

Since the various descriptions of the levels are self-explanatory, these will not be described in greater detail. The exhibit space in each instance was 20 feet in width. At the American Medical Association exhibit 245 signed in for more information, 100 at the psychiatric exhibit.

Comprehensive Environmental Control

As mentioned earlier, the first attempt to fast patients at the St. Francis Hospital in 1950 preparatory to testing single foods by ingestion had failed. The apparent reason for the failure, unrecognized at the time, was the willful accentuation of associated withdrawal effects. Not realizing that I was being manipulated, the fasts were terminated on the third day by feeding all six patients.

It was not until 1956 that the fasting program was tried again at the same hospital at the suggestion of Dr. Donald S. Mitchell of Montreal, who had tried it successfully. There were two main reasons for its success this time. Patients were placed in different parts of the institution and the fasts were started at different times. Instead of patients being allowed to reinforce their withdrawal manifestations by talking with each other, I used the experience of the previously fasted patients to help me tide the currently fasting individuals over their withdrawals. This general program has been the rule ever since.

The real reason for reinstituting the fasting and feeding diagnostic technique was the fact that many chemically susceptible patients could not be diagnosed specifically while maintaining environmental chemical exposures in their jobs and/or homes. Combining fasting with isolation in a controlled environmental setting promised to solve both of these problems. Moreover, such a program effectively removed a patient from the chemical exposures associated with his work and home to permit recovery from their cumulative effects. Then when the patient was returned, first to his home and then to his job, clear-cut convincing test reactions were often demonstrated.

This program of fasting under environmentally controlled conditions, started at the St. Francis Hospital, was continued at subsequent hospitals as will be described later. At this time, individual patients were isolated in either single or two-bed rooms. Two patients per room had certain advantages in that they aided in observing each other's reactions. With time, the disadvantages of doing this work in different parts of the hospital with nursing staffs relatively unfamiliar with these routines became increasingly apparent. It was also difficult to control tobacco smoke or to use rooms previously employed for isolation of patients for control of infection because of chemical contamination. Moreover, the need to maintain special cleaning techniques, special bed linens, mattresses, pillows and furniture without sponge rubber padding and plastic upholstery, became increasingly apparent. And most importantly, the necessity for nurses, technicians, janitors and other personnel entering the unit to abstain from the use of odorous cosmetics and tobacco became mandatory. Optimally effective techniques will be described later.

1954-1956

Chapter VI — 1957-1959

A Period of Transition

Arthur F. Coca, M.D. — Continued

The last red ink letter that I received from Dr. Coca was in 1957. Since this summarizes differences in our points of view as seen by him, I will quote parts of it.

February 7, 1957
Dear Dr. Randolph:

I am embarrassed by your invitation to "comment" on your large essay on "Allergy and Specific Adaptation." I can avoid some controversy, as you have politely done, by freely paraphrasing the fourth paragraph of your letter.

"The essay is well done, as well done as it is possible to present an abstruse subject in the language that an extraordinary reader (not me) might be able to get something out of."

But I am impelled, with little expectation of influencing you, to lament your preoccupation with the symptomatic *effects* of the idioblaptic constitutional disease-entity and their alleviation — by whatever therapeutic means — *and* your total disregard of the existence of that disease-entity.

Not having mastered the time consuming art of interpretation of the objective pulse-record, your investigation of the disease is limited. Your procedure, for example, has nothing to offer to the symptom-free allergic person approaching the first heart attack or in the interval between heart attacks.

Selye's imaginative contributions concern effects, not specific causes; he has no method. His misused term "adaptation" is sterile; it cannot survive the inevitable recognition of the existence and nature of idioblaptic disease in man *and* lower animals.

I doubt it will change your thinking (you're too far up a blind alley), but

I should let you know that investigations of idioblaptic allergy are well underway in animals with electrical recordings of the pulse-rate. The published results of these studies in reputable institutions will not allow further question of the existence of the disease-entity "idioblapsis."
    Friendly regards from us both to you *two*
    in our uncompromising disagreement,
        Most sincerely,
        Arthur F. Coca

I will try to answer some of Dr. Coca's criticisms on the basis of knowledge available to me at that time. In my defense, I tried diligently to master the pulse acceleration technique over an extended period of time. I agree that this technique is helpful to the clinician, but it is far less convincing to the patient than the Rinkel technique of preliminary avoidance of one or more foods followed by test re-exposures, noting both the initial withdrawal accentuation of symptoms and, especially, the precipitation of an acute reaction associated with a test re-exposure. Moreover, I incorporated measurements of the pulse rate for a full minute before and repeatedly following exposures to single foods as an integral part of the program of comprehensive environmental control, as will be described in more detail later. As I see it, my day to day job was convincing the patients sufficiently that they would follow specific dietary instructions. And once a sufficient number of patients were convinced, hopefully this might be convincing to the medical profession. I was not disinterested in mechanisms, as witnessed by the development of the acid-anoxia-endocrine concept of allergic inflammation and the therapeutic use of the alkaline salts of sodium and potassium bicarbonate in its treatment.[68,69,76,86,105]

I regarded tachycardia as another cardiovascular manifestation of reaction, agreeing that it was present more frequently than certain other cardiovascular effects, such as erythema of the face or ears, chilling, variations in blood pressure, arrythmia (irregularity of the pulse) and others. I simply could not agree with Dr. Coca's statement in respect to an "idioblaptic constitutional disease entity." To me, this was a magnification of one of the effects of the allergic reaction, but it was not as characteristic of the allergic response as claimed. It was my impression that a pulse acceleration was no more characteristic of an allergic response than intravascular agglutination of erythrocytes. Both occurred as a result of infection and tachycardia was often a response to exercise and otherwise. Neither did tachycardia occur invariably in allergic responses. For instance, bradycardia was by no means rarely observed. When I asked Dr. Coca about this after the publication of the third edition of his book in 1956 and the book, *The Pulse Test,* written in conjunction with Lyle Stuart, and also published in 1956, he admitted having observed this. But he and Mr. Stuart thought that including this observation might be misleading.

Dr. Coca also commented that Selye has no method. I pointed out that I am extending Selye's concept of adaptation in a general sense, emphasizing its specific features which is in keeping with Rinkel's views and provides a method for identifying the causative role of given environmental exposures in numerous chronic illnesses.

I also discussed with Dr. Coca the range of the pulse changes that he considered significant. Although not answering precisely, he said that he had seen instances in which an increase of two beats per minute after exposure had been consistently significant. This point will be taken up later as a result of Dr. Corwin's statistical analysis of Dr. Coca's data.

Finally, Dr. Coca's allegation of my total disregard of the existence of the entity of "idioblaptic constitutional disease" was not entirely justified. It is true that I never used this term which seemed to me to be more confusing than helpful, and unnecessary. Neither was I as convinced as Dr. Coca about the hereditary influence controlling its occurrence. My observations in this regard were admittedly preliminary. I simply was not sufficiently interested in making an adequately detailed study of this aspect of the problem. It had been my impression that reactions to foods or environmental chemical exposures might be induced in anyone as a result of sufficiently great exposures, either massive in the sense of a single exposure or more commonly as a result of protracted exposures. The reason some individuals failed to develop allergic responses has simply not been determined. Neither have the distinctions between reactions on the basis of individual susceptibility and toxicity been clearly demarcated. These two types of response often overlap, interdigitate and under propitious circumstances may occur in most individuals. This will also be presented in more detail later.

The last green ink letter that I received from Dr. Coca, March 27, 1958, starts off as follows: "You will not be surprised to hear that I am again (or still) having trouble with orthodox medical authority in my very unconventional management of the chronically ill."

The last communication from Dr. Coca was a copy of a letter and discussion of the subject of alcoholism written to Dr. Ruth Fox, president of the National Council on Alcoholism, of which copies were sent to me. I had previously suggested to Dr. Fox that she contact Dr. Coca in respect to discussing my presentation of alcoholism in New York City.

November 23, 1959
My Dear Doctor Fox:
  Since your most enjoyable and instructive visit here, I have come to realize that you were really the first to teach me the truly practical problem of alcoholism. My previous interest had been in ascertaining the nature and cause of the phenomenon, thinking that after determining the cause of alcoholism, one should have only to avoid it.
  But after Dr. Randolph's reprint (which you kindly sent me) and after I had prepared my "discussion" of his work, it became clear: 1) that the cause of alcoholism *has* been ascertained by him, 2) that he has

actually been able to abolish the "craving" for alcohol by avoidance of the cause(s), and 3) that the one case that I was to report merely confirms, in a striking way, what he has already discovered. However, the practical problem of the alcoholic remains; and that is what you taught me — the danger of the "slips"!!

I have had plenty of difficulties in the past 25 years applying the tedious self-denying technic of avoidance of food and inhalant allergens; not to mention the 15-20 percent of the cases that require limited sympathectomy on account of their *multiple* food sensitivities. But the average alcoholic, even after having submitted to such a self-denying course, can have no assurance that this restauranteur or home cook has not innocently "slipped" one of his food-poisons into some item of his diet. In such event, the food-allergic person who is *not* an alcoholic may get off with a headache or indigestion. Whereas the victim of alcoholism, especially the "social drinker," may be unwittingly started on one of his devastating seizures.

The safest "alcoholic" is one who takes out complete "health-insurance" through the preventive method of the pulse test, which in many or most alcoholics will *abolish the alcoholic craving;* and who also avoids "social drinking" thus insuring himself against the "slips."

My inhalant allergies are making me unhappy. Added to those, I have recently had a recurrence of a viral infection which cost me a lot of gastrointestinal blood. In these circumstances I must deny myself the Randolph meeting December 1.

Do you think any good could come from your reading the enclosed "Discussion" with any comments *you* care to make about it — you are well informed.

With apology and kindest regards.

Most Sincerely,
Arthur F. Coca, M.D.

The "enclosed discussion" follows:

I have been invited by Dr. Fox to "discuss" Dr. Randolph's paper on Alcoholism, not as one experienced in that manifestation but because of my special interest in the manifestation of specific sensitivity. As an observer of alcoholism I am decidedly a Johnny-come-lately and my personal information concerning that condition is drawn from a single case. However, that case seems to me to contribute irrefutable support to Dr. Randolph's concept as to the nature and specific cause of alcoholism.

This man was not specifically sensitive to any food nor to any constituent of any alcoholic beverage; he was allergic only to house-dust. So long as he avoided house-dust through the treatment of his house furnishings with Dust-Seal, he experienced no *craving* for alcoholic beverage and the abnormal compulsion did not recur when he drank socially.

The following statement published by Dr. Randolph in 1956 generalizes the lesson so clearly evident in the uncomplicated facts of my case; he writes:

"Abstaining addictive drinkers have been diagnosed and treated by means of avoidance of foods and other environmental excitants to which they are *specifically sensitized.* Patients so handled have reported a general symptomatic improvement as well as having a lessening-to-an-absence of their craving for spiritous beverages. This emphasis carries the connotation that problem-drinkers may become *social drinkers* provided they eat and drink only compatible foods and their derivatives, or remain controlled in respect to house-dust or other specific environmental materials."

Dr. Randolph points out that the greatest drawback of such a program of prevention is the *"danger of a specific dietary 'slip' reactivating the craving."*

Such a "slip" has occurred on several occasions in my one case, when he has been exposed to house-dust in homes that have not been Dust-Sealed. To be sure, each such "slip" gave added evidence of the correctness of Dr. Randolph's thesis.

This meeting would seem to be one occasion on which to extend recognition to Dr. Randolph for his single-handed struggle with the heartbreaking problem of alcoholism; but like all the pioneers in the unorthodox advancement of knowledge he is still being punished by his colleagues for his use of the very technics and principles with which he has been able to clarify that problem.

Now I should like to offer for your consideration some thoughts concerning the broader field of the automatic internal compulsions which influence all animal life including, of course, human existence.

The Creator put the normal internal compulsions and regulators into man far in advance of the intelligence and will. Long before he knew *why,* man was inwardly advised as to the vital necessity of the "balanced" diet, the minerals and the vitamins; he never needed a timetable for breathing, nor a time schedule for "planning" his family. The automatic internal compulsions and regulators took care of all those matters.

Then Nature in her apparently blundering course through trial and error committed her most *patent* mistake in teaching animals to protect themselves against "infections" through the production of antibodies and other specifically reacting substances. The "mistake" became apparent when it was found that some of the antibody reactions released or produced sometimes dangerously injurious substances causing disease or even death. This condition of specific sensitivity has been given the general name of allergy, which is in popular use.

You have heard from Dr. Randolph that the disease of specific sensitivity (allergy) is capable of generating the abnormal compulsion known as alcoholism. But notice the astonishing nature of that compulsion — namely, that it is not caused by the alcohol but by another substance (in my case, by house-dust). This case alone would seem to exclude the principle of drug-addiction in explanation of alcoholism — a broader view of the situation seems called for.

Dr. Randolph has called attention to other abnormal compulsions; especially the tendency of some food-allergic persons to crave their food-allergens, which he considers as specific. I have not encountered such a case, but I have never tried to identify one such.

Other abnormal compulsions come to mind which may belong in this category: drug-addiction stands out; then I should suggest kleptomania and homosexuality.

In closing, I see the problem of alcoholism as follows: The dearest wish of the victim of alcoholism is to be able to drink alcoholic beverage in unlimited quantity without suffering the dreadful physical and other consequences. It is not the compulsion of a normal appetite that he wants to satisfy but an abnormal craving which is a manifestation of an inherited constitutional disease.

Dr. Randolph's demonstration of the fundamental cause of the abnormal craving does not enable the alcoholic to satisfy that craving as he wishes; it only teaches him how (with constant self-denial of many of the pleasures of life) *to get rid of the craving.*

The courageous course would still seem to be that urged upon the subjects of alcoholism by the responsible leaders of the organized preventive societies; namely, the complete avoidance of alcoholic beverage. The struggle against the other manifestations of "nonreaginic food allergy" will of course continue on its own important account.

Arthur F. Coca, M.D.
November 21, 1959

On December 13, 1959, I wrote the following long hand note on the margin of this discussion: "This is one of the last things written by Dr. Arthur Coca. It was written, apparently, after the onset of his terminal illness resulting in his death, December 12, 1959. When I was in New York City December 1, he was unable to talk, according to his wife. He apparently continued on a downhill course, according to a telephone call which I had this morning with Leonard Green, the manufacturer of Dust-Seal and a great friend of Dr. Coca."

In view of Dr. Coca having mentioned the product, Dust-Seal in this and certain other letters, I wish to record my experience with this simple hydrolysate of mineral oil. I found that applying Dust-Seal to fabrics, especially

A Period of Transition

bedding, upholstery and carpeting, was helpful in reducing dust exposures in selected dust sensitive patients. But many patients, who were also susceptible to environmental chemical exposures, reacted much more acutely to sitting in a "Dust-Sealed" chair than sitting in a chair which had not been Dust-Sealed. Although I reported these observations to Dr. Coca, he apparently did not believe the reliability of these observations. I failed to use the product, Dust-Seal after these observations.

I wrote the following letter to Mrs. Coca:

December 26, 1959
Mrs. Arthur F. Coca
425 Grant Avenue
Oradell, N.J.
Dear Mrs. Coca:

I am so sorry that I missed seeing Dr. Coca on my recent trip to New York City. I appreciate the time and effort that he spent in writing the discussion of my alcoholism presentation. This must have been one of the last of his amazingly long list of medical contributions.

I regret that I was unable to attend his funeral.

Dr. Coca not only enriched the lives of many people, but added tremendously to medical knowledge. His contributions as far as food allergy and related fields are concerned seemed to be, at times, against an almost organized resistance. Nevertheless, he kept at it. Since his views are basically correct, someday he will be given full credit. Although it is unfortunate that this did not come sooner, the success of his recent book must have been a great source of satisfaction to him.

Leonard suggested the possibility of a letter to the letter section of the New York Times about Dr. Coca's work. I tried to write such a letter but did not come up with anything with which I was satisfied.

I hope that you remain well. Please let me know if at any time I may be of help to you.

Very sincerely,
Theron G. Randolph, M.D.

Max Gerson

In a somewhat similar manner, Dr. Max Gerson remained the center of a raging controversy in regard to the role of environmental exposures in the etiology of cancer and many other chronic degenerative diseases, including allergies. (Gerson M. *A Cancer Therapy, Results of Fifty Cases*. Del Mar, California: Totality Books, Publishers, 1958) A review of Dr. Gerson's point of view is beyond the scope of this presentation, except to say that his overall approach to degenerative diseases was detoxification — the removal of toxins and poisons from the body, on the one hand, and minimizing their intake, on the other. He said: "We cannot detoxify our bodies when we add poisons through our foods which is one of the reasons why cancer is so much on the increase." Gerson's basic therapeutic program was to use a diet of fresh vegetables and fruits, organically grown as much as possible. Despite continued criticism of Dr. Gerson's point of view, there has been increasing evidence in recent years that the chemical contamination of the human environment is a major contributing fact to the current level of degenerative diseases, including cancer.

Both Dr. Coca and Dr. Gerson encountered the organized opposition from the medical profession to their beliefs and the medical applications of these beliefs. It is significant that their deaths, both in 1959, mark the end of one environmental-medical era, and 1960 the beginning of the third great thrust of environmental-medical interrelationships. With this transition, the burden of defending the role of the environment in medicine passes to a younger generation.

Environmental-Medical Interrelationships — Further Developments

The history of environmental-medical thrusts in America is conveniently divided into three major eras. As previously mentioned, the first major environmental thrust having profound medical implications started with acceptance by academia of the germ theory of disease in 1880.

In contrast to infectious diseases, which may involve all persons to various degrees at different times, the second great environmental thrust — allergic illnesses — involves persons more selectively, at least as far as more overt clinical manifestations are concerned. This selectivity is based primarily on the presence of individual susceptibility and, secondarily, upon exposure to the environmental excitant(s) in question. For instance, only pollen sensitive persons are subject to hay fever or asthma on that basis when specifically exposed. In contrast, the majority of people who are not specifically sensitive are unaware of a given pollen season. As also previously noted, Englishmen identified grass pollens and animal danders responsible for rhinitis and asthma in the 1800s. M. Wyman, in America, was the first to recognize ragweed pollen as a cause of ragweed hay fever in 1872 (*Autumnal Catarrh*. New York: Hurd Houghton). However, the main thrust of clinical allergy developed after the turn of the century, following the description of anaphylaxis in laboratory animals.

Also, in contrast with infectious disease which has become widely accepted and applied, this is not the case with the allergic concept of disease. There are several reasons to account for this disparity. One of these reasons is the prolonged controversy involving this subject. Clinical allergy, developing on the heels of infectious diseases, was also interpreted on the basis of immunologic mechanisms by some, and finally accepted by medical academia as limited to antigen-antibody relationships in respect to its underlying mechanisms, despite strenuous objections by Coca, the leading theoretician in immunology in the United States at the time and editor of the Journal of Immunology. Moreover, the unfortunate acceptance of allergy as an immunologic response prior to the description of several major clinical manifestations demonstrated to be on the basis of individual susceptibility — many of which are not characterized by antigen-antibody responses — has further confused this subject.

Differences of opinion within the medical profession are sometimes referred to as the town versus gown debate. Whereas medical academia in America had plunged precipitously in favor of the germ theory of disease in 1880, while practicing physicians defended the status quo, these roles in respect to environmentally related medical care of allergic diseases are reversed. Academicians, including those responsible for the medical curriculum and physicians dominating concepts and views in the national allergy societies, are presently defending the bodily and immunologically focused status quo in medicine, in respect to allergy, against the environmentally focused and nonimmunologically explained views of a growing minority of physicians and a growing number of patients. Moreover, these differences are sharpening as full-time faculty members are becoming increasingly dominant in medical schools.

The year 1957 also marked the transfer of my hospital from the St. Francis in Evanston, Illinois to the Swedish Covenant Hospital in Chicago. The principal reason for this change was the increasingly widespread nature of my practice and the fact that admission policies of the St. Francis Hospital were largely geared to the hospitalization of local patients. The concept of comprehensive environmental control, first appreciated at St. Francis, was continued at the Swedish Covenant Hospital — in both instances observing patients in single rooms or two-bed rooms with both patients being on this program. As a result of several opportunities of presenting this concept of illness and techniques to demonstrate these environmental relationships to members of this hospital medical staff, a better rapport with other physicians developed here than in any previous or subsequent staff affiliation.

Several of us, interested in the application of this medical point of view in respect to mental and behavioral aspects of illness, organized the first section of the American College of Allergists called Allergy of the Nervous System in 1957. I was elected the first president of this group. Yearly half-day meetings were held in conjunction with the annual meeting of this parent organization for the next several years.

The first time that I used the word, ecology, in a presentation was in conjunction with Dr. Donald S. Mitchell in 1958 in the title of a program entitled, Specific Ecology and Chronic Illness.[117]

In 1959, I was invited by Norris Nesset, the chairman of the board of trustees of the Lutheran General Hospital, in Park Ridge, Illinois, to join the staff of that new hospital. I had had a close association with Dr. Nesset in his former position with Baxter Laboratories, in that this institution had financed a grant which defrayed the expense of making the first recorded motion picture of an acute psychotic episode following the intubation of an allergenic food (beets) in 1950.[79] The diagnostico-therapeutic program of comprehensive environmental control was also carried out at the Lutheran General in individual rooms occupied by either one or two patients, pending the construction of the first comprehensive environmental control unit at that hospital. More will be said about this later.

In 1959, a scientific exhibit, in conjunction with Doctors Donald S. Mitchell of Montreal, Canada and Richard Mackarness of Kew, England, entitled, "A Practical Nonpsychiatric Approach to Mental and Related Physical Ills," was presented at the Joint Meeting of the British Medical Association and the Canadian Medical Association in Edinburgh, Scotland. This exhibit, essentially the same as shown earlier, follows the same V-shaped format, but by this time consisted of four instead of five levels.

A photograph taken at the time of this scientific exhibit (Figure 19) shows Dr. Mackarness (left) demonstrating the "Pick Up" Intensities and the Depths of Hangovers in the "V" shaped area of the exhibit at the right. I am standing at the right. The actual levels of reaction displayed at this time are shown in Figure 20.

I had the opportunity of speaking several times in both England and Scotland including one presentation before the Royal College of Physicians in London. These presentations on the subject of clinical ecology prepared the way for the formation later of the British Society for Clinical Ecology.

Figure 19. Scientific exhibit in Edinburgh, Scotland with Dr. Richard Mackarness of Kew, England and Dr. Theron G. Randolph.

A Period of Transition

LEVELS OF REACTION

Minor symptoms precede major disease,
as will be illustrated in the following chart

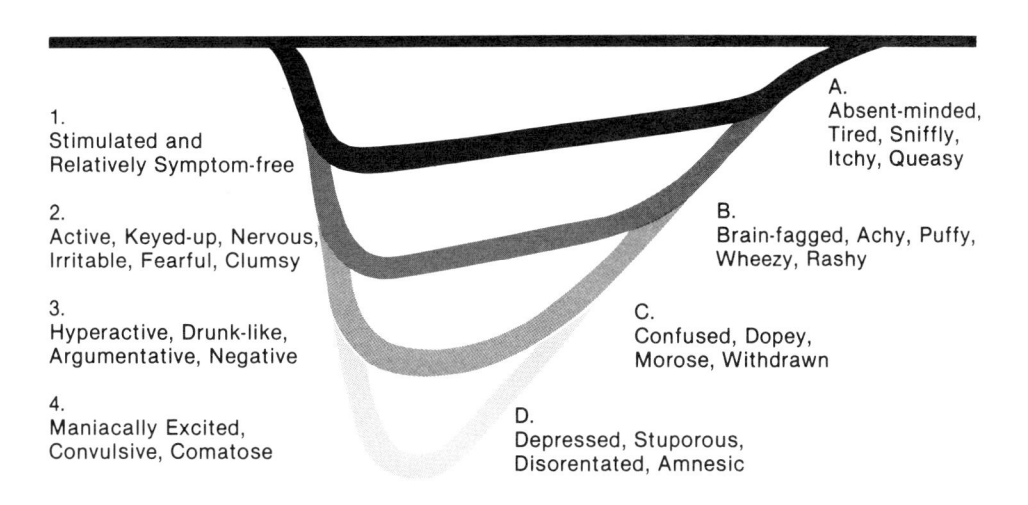

"PICK UPS"   **"HANGOVERS"**

1. Stimulated and Relatively Symptom-free

2. Active, Keyed-up, Nervous, Irritable, Fearful, Clumsy

3. Hyperactive, Drunk-like, Argumentative, Negative

4. Maniacally Excited, Convulsive, Comatose

A. Absent-minded, Tired, Sniffly, Itchy, Queasy

B. Brain-fagged, Achy, Puffy, Wheezy, Rashy

C. Confused, Dopey, Morose, Withdrawn

D. Depressed, Stuporous, Disorentated, Amnesic

Figure 20. Levels of reaction in stimulatory-withdrawal phases.

Chapter VII — 1960

The Environmental-Medical Watershed

General Review

The year 1960 might be regarded as something of a watershed or dividing line between the relative absence of environmental concern, generally, and a period of increasing interest in the environment in respect to its pollution and effects on human health. For instance, prior to 1960 the term "ecology" had little meaning for most persons, but after 1960 the term ecology and what it stands for — interrelationships between an individual and that person's intake and surroundings — became a household word.

As a brief review of the earlier part of this book, as well as a foretaste for the remainder of it, the following table summarizes environmental and health interests in the United States by double decades before and after 1960.

Dates in Scores of Years	Environmental-Medical Interrelationships	U.S.A. Presidents
1861 - 1880	Gradually increasing acceptance of the germ theory of disease in Europe	Lincoln, Johnson, Grant, Hayes
1881 - 1900	Precipitous acceptance of the germ theory of disease by academia in United States	Garfield, Arthur, Cleveland, Harrison
1901 - 1920	Allergic concept of disease (descriptively) — gradually increasing environmental pollution	McKinley, Taft, Roosevelt, Wilson
1921 - 1940	Allergic concept of disease (immunologically) — progressive environmental deterioration	Harding, Coolidge, Hoover, Roosevelt
1941 - 1960	Environmental Pollution debacle — with rapidly increasing popular concern	Roosevelt, Truman, Eisenhower
1961 - 1980	Public awareness of environmental deterioration. Enactment of federal regulatory statutes	Kennedy, Johnson, Nixon, Ford, Carter
1981 -	Politicization of environmental movement; De-emphasis of environmental regulation	Reagan

Coincident with this interest of the public in the environment and its medical effects, both Democratic and Republican administrations enacted environmental protective legislation. This started with the Environmental Policy Act (EPA) and ended with the Toxic Substances Control Act (TSCA). Indeed, one of the last functions of retiring President Ford was signing this latter legislation. But when President Carter brought many environmentalists into his administration, this so alerted corporate and industrial interests that they opposed much of this activity, with the result that the environmental record of the Carter administration was essentially an environmental stalemate. Much of the favorable environmental policies enacted during the score of years 1961-1980 has since been significantly modified by the Reagan administration.

Environmental pollution, starting with World War I, has increased gradually, especially during and following World War II. This justifies reference to the years between 1941 and 1960 as an environmental pollution debacle.

Also coincident with these changes, medicine became increasingly bodily centered, excessively analytical, environmentally alienated, progressively dependent on drug suppression of symptoms and increasingly costly. Economic measures to hold down costs during the present administration seem to be interfering with the quality of medical care.

Other Activities

My own activities in 1960 will now be described against this backdrop. To meet the demand for organically grown produce for patients diagnosed as having the chemical susceptibility problem, I organized the organic growers of northern Illinois to provide this type of food in 1960. Also, to meet the requests of this group of patients for others with whom they might fraternize, the Chemical Victims organization was set up. This support group, later called the Human Ecology Study Group, has held regular meetings since this time.

In keeping with the general direction of physicians practicing this type of medicine, the Board of Directors of the Rockwell M. Kempton Medical Research Foundation changed the name of this organization in 1960 to the Human Ecology Research Foundation.

Finally, by the year 1960, the general direction of environmentally focused medical care had reached much the same overall status as it enjoys today. Since this was expressed in a short article entitled, "A Third Dimension of the Medical Investigation," published by me in Clinical Physiology in 1960,[125] this is being reproduced herewith.

> The investigation of the patient's sickness needs a third dimension. The patient's medical workup — consisting of a history, physical examination and symptomatic treatment — originated when medicine was dominated by descriptive concepts of illness. But now that descriptive medicine is giving way to medical management, based on

mechanisms and causative factors in disease, another tool is needed.

The planned maneuver of the patient in respect to his diet and environment is neither purely diagnosis nor treatment, but bridges the two with features of both. Medicine has been approaching this course for some time. Although aspects of this third dimension of the patient's investigation may be recognized, the totality of the maneuvers to be described is relatively new. It is best presented from an historical point of view.

Whereas formerly the diagnosis of a case rested principally on the history and physical examination, in recent years laboratory procedures have become increasingly important diagnostic aids. Treatment, no longer entirely symptomatic, has been more concerned with the correction of faulty mechanisms as determined by laboratory diagnostic procedures. This increasing emphasis on finding and correcting abnormalities of physiology and chemistry has resulted in great medical strides. But, unfortunately, certain relatively undesirable changes have accompanied these advances.

Associated with this tremendous increase in detailed knowledge has been a progressive tendency to consider the patient as well as the diseases under investigation as collections of constituent parts. Anatomically and technologically demarcated medical specialties and subspecialties have changed long established physician-patient relationships. Indeed, there is some basis for the complaints that doctors are becoming technicians interested only in parts of patients.

Neither has the investigation of medical problems by an analysis of them provided satisfactory answers to many chronic illnesses. Although various limited mechanisms in medicine may be understood — even to the point of their chemical makeup — more is required than a mere summation of analytically derived details. Drug treatment, based on limited mechanisms in defiance of the body's dynamic response to the introduction of foreign materials, often defeats a practical application of such knowledge. Interactions between isolated parts and induced reactions to chemical drugs have very often introduced new complications of old illnesses.

The body as a whole is known to respond to a wide range of environmental exposures in a basically similar manner — that is, by adapting to them. The pattern of these responses in animals has been described by Selye as the "general adaptation syndrome." Although the same merging stages of adaptation occur in humans, the response in man seems to be more specific to the excitant, more individualized to the person and associated with greater degrees of susceptibility than is generally recognized.

A person adapting or partially adapting to one or more common dietary or other environmental materials to which he is susceptible and which are responsible for his chronic illness rarely ever suspects them. But if oft-repeated exposures are avoided until such an individual has recovered from the effects of the most recent specific dose, an isolated re-exposure then induces an acute immediate reaction. This maneuver, which causes an adapted or partially adapted stage to revert to a nonadapted stage, not only changes chronic illness to acute illness but also clearly demonstrates its causation. When chronic illnesses are investigated in this manner, exciting and perpetuating factors of a surprisingly large number may be demonstrated.

Since maladaptation and susceptibility to a wide range of allergens, irritants, and lesser doses of toxins and poisons tend to occur in a given person, and as these are not ordinarily detected from the history, physical examination, X rays, skin tests and other laboratory procedures, some more fundamentally sound medical approach is needed. And since treatment aimed at the control of limited mechanisms is at best only palliative and often compounds the original illness, a more rational basis of therapy is also required.

In addition to the time-tested tools of the history, physical examination and laboratory tests, a more accurate method of demonstrating causative factors of chronic illness is to observe the patient while multiple suspected and probable environmental exposures are being avoided, then subsequently returned one at a time. By use of this method, the following program of comprehensive environmental control and subsequent test exposure gradually evolved.

Patients are fasted in hospital quarters relatively free of chemical odors and fumes as well as pollens, spores, danders and dusts. Indoor air pollution is controlled as far as possible by means of passing incoming air through activated carbon and mechanical filters. Patients also avoid all drugs, cosmetics, tobacco and synthetic wearing apparel and drink only spring water.

Symptoms tend to be accentuated at first, usually reaching a peak on the third or fourth day. Chronic manifestations resulting from foods, previously eaten regularly, tend to improve; those from susceptibility to chemical additives and contaminants of foods, air, water, synthetic drugs and related chemical exposures subside more slowly. Although the average duration of the fast is five days, it is sometimes continued seven to ten days and occasionally longer.

Indications for breaking the fast are a relative absence of or a marked decrease in previous chronic symptoms for a period of at least 24 hours; satisfactory sleep, and a stabilization of the pulse rate at significantly lower levels. The fast is broken by a meal of unsuspected and chemically less contaminated specific food. Other chemically uncontaminated foods are returned one at a time, preferably not more frequently than twice or thrice daily, which favors the observation of both immediate and delayed objective and subjective effects. This routine incorporates the basic principles of Rinkel's individual food ingestion tests. Pre and post ingestion pulse determinations, as recommended by Coca, help in interpreting borderline reactions.

After all foods used in some form once in three days or more frequently have been tested in this way, susceptibility to the chemical additives and contaminants of the diet is investigated. This is done by feeding seldom eaten foods known for their chemical additives and/or contaminants in three daily feedings for at least two days. These menus usually consist of canned blueberries sweetened artificially, canned peaches, canned dark red cherries, canned salmon, canned tuna, frozen broccoli, frozen cauliflower, frozen spinach, raw apple, raw celery and the outside leaves of head lettuce. These foods are selected because of their spray residues and chemical contamination arising from can linings (the golden brown inner lining of metal tins, usually a phenolic resin) and because of their availability in chemically less contaminated form. In addition to these two major sources, foods are also frequently contaminated chemically by fumigation, fungicides, sulfur treatment, artificial coloring and sweetening agents, artificial ripening procedures, protective waxes and certain packaging materials.

It should be emphasized that the clinical effect of chemical additives and contaminants of the diet not only involves a highly variable factor of individual susceptibility but is also cumulative. At least ten days may be required for the first evidence of recovery after their removal from the diet and surroundings. In instances of extreme susceptibility, acute reactions may follow the first ingestion of chemically contaminated foodstuffs, but at least two days of continued use may be necessary before a susceptible but relatively symptom-free patient may manifest convincing symptoms. At other times, the immediate effects of a chemically contaminated diet may be stimulatory. This phase is then followed by the recurrence of any allergic-type response or by a wide range of constitutional symptoms not ordinarily regarded as allergic. The most common of these are chronic fatigue, headache, myalgia, arthralgia, arthritis, depression, and certain other musculoskeletal, neurological and/or mental syndromes. Susceptibility to foods per se or to their chemical additives and contaminants may give rise to identical symptoms. Although either may occur alone in problem patients, they most commonly coexist.

Acute ingestive reactions resulting from test doses of foods (or their chemical contaminants) are treated by emptying the gastrointestinal tract as soon as possible after cause and effect relationships have been demonstrated. This is best accomplished by administering 10-15 grams of mixed alkali salts (2/3 $NaHCO_3$ and 1/3 $KHCO_3$) orally in approximately a quart of spring water. This not only induces a saline laxative action but also combats the apparent mechanisms involved in keeping with the acid-anoxia-endocrine-enzymatic concept of it.[68,69,105]

Then with the patient relatively symptom-free on a compatible intake of food and water, he is returned successively to his home, his water supply, work and avocations. A recurrence of reactive symptoms during the first 48 hours at home usually indicates susceptibility to indoor chemical air pollutants — house dust, animal danders, silk or other home exposures to which susceptibility exists. Chemical air pollutants in the home have been demonstrated in the following order: odors of home utilities — especially those of gas kitchen ranges, gas or fuel oil space heaters and water heaters; warm air heating systems; sponge rubber bedding, padding and upholstery; odorous plastics, insecticides, perfumes and certain others.

The writer has found that the planned maneuver of the patient in respect to his total environment (as far as this is possible) answers questions of the etiology and treatment of certain chronic illnesses more accurately than other available programs. Indeed, now that he is accustomed to this routine, he would be relatively ineffective in the practice of medicine without this third dimension of the medical investigation.

The name "third dimension" is well chosen. In comparison, the standard diagnostic and therapeutic approaches to illness may be likened to lines at different angles on the same plane. The experimental maneuver of the patient in respect to his diet and environment not only adds a specific element to medical management, but also provides an infinitely greater depth and understanding of the clinical pictures presented by patients.

The addition of the third dimension to the medical investigation of patients needing it causes certain chronic disease syndromes to revert to more readily observable acute reactions and demonstrates their etiology. This enables a physician to instruct his patients in how to avoid the causative factors of illness — a program which is generally preferable to merely treating the effects of illness.

Limiting interpretations of etiology to demonstrable cause and effect relationships has the additional advantage of minimizing psychogenic and other alleged inner "causative" forces neither measurable nor demonstrable.

Another desirable feature of this program is that a patient is regarded as a whole person, responding to his or her environment as an integrated unit, rather than as a collection of fractionated parts. Being a joint endeavor, it is based on the physician's know-how and the patient's compliance with recommended instructions; its effectiveness depends upon the observations of both. Not being done on, to, or even for the victim of the illness, but with him, it tends to bring the patient into the routine of the medical investigation. Changing the status of the patient from that of a passive observer to an active participant, as well as the overall effectiveness of the program and absence of overtreatment by drugs, eliminates many of the criticisms of present-day medicine.

Finding and avoiding causes of chronic illness by means of the planned maneuver of the patient in respect to his or her total physical environment is infinitely more satisfactory than a patchwork treatment of effects and/or of limited mechanisms of many chronic diseases.

The planned maneuver of the patient is a long-neglected phase of the medical investigation. It supplements rather than supplants the history, physical examination, diagnostic laboratory procedures and standard therapeutic approaches. Being neither primarily diagnostic nor primarily therapeutic, it is related to both in the sense of being a common extension of both.

Chapter VIII — 1961-1964

The Beginning of an Ecological-Medical Focus

Appraising the Totality of the Chemical Environment

After trying unsuccessfully for a decade to integrate the various facets of the chemical susceptibility problem into a comprehensive presentation, this was finally accomplished by 1961 with the publication of four manuscripts in successive issues of the Annals of Allergy.[129,132,133,134] Each time that this had been attempted earlier, some relatively new clinical observation developed which not only required inclusion, but also an evaluation in terms of its relative importance. For instance, the first patient in whom this clinical syndrome had been recognized had been living for several years in an all-electric home and did not complain of reactions to the fumes of gas utilities. It was not until several other patients had been diagnosed did the significance of fumes from gas burning utilities in the home become apparent. And only later did the preeminence of pesticide exposures become apparent and fumes from gas burning home utilities become of secondary significance.

One of the most difficult concepts to grasp was that the so-called chemical susceptibility problem was more than the sum of its constituent pieces. Although this medical problem might be initiated as the result of a single massive or relatively sustained environmental exposure, once this had occurred individual susceptibility usually had spread to involve other combustion products and derivatives of gas, oil and coal. By the time that a person's health and/or productive capacity became sufficiently impaired for one to seek medical advice, this problem had usually become widely spread and complicated.

Although the chemical susceptibility problem presents as both localized and systemic syndromes, it is especially apt to manifest in such advanced constitutional illnesses as headache, fatigue, muscle and joint aches and pains (myalgia, arthralgia and/or arthritis), "brain-fag," depression and a wide range of cardiovascular manifestations.

It is relatively rare for the chemical susceptibility problem to exist as the only environmentally related illness. At least by the time when seen medically, the majority of cases are also susceptible to foods, and many also are sensitive to pollens, house dusts, animal danders and/or other biological inhaled particles.

From the standpoint of the type of environmental chemicals incriminated, pesticides seem to be the most hazardous, followed closely by odors and fumes from solvents and fuels, including their combustion products, synthetically derived drugs, textiles, dyes and many other food and water additives and contaminants.

From the standpoint of the portals of entry involved, inhalation, ingestion and contact seem to be incriminated in that order of relative importance.

Although initial manifestations were largely localized to occupational exposures and confined largely to the urban work place, chemical contamination of the environment now occurs about equally as a rural and urban problem. Indeed, with the direction of modern agriculture, there seems to be little difference between industrial and agricultural pollution. Housing, whether urban or rural, tends to be chemically contaminated to about the same degree as far as inherent sources of pollution are concerned.

From the standpoint of presentation, the chemical susceptibility problem is best divided into indoor (domiciliary) and outdoor (ambient) air pollution, food and water additives and contaminants, synthetic drugs as well as synthetically derived clothing, toiletries, etc. One must recognize, however, that these subdivisions are arbitrary and tend to detract from the way the problem manifests clinically.

It must be recognized that the chemical susceptibility problem overlaps both toxicity and individual susceptibility. These relationships will be discussed in greater detail later.

Another equally important consideration is in respect to the acquisition of the chemical susceptibility problem that constitutes an injury, from which a person so injured never completely recovers. For instance, never again will a highly susceptible person ever tolerate exposures to incriminated environmental agents as formerly. In short, these materials referred to as GRAS (generally regarded as safe) cannot ever be so regarded by such a person in the future. Although avoidance and improvement in general health may enable one to tolerate greater amounts with less reactions or the responses induced may be less severe and shorter in duration, complete reversion to the normal GRAS status cannot be expected.

These four chemical susceptibility manuscripts were indexed and republished in book form in 1962.[137]

Organic Growers

Although organizing the organic growers of northern Illinois was helpful in the production of chemically less contaminated produce, it did not solve its distribution. Mr. and Mrs. Thomas Barry, whose hyperactive son required this type of chemically less contaminated food, had been encouraged by other patients to double the size of their garden each year since 1960, as relatively pesticide-free foods were still insufficient to meet demands for this type of produce. Consequently, a meeting of representatives from 80 families requiring such food was held at the Lutheran General Hospital in 1962 for the purpose of raising funds to set up Mr. Barry in the full time production of organic garden produce on 12 acres of land. Mr. Barry resigned his foreman's job. His life was insured. Both Mr. and Mrs. Barry spent full time in raising and selling this produce in the store on their farm located 60 miles west of downtown Chicago. Ten-thousand dollars were contributed for the purpose of buying a tractor, truck and other farm equipment for growing garden produce. An account of this venture in growing natural food on a cooperative farm set up by consumers was published in 1962.[141]

Although this solved the seasonal availability of organically grown food, we were unable to finance the freezing, canning and drying of seasonally available supplies. This being left for the consumers was unsatisfactory, in that many were unable, unwilling, did not know how or did not have adequate space and storage facilities to do this. Gradually, this retail store was able to obtain year-round supplies of fresh produce from southern California to fill this need.

It should be emphasized that within our group of purchasers we had a built in system to determine whether or not the produce was or was not reliable. On each of several different occasions we were able to detect chemically contaminated produce on the basis of epidemics of illness amongst our consumers. Since most of the purchasers were also members of Chemical Victims, this type of multiple reaction was spotted very quickly. For instance, I learned that four different patients had become sensitive to chicken for the first time within a week. Checking back, the organic farmer had made no changes in growing and selling the chickens, but the slaughter house found that if the chicken carcasses were swabbed down with detergents they looked better. This was changed immediately and all of the "new chicken sensitive patients" recovered promptly! Other examples could be cited in which lying producers and fudging middle men have been detected by organized sleuthing consumers whose decisions have been based on "inside" tips.

Pesticide Hearings

At the suggestion of Senator Hubert Humphrey of Minnesota, I submitted testimony at Hearings of the Interagency Coordination in Environmental Hazards (Pesticides); Of The Committee on Government Operations, U.S. Senate, 88th Congress, Part II, July 17, 1963, under the Chairmanship of Senator Abraham Ribicoff of Connecticut. The gist of this testimony was the statement that pesticide exposures are the single most important part of the chemical susceptibility problem — essentially the crux of that clinical problem — as these combustion products and derivatives of gas, oil and coal are encountered in daily life.[150]

The National Conference on Air Pollution

Supported by a ground swell of public interest and indignation, the National Conference on Air Pollution in Washington, D.C. on December 10, 1962, based on this public concern, initiated a wave of environmental regulations lasting for the following one and one-half decades. President Kennedy grasped the significance of this meeting with the following announcement:

THE WHITE HOUSE
Washington

I am pleased to send greetings to the delegates and participants attending the National Conference on Air Pollution.

It is imperative that this nation act to preserve now, and for the years ahead, the purity of its air. The pollution of this priceless resource continues to jeopardize the economic vitality of our nation and the health of millions of our citizens. This fact we can neither condone nor tolerate; we can and must, instead, resolve to use every appropriate means in a concerted effort to clear the air of its burden.

It is my hope that this Conference will provide the wise counsel and informed guidance that will hasten the needed action on this problem. I urge every American to lend his full support to this endeavor.

With all good wishes for a highly successful meeting,

John F. Kennedy

Although this meeting was well publicized in advance, there was no way to arrange to participate in it because the agenda and participating personnel had been "canned in advance," as is so common in such sessions organized by governmental agencies. Neither was there any provision made for spontaneous discussion from the floor. The only time allowed for discussion in the preplanned agenda was one and one-half hours at the end of the session!

I was the second person from the floor recognized to speak. I made the following comment as published in the Proceedings of the National Conference on Air Pollution, 1963, page 517, as follows:

Theron G. Randolph. I am speaking as a practicing internist and allergist.

Attempts to appraise the clinical significance of the air pollution problem, as far as chemical exposures are concerned, must begin with the realization that there are two major divisions of the subject. One has to do with outdoor chemical air pollution, which you have heard discussed from various standpoints. Equally important is the subject of indoor chemical air pollution. In my experience, this indoor air contamination is such a frequent cause of chronic illness in susceptible persons that it must be evaluated and preferably controlled before one is justified in drawing deductions as to the clinical significance of chemical contamination of the outdoor atmosphere.

Although outdoor chemical air contaminants enter homes located in contaminated areas, chemical air contamination arising within a patient's living quarters may occur irrespective of the location of such dwellings. The major sources of such contaminants are odors and fumes arising from leaking utility gas or the combustion products of gas, oil, or coal. The gas kitchen range, gas-panel heating units, and fuel oil space heaters are the most significant, although appreciable air contamination is also derived from hot air furnaces (irrespective of the type of fuel used), sponge rubber padding, bedding and upholstery, insecticides, paint odors, disinfectants, and various other odorous household materials. Contaminated air from the garage also often enters the house.

These relatively constant sources of indoor chemical air pollution are rarely suspected as inciting and perpetuating causes of chronic illness. Only as patients are maneuvered in respect to these exposures may acute reactions be observed which demonstrate cause-and-effect relationships. For instance, on the basis of observing clinical effects in moving patients in and out of their homes and, later, of maintaining the patient in his home but removing and then replacing his gas kitchen range, over 800 such devices have been removed permanently from the homes of highly susceptible persons.

A wide range of chronic illnesses result from such day-in and day-out hydrocarbon exposures. The most serious are depressions and other advanced psychotic states.[131] Lesser grade cerebral reactions manifest as mental confusion and "brain-fag" as well as physical fatigue. Closely related manifestations are rheumatism, arthritis, myalgia, neuralgia, headache, and related musculoskeletal and neurological syndromes. Any of the responses ordinarily considered as allergic, especially stuffy nose, coughing, bronchitis, bronchial asthma, are also commonly on the basis of susceptibility to airborne chemical contaminants.

Finding and avoiding these home environmental incitants impinging on the physical and mental health of susceptible persons is opening a new experimentally orientated medical approach to many chronic illnesses.

Dr. Goldsmith. I should like to say one thing about Dr. Randolph's very interesting observations, which doesn't in any sense conflict with them. That has to do with the value of staying at home during periods of severe air pollution. It's important, as Dr. Randolph emphasized, from all the facts we know, and they of course are not adequate, that in a severe air pollution episode in any major metropolitan area, including London or Los Angeles, there is every reason to recommend as a public health measure that the people who may be most susceptible to air pollution do in fact stay at home. Perhaps they should keep their ranges off, but at least they should remain indoors.

It should be mentioned that Dr. Goldsmith was regarded at this time as the most knowledgeable spokesman of the Los Angeles "smog" problem emerging as a major concern.

Then from the other side of the room Frank Silver from Martinsburg, West Virginia, identifying himself as a gas engineer, agreed wholeheartedly with my interpretation of the significance of indoor air pollution. This was the first time that I had ever seen or heard of Frank Silver, who has since become a steadfast friend and a major contributor to our soon to be formed Society for Clinical Ecology. This discussion initiated a series of conferences during the next few years on the subject of indoor air pollution.

Herbert J. Rinkel, M.D. — Continued

In 1959, Dr. Carlton Lee of Joplin, Missouri, observed that allergic reactions to foods could be treated or "neutralized" by appropriately dosed injections of specific food extracts. Rinkel postulated that, if true, it might also be possible to provoke acute reactions diagnostically. The two of them worked out the details of the provocation-neutralization technique for food allergy. This was published in the Archives of Otolaryngology in January 1962 (see Appendix page 335).

Herbert Rinkel, whose picture is shown in Figure 21, had been diagnosed as having an inoperable abdominal malignancy in 1962. At about this time, Herb turned his practice over to Dr. James W. Willoughby and spent most of his time preparing a final comprehensive presentation of his work, which was published serially in the Archives of Otolaryngology (see Appendix, page 335).

Figure 21.
Herbert J. Rinkel, M.D.

Ecological-Medical Focus

In view of Dr. Rinkel's terminal illness, four of us visited him on May 12, 1963. The letter to Dr. Rinkel proposing this visit is as follows:

May 7, 1963
Dear Herb:

Just to confirm our telephone conversation. We are descending on you at your home Sunday morning, May 12.

Since you said that it would not tire you out any more talking to several than to one, we are sending quite a delegation. In addition to myself, it consists of:

Dr. George S. Frauenberger — an Evanston pediatrician and long-time friend of mine with whom I work closely and who is using your titration technique and has reported on its use to the College in the past.

Dr. Richard Mackarness of Kew, England, a suburb just to the west of London on the Thames. He is a general practitioner half-time and a medical writer half-time. I have known him for five years and did a joint exhibit with him four years ago at the joint Canadian-British Medical Association Meeting in Edinburgh. He is in the midst of writing a popular book in which our various views are brought up-to-date.

Dr. Marseille Spetz, a woman internist from northern California who worked with me in the Allergy Clinic at the University of Michigan Medical School, just before I came to Chicago. She is currently working for Bob Godlowski, checking manuscripts and helping him generally with his writing.

Then there is Bob Godlowski.

We leave here on the San Francisco Chief and arrive in Kansas City at 10:25 p.m. We have reservations Saturday night at the Muehlebach. We will have breakfast at the hotel and come out. Please leave word at the hotel when we are to come, or if you would like to come to the hotel. We wish to make this as easy for you as possible. In the event that we do not hear from you, we will call your house Sunday morning.

We leave that evening (Sunday) at 10:55 p.m.

Looking forward to a good visit,

Sincerely,
Ted

Extensive motion pictures were taken in the basement of Dr. Rinkel's home — his work space — as well as in his office. These have been projected a few times at medical meetings. The following letter of acknowledgement was written to Mrs. Rinkel after Herb's death:

June 22, 1963
Mrs. Herbert J. Rinkel
6405 Sagamore
Prairie Village, Kansas
Dear Marie:

It was my good fortune to have worked closely with Herb for twenty years. During this interim, I had confirmed many of his important medical observations. It is most regrettable that this genius did not quite make the trip to France, to give his important paper and to receive the plaudits which he so richly deserved.

Since hearing the news of his death from Lucille Zeller, I have thought many times how fortunate we were in having the opportunity of spending the day with you both in mid-May. May I express my somewhat delayed appreciation to you for making this a long-remembered pleasant and instructive occasion.

I regret that I was unable to attend the funeral. In lieu of flowers, I am enclosing a check for $25 made out to the Herbert J. Rinkel Fund. Although I have not been told that such a memorial fund will be established, this must be the case. In the event that it is named in some other way, I will gladly make over the check accordingly.

Very Sincerely,
Theron G. Randolph, M.D.

P.S. Fortunately, the motion pictures came out very well. I am taking these to France tomorrow and hope to be able to project them in some way as a part of a memorial to Herb. I will let you know the outcome of this. I also hope to have the opportunity of showing these motion pictures to you at some future time. In the event that you should visit or pass through Chicago at any time, please let us know your plans so that arrangements may be made to see you.

The first International Congress of Food and Digestive Allergy was scheduled for Vichy, France, in June 1963. According to plans, Dr. Albert H. Rowe and Dr. Herbert J. Rinkel were to be honored for their pioneering work in food allergy. Herb was looking forward to this meeting, where he had hoped to present Lee's and his observations on the provocation-neutralization technique in food allergy. Unfortunately, he died ten days before the meeting. His paper was presented by his associates, Dr. James Willoughby and Dr. Dor Brown.

Fortunately, however, the Rinkel point of view in allergy has been carried on, as far as yearly instructional courses are concerned, by three independent programs under the direction of Drs. Russell Williams, James Willoughby and Dor Brown. These three men were the heirs designated by Rinkel to continue this point of view in clinical allergy. At first I wondered if it was wise to have three instructional courses in clinical allergy but, since Rinkel's death, it has become increasingly apparent that this course of events has been highly desirable. These courses have been justified on the basis of numbers alone, since it is difficult to arrange the logistics of such an instructional course with more than 300 in attendance. Currently, the Rinkel point of view in food and inhalant allergy is being presented annually not only in these three courses, but also in a course for otolaryngologists sponsored by Dr. Sam Sanders of Memphis, Tennessee. On a broader scale, including instructional courses attended by many allergists, pediatricians, internists, physicians in general practice and others in addition to otolaryngologists, yearly courses have been conducted under the supervision of Dr. Joseph B. Miller of Mobile, Alabama.

Over the years, there has been an increasing activity of university medical and teaching centers in sponsoring educational efforts in this area. Particularly active have been the University of Texas, University of Florida, University of Hawaii, Tulane University, University of Alabama and others. Some of the essentially empirical nature of Rinkel's clinical observations are gradually being supported by measurements of immunologic mechanisms. For instance, Dr. Richard G. Fadal and Donald J. Nalebluff recently held an instructional course in Pennsylvania in which documented measurements of immunoglobulins supported the relative superiority of the Rinkel approach to the management of inhalant allergy, as compared with conventional approaches. At the present level of postgraduate medical education, probably 1500 to 2000 physicians have received or are receiving instruction in aspects of clinical allergy which Herbert J. Rinkel originated or to which he made significant contributions.

It was to be expected that these major changes in approaches to the subject of clinical allergy, in competition with long-standing traditional concepts and practices, would be resisted by advocates of the latter. This resistance has arisen largely from academic medical centers and allergists certified by the American Board of Allergy and Immunology (a conjoint board of the American Board of Internal Medicine and the American Board of Pediatrics). Although several of us have been disappointed that the arbitrary decisions of this certifying agency have excluded otolaryngologists because of their lack of training in internal medicine and pediatrics, there have been advantages of this discriminatory policy. One of the chief benefits has been the independence of the American Society of Ophthalmologic and Otolaryngologic Allergy, its transactions and other journals in the field of otolaryngology. These medical meetings, instructional courses and publications, along with the Annals of Allergy, have been the major avenues through which the Rinkel school of thought in allergy has been presented over the years. Not only has substantial headway been made, but the speed of acceptance of these views is increasing rapidly. The program of the Society to set up a comparative study of the Rinkel and conventional approaches to inhalant allergy, if conducted impartially, might be helpful in settling this continuing controversy.

In addition to being a photographer, Herbert J. Rinkel was also an outdoorsman as shown by his picture in Figure 22.

Figure 22.
Herbert J. Rinkel, M.D.

In summary, Herbert J. Rinkel was a technological genius, an innovator and an inventor with a passion for making cause-and-effect observations of patients and, especially, for measuring them. Under these circumstances, it is not surprising that he should come up with unusual and unique clinical observations. He chose as his main subjects for medical observation, allergic responses to foods, pollens, molds, danders, dusts, insects and other inhaled materials. He is to be credited with having changed the course of both food and inhalant allergy. Although this change in the direction of these fields is still being debated, his basic observations have been widely confirmed and ever more widely accepted. Without doubt, he was the outstanding clinical investigator of his day as far as the field of allergy is concerned.

ANALYSIS OF COCA'S PULSE RATE STATISTICS
Bio-Assay of Food Allergens

Alsoph H. Corwin, Ph.D, Maravene Hamburger and
Francis N. Dukes-Dobos, M.D.

I. Statistical Examination of Daily Ranges of the Human Heart Rate as Influenced by Individually Incompatible Foods

In many cases of food allergy, serological and skin tests fail. A reliable procedure for the bio-assay of food allergens should involve measurements of some physiological variable which could be analyzed by statistical methods. Coca's pulse acceleration method satis-

Ecological-Medical Focus

fies this criterion. However, it was based upon clinical observations only, and before his method can be adopted for objective assay, several pertinent questions must be answered. For example: (1) In a given individual, are the accelerations which are caused by incompatible foods statistically significant or are they within the limits of normal variation of the daily pulse range? (2) How consistently is the ingestion of an incompatible food accompanied by pulse acceleration in a given individual and in a population? (3) Does this pulse reaction occur similarly in laboratory animals? (4) Is this pulse acceleration a consequence of antigen-antibody reaction, either direct or indirect, or is it produced by other physiological or psychological mechanisms?

Conclusions

Our analysis of Coca's data permits the following additional conclusions to be drawn concerning the group of subjects on which his observations were made: (1) All four sets of data, minimum and maximum pulse rates, before and after treatment, fit normal distribution curves satisfactorily, indicating the absence of systematic bias in the selection of data. (2) Avoidance of pulse accelerating substances significantly lowered daily maximum pulse readings. (3) As a rule, persons with low daily minimum pulse readings will have low daily maxima and vice versa. (4) The range between daily minimum and maximum pulse rates decreases as the value of the minimum increases. (5) Table II provides a means for the estimation of the normal daily range of the pulse rate of an individual if the daily minimum is established, thus supplying the clinician with a criterion for the better evaluation of a daily pulse curve in cases in which the diagnosis is difficult.

TABLE II. STATISTICAL PATTERN OF DAILY PULSE RATES RELATIONSHIP BETWEEN MINIMA (X), MAXIMA (Y) AND DAILY RANGES

x	$y-2s_y$	y	$y+2s_y$	Daily Ranges
50	56	63	70	6 to 20
55	60	67	74	5 to 19
60	64	71	78	4 to 18
65	68	75	81	3 to 16
70	72	79	85	2 to 15
75	76	83	89	1 to 14

(6) Daily minima below 49 should not be considered "normal." (7) Daily maxima above 85 are not "normal" except in the special cases in which the minima are above 70. (8) Daily minima above 76 are not "normal." (9) Examination of the low pulse values indicates that disturbances due to emotional excitement of the observers were probably not a critical factor in the observations. (10) The standard deviation of the observations could be diminished if the minimum counting intervals were consistently raised to one minute. (11) Evidence exists that Coca's observers were affected by a bias for even numbers, especially those divisible by four, and by a special bias for the number 100.

(12) The number bias can be essentially eliminated in comparing groups of observations statistically, even though it does increase the standard deviations of the averages. (13) By indicating that statistically significant changes take place on dietary control, this study provides a justification for the further study of the pulse rate as a means of bio-assay in food-induced reactions that do not respond to skin testing or to immunological assay procedures.

It seems probable that the general technique proposed by Coca of reading the pulse systematically throughout the day and controlling the high excursions by means of dietary and environmental control is a valid one for the study of the "normal" human heart. The deficiencies of this study are mainly technical. With better counting techniques, and particularly with counting intervals no shorter than one minute, it should be possible to reduce the standard deviation of the observations. These improvements would alter the numerical details of the results and narrow the limits to be regarded as "normal." In future studies on the "normal" human pulse, the refinements in technique suggested by these studies should be taken into account. (Annals of Allergy. Vol. 19:1300,1309,1310, 1961)

II. Food-Induced Reactions of the Heart Rate of Guinea Pigs

Our first paper on bio-assay of food allergens dealt with A.F. Coca's data on pulse acceleration in humans elicited by ingestion of individually incompatible foods. We found that the changes in the heart rates of sensitive patients follow a regular pattern and their magnitudes satisfy the requirements of statistical significance. Some problems which arise in the use of the "pulse-dietary" method as a clinical diagnostic tool have been studied recently by Ettleson and Tuft.

Our long-range objective is the identification of the pure chemical substances which are responsible for these food reactions. Thus for our purposes the major importance of our previous results is that these methods open up a new procedure for the bio-assay of food allergens which cannot be identified by skin testing or by immunological assay procedures. To establish whether these "allergens" are full antigens, haptens, atopens or other biologically active substances would be the next step. The achievement of this goal would help to discriminate between food reactions of true allergic etiology and other food incompatibilities.

Summary

1. Approximately 17,000 electrocardiographic and piezo-electric recordings of heart rates on the same nine guinea pigs have been made over a period of 29 months, the majority of them before and after feeding single foods.

2. Environmental factors such as cage material and cleanliness, atmospheric pollution, startling sounds, unaccustomed conditions,

variations in temperature and fasting time, which affect the heart rates of guinea pigs under experimental circumstances, have been controlled to secure a significant diminution in the standard deviations of pulse rates in fasting animals. In the absence of such disturbing factors, the heart rate of a normal fasting animal was found to be approximately 200 beats per minute at 74-76 degrees F. This value is dependent upon temperature and laboratory temperatures around 85 degrees F. give lower readings.

3. Variance analysis shows that foods may be divided into three categories for each guinea pig individually according to their effects on the heart rate: (a) "Normal," producing an average elevation of 15 beats per minute in heart rate; (b) "Accelerating," producing an average elevation of 36 beats in rate; (c) "Decelerating," producing either an abnormally low acceleration or an actual deceleration in rate, on the average, five beats per minute deceleration.

4. The most consistent accelerator observed was lettuce; the most consistent decelerators were grains.

5. No correlation was found between acceleration and either raw weight, dry weight, or caloric value of the foods.

6. Our results suggest that the effect of foods on the heart rate of guinea pigs which we have observed are not consequences of the specific dynamic effect.

7. Our observations on the deviations in guinea pig heart rates after feeding can be used as a baseline in future investigations for identification and isolation of pure chemical substances which are responsible for these abnormal reactions. (Annals of Allergy. Vol. 21:547,560, 561, 1963)

By the mid-1960s, details of the comprehensive environmental control unit had been worked out, both from the standpoint of the hospital administration and participating physicians, and had been published.[155,156,166] It should be emphasized that experience with this type of diagnostico-therapeutic program is so important, in appreciation of the existence of the chemical susceptibility problem, that the full implications of this problem are rarely appreciated by physicians unless they have the disease personally or have had the clinical experience in working in such a unit.

In 1964, in searching for a simpler way of diagnosing the existence of the chemical susceptibility problem than observation during environmentally controlled circumstances, I began to think of subjecting patients to synthetically derived chemicals whose pharmacology and toxicology had been determined to be safe. Synthetic ethyl alcohol came to mind as it met these requirements. I then remembered that when I had previously attempted to use this substance as a control in testing alcohols derived from specific foods that selected patients had become acutely ill for no apparent reason. In checking out this old protocol, I found that those reacting to the ingestion of synthetic ethanol did so in proportion to the relative severity of their chemical susceptibility problem as this was later appreciated. Testing new patients sublingually into the positivity of the way they replied to the chemical questionnaire ([137] page 25) offered a safe and effective way for testing for the existence of the chemical susceptibility problem. Since this subject has only been reported preliminarily, I will cite that abstract.[165]

The Provocative Hydrocarbon Test, Preliminary Report

Susceptibility to a wide range of commonplace chemical exposures (synthetic drugs and chemical additives and contaminants of air, food, and water) has been demonstrated as a common cause of many chronic physical and mental illnesses.

Although the ingestion of synthetic alcohol (derived from ethylene gas) induced acute reactions in selected cases, this response was too variable to be used diagnostically. Consequently, the chemical reaction syndrome has been detected by history (questionnaire) and by response to single exposures following simultaneous avoidance of multiple probable incitants. This preliminary report, correlating results of questionnaires and injections of synthetic alcohol in 50 cases, suggests the latter as a safe, simple diagnostic procedure.

The provocative hydrocarbon test is patterned after the provocative food test of Rinkel and associates. Serial 1:5 dilutions of absolute synthetic ethanol are called: "1," "2," "3," etc. Subjects are observed following intradermal injection of "3" in 4 sites of 0.025 ml. each on the forearm. If reactive symptoms have not developed after 10 minutes, a "kicker" dose of 0.02 ml. No. "5" is injected similarly above the previous sites. If there is no symptom response after 10 minutes, the test is negative. Positive responses may be systemic, localized, objective, or subjective. They consist most commonly of headache, fatigue, sleepiness, flushing, chilling, hyperactivity, and sometimes aching, mental confusion, depression, and nasal, bronchial, dermatologic, and gastrointestinal symptoms.

A highly individualized smaller intradermal injection "neutralizes" 90 percent of the induced reactions. Start with 0.01 ml. of No. "10." If symptoms are unchanged, apply at successive 10 minute intervals, 0.01 ml. No. "9," "8," etc., until relief is obtained. If No. "10" is followed by increased symptoms, and No. "12" also, apply No. "9," "8," etc., similarly. If No. "12" is ineffective, No. "11" is most likely the "neutralizing" dose.

Initial test doses apparently provoke acute nonadapted responses. "Neutralizing" doses seem to reinduce immediate relatively symptom-free adapted responses.

We now use only one injection or sublingual dose. Determining the treatment dose has been helpful in the management of chemically susceptible persons unable to avoid exposures to perfumes, solvents and certain related environmental chemical exposures.

Chapter IX — 1965-1970

Era of Ecological-Medical Enlightenment
Favorable Governmental Environmental Protection

Increasing Medical Resistance
Formation of Society for Clinical Ecology

Inasmuch as resistance to acceptance of abstracts for program presentations and manuscripts for publication had been increasing as problems by this time, I complained about this increasing resistance on the part of traditional allergists, other physicians and, especially, their organizations to the Board of Directors of the Human Ecology Research Foundation. In their opinion, what was needed was our own national organization to serve the following functions:

1) Regularly scheduled meetings to act as a forum for criticism of developing concepts and techniques before these were presented to the medical profession at large, and 2) Eventually, developing our own national publications. Indeed, the Directors felt so strongly about this that they raised $10,000 for this purpose.

Since my previous threat to form a competing clinically oriented organization had not been taken seriously, and attitudes between several other members of the Editorial Board were hardening, with the exception of the Editor of the Annals, Ethan Allen Brown, those of us attending the Annual Meeting of the College of Allergists interested in this had a meeting April 7, 1965 and founded the Society for Clinical Ecology at the Dunes Hotel in Las Vegas, Nevada.

Despite unanimity amongst us for the need of such an organization, most of the evening was spent in discussing its name. We agreed with my suggestion that the name should include clinical ecology, but debated whether it should include American. In view of the increasing interest of Canadians, we finally decided on the Society for Clinical Ecology. Clifton Brooks was elected president, (since I was regarded as too controversial) and Marseille Spetz as secretary-treasurer.

An announcement of the formation of the Society for Clinical Ecology by Clifton R. Brooks, M.D., Marseille Spetz, M.D., George S. Frauenberger, M.D.,

Theron G. Randolph, M.D. and Jonathan Forman, M.D. was written by Dr. Forman, Editor of the 28th Series, Letters of the International Correspondence Society of Allergists, Worthington, Ohio, 1965.

A Clinical Ecology Study Club had also been started in February 1965 following an instructional course in clinical allergy. This is described in a letter from Thea Hennecke of Windsor, Ontario, Canada to Dr. Lawrence D. Dickey in connection with his account of the History of the Society for Clinical Ecology, published as Part I of the Appendix of the book, *Clinical Ecology*, which he edited and which was published by Charles C. Thomas, Springfield, Illinois in 1976.

The first Annual Scientific Meeting of the Society for Clinical Ecology was held at the Palmer House in Chicago, April 24, 1966. An announcement was published in Science in 1966,[182] the program of which follows:

Program:	Moderator: Jonathan Forman, M.D., Columbus, Ohio
10:00 A.M.	Introductory Remarks, Clifton R. Brooks, M.D., President, Bethesda, Maryland
10:15	Biologic Basis of Medical Practice, Donald Mitchell, M.D., Montreal, Canada
10:45	Role of Specific Adaptation in Chronic Illness, Theron G. Randolph, M.D., Chicago, Illinois
11:15	Ecologic Implications of Individuality, Frederick Sargent II, M.D., Director, Center for Human Ecology, University of Illinois, Urbana
	Moderator: George A. Zindler, M.D., Battle Creek, Michigan
2:00 P.M.	Objective Measurements of Typical Chemical Reactions, Eloise W. Kailin, M.D., Washington, D.C.
2:30	Carbon Monoxide Poisoning, Frank Silver, P.E. (Consulting Environmental Engineer), Martinsburg, West Virginia
3:00	Intermission
3:15	Food and Inhalant Allergy in Guinea Pigs, Alsoph H. Corwin, Ph.D., Professor of Chemistry, Johns Hopkins University, Baltimore, Maryland
4:00	Tissue Eosinophilia of Gastrointestinal and Genitourinary Lesions of Food Allergy Patients, Lawrence D. Dickey, M.D., Fort Collins, Colorado
4:30	Business Meeting

There will be a $5.00 registration fee, payable at the meeting.
*This is immediately before the 22nd Annual Congress of the American College of Allergists, Palmer House, Chicago, Illinois, April 24-29, 1966.

Background:
The Society was organized preliminarily at Las Vegas, Nevada, April 7, 1965, to advance the medical aspects of human ecology.

Whereas traditional approaches in this field have dealt largely with epidemiologic and demographic studies, this organization is interested in the interactions between a person and his or her constant intake and immediate surroundings, as these may be demonstrated to be reflected in his or her total health.

Interest in these studies developed from the observation and recording of cause and effect relationships between given environmental exposures and particular illnesses. Techniques for demonstrating these relationships have evolved. Underlying adaptive mechanisms have been revealed. Thus the inciting and perpetuating causes of many physical and mental illnesses have become apparent and therapeutic and prophylactic procedures clear.

The Society hopes to make information now at hand more generally available while promoting the application of sound ecologic techniques and principles to a wider sector of medical practice.

Clifton R. Brooks, M.D.
Marseille Spetz, M.D.
George S. Frauenberger, M.D.
Theron G. Randolph, M.D.
Jonathan Forman, M.D.

Subsequent meetings of the Society for Clinical Ecology and the Clinical Ecology Study Club have been described by Dr. Dickey in the book, *Clinical Ecology*. On October 26, 1967, the two organizations were combined, with the Clinical Ecology Study Club becoming a committee of the Society for Clinical Ecology.

Since there were wide differences in the training and abilities of the membership, the Grand Cayman Meeting of the Study Club in November 1966 has been considered as the First Seminar of the Society for Clinical Ecology. The purpose of this meeting was to update all those attending to the general level of the most experienced members of the group. I recall lecturing repeatedly for two days. Proceedings of this and all previous and subsequent early meetings were tape recorded by Dr. George S. Frauenberger. I also have taken motion pictures of all meetings.

All expenses incurred for the meeting place, announcements, mailings and special speakers for meetings held during 1966 and 1967 were defrayed from the $10,000 provided for this purpose by the Human Ecology Research Foundation.

In 1968, Dr. Jonathan Forman established the Jonathan Forman Gold Medal Award, honoring those who had contributed to the field of clinical ecology. This gold medal is presented at the annual dinner meeting of the Society. It is awarded to an individual, not necessarily a physician, who, in the opinion of the Society, has made an outstanding contribution to the field of clinical ecology.

I was the first recipient of this award in 1967. I will quote from Dr. Dickey's account on this occasion.

Dr. Randolph received the Jonathan Forman Award because of his development of the concept of an ecological orientation in medicine which was set forth in his 1962 book, *Human Ecology and Susceptibility to the Chemical Environment;* secondly, for his development of the comprehensive environmental control program as a technique of demonstrating cause and effect relationships from exposures to common foods and environmental chemicals; and thirdly, because of his role in founding and encouraging the development of the Society for Clinical Ecology.

The two sides of this medal are shown in Figure 23. My picture at this time, age 60, is shown in Figure 24.

John Maclennan was elected president of the Society in 1968. I was elected president in 1969.

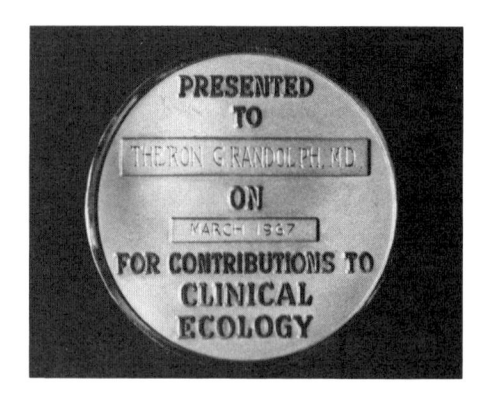

Figure 23. The Jonathan Forman gold medal, front and back.

Figure 24. Theron G. Randolph, M.D.

Hospital Staff Appointments

As my staff appointment at Lutheran General Hospital developed well, this seemed to be a logical place for setting up a comprehensive environmental control unit.[155,156,166] Consequently, detailed plans for such a unit were included in the expansion of the hospital. In addition to knowing the chairman of the board of directors of the hospital and having worked with him for several years, another director, who had given the ground for the hospital, had also been a rehabilitated asthmatic patient. When this institution was looking for a name for the holding company incorporating all of the institution's activities, I had suggested the name: Lutheran Institute for Human Ecology. As this developed, sociological features of ecology were emphasized over biological aspects.

In the meantime, I had been invited by the Food and Drug Administration to participate in the Peanut Butter Hearings (Testimony in the Matter of a Definition and Standard of Identity for Peanut Butter, Docket No. FDC-76) February 17-18, 1966, p. 6511.[177] I again emphasized the desirability for *all* ingredients of processed food, especially specific sugars, to be listed on the labels of processed foods. Of course, these recommendations were resisted by manufacturing interests, including a concerted effort to keep me from testifying. This is all a matter of public record.[177]

After the ecology unit had been completed I inquired, December I, 1966, when patients might be moved in and the area used for its intended purpose, only to be told that it would be better to admit other patients first, because of trouble that I had had with Blue Cross. Two weeks later, I was called in and told that my staff appointment had been terminated, to be effective in two weeks — January 1, 1967. When I inquired for the reasons for this decision, absolutely none were given! The fact that this occurred within a year after my testimony on the Peanut Butter Hearings in February 1966 could not be dismissed, in view of the opposition testimony submitted from food manufacturers at the time of this hearing.[177]

Since I had maintained my former hospital staff appointment at the Swedish Covenant Hospital, I carried on the same type of environmentally focused medical care there temporarily. But because of a shortage of bed space, I applied and received a hospital staff appointment at the Henrotin Hospital in Chicago. Unfortunately, however, it was necessary to hospitalize my patients in single or two-bed rooms in both institutions rather than in comprehensive environmental care units.

Having been asked by the authors of the book, *Interactions of Man and His Environment, Proceedings of the Northwestern University Conference,* held January 28-29, 1965,[173] to write a chapter on "The Toxic Environment," this relatively short manuscript is here reproduced, inasmuch as this subject of interrelationships between toxicity and individual susceptibility to environmental chemicals had not been presented previously:

The Toxic Environment

The greatest single change in man's surroundings during the past century has been the ever-increasing chemicalization of his environment. Although this process started off with coal burning home heating units, kerosene lamps, and fuel oil and gas burning household utilities, it remained for the modern petrochemical revolution to stamp the imprint of the chemical environment on the present generation. In addition to automotive products, the chemical environment also includes the widespread use of solvents, synthetic drugs, dyes, plastics, and fabrics, as well as detergents, pesticides, and other materials. Unparalleled chemical contamination of air, food, and water has resulted.[137]

As judged by this ever-increasing rate of chemicalization of the environment in recent years, the habitat is becoming more and more menacing. It is already hostile to the point of impaired health and productivity for an expanding sector of the population.

Many questions arise in a discussion of this subject. For instance, how does the body react to lesser exposures of substances known to be toxic in greater concentrations? What are the distinctions between reactions on the basis of susceptibility (allergy) and toxicity? Are individuals capable of adapting to this relatively new chemical environment? Are we protecting ourselves adequately against our present chemical surroundings? Are our techniques of testing the safety of new chemical compounds sound? Do present practices increase the susceptibility of individuals, thereby enhancing the impact of the chemical environment, and threaten their future health and behavior?

It is well known that man possesses an amazing ability to adapt to changes in his intake and surroundings, provided such variations occur slowly. But ecologically speaking, the chemical environment has been an explosive development. Since a large sector of the population is already believed to be maladapted to various environmental chemicals, what may be expected of a future which promises greater concentrations of present materials plus many additional ones?

Descriptive features of these maladapted responses will be outlined. Interrelationships between reactions on the basis of individual susceptibility and toxicity will be emphasized to serve as a background for proposed changes in testing and regulatory functions.

The development of adapted responses to chemically derived materials is exceedingly subtle and relatively unappreciated either by victims of this process or their physicians. When such relationships are demonstrated, the startled individual often asks: "Why did this just suddenly start to bother me?" Actually, the process had been building up for some time as a result of the hammering impact of frequently repeated small exposures.

When a previously well person is first exposed to this type of chemical impingement, there is usually no obvious reaction. A particular agent, not being suspected, often results in an "I can take it" or "That does not bother me" attitude. Neither is adaptation to continued regular exposures in the presence of individual susceptibility apt to be recognized.

Instead of feeling worse, a person who is specifically adapted tends to be relatively stimulated immediately following each re-exposure. These oft-repeated or constant-sized doses not only tend to maintain a relatively stimulated state, but some persons actually resort to such items as often as necessary to remain "normal." Most commonly, however, a specifically adapted individual merely carries on with his accustomed routines, oblivious of any type of response. The housewife reacting to the fumes of her gas kitchen range rarely ever suspects them.

But sooner or later, specific adaptation tapers off. The impact of exposure, enhanced by increasing individual susceptibility, gradually breaks through bodily defenses. Postexposure "pickups" then become relatively less beneficial and shorter. Delayed "hangovers" are manifested as increasingly chronic physical and mental illnesses.[131] Localized effects such as conjunctivitis, nasal disturbances, frequent colds, mild gastrointestinal distress, itching, or other lesser regional symptoms are first apparent. More advanced localized responses include asthma, hives, and eczema and musculoskeletal, gastrointestinal, genitourinary, and neurological manifestations. Closely related are general effects such as chronic fatigue, headache, and impairment of higher integrated cerebral functions. The latter also include cutbacks in humor, initiative, and ambition, as well as forgetfulness, incomprehension in reading, inability to make decisions, and faulty concentration. Confusion, depression, and other psychotic behavior collectively referred to as ecologic mental illness may eventuate.[131]

Such maladapted stages of chronic reactions may persist for many months or years, as long as accustomed routines are continued. Indeed, life tends to become something to be endured in sick boredom, instead of a challenging and exhilarating experience. But if the responsible agent or agents are avoided sufficiently long to permit "hangovers" to subside, recovery recurs. Re-exposure to such previously avoided incitants then induces acute immediate reactions. This course of events — specific avoidance followed by re-exposure — changes chronic illness of obscure causation to acute illness and demonstrates its inciting and perpetuating environmental causes.

Susceptibility and maladaptation to multiple chemical and other environmental exposures are the rule, due to the tendency for cross-reactions to occur to materials of common genesis. Therefore, the simultaneous avoidance of probable, potential, and suspected agents prior to testing singly is most rewarding. This program, called comprehensive environmental control,[166] has been carried out in an ecologic unit.[155,156] The effects of single re-exposures are first observed in the hospital setting, subsequently as subjects are returned to their homes and to their work.

With the application of this technique, the significance of chemical exposures as causes of chronic illnesses in the average householder gradually emerged.[137] One by one, the effects of synthetic drugs, pesticides, and certain other chemical additives and contaminants of the diet and indoor chemical air contamination were incriminated. Of the latter, fumes from gas utilities, sponge rubber padding, plastics, and airborne insecticides were demonstrated to be most troublesome.

Although any person may adapt for a time and then maladapt and sicken as described, this exposure-reaction pattern occurs more readily in selected individuals. Such a selective susceptibility has long been noted in occupational situations, even though dealing with allegedly nontoxic levels of exposure. Greater quantities, known to induce reactions in most organisms, are usually referred to as toxic or poisonous exposures.

The important point to emphasize is that this exposure-reaction pattern, as it now occurs in the householder, is appearing as an ever-increasing factor in chronic illness. Regular repetition of minimal chemical exposures induces adapted responses readily confused with apparent tolerance. Relatively larger or more rapidly absorbed doses to which one had been adapted in smaller quantities, might break through and precipitate acute immediate reactions. When such exposures occur by chance, they are apt to be confused with those of acute toxicity. But when these governing circumstances are controlled, this technique of avoidance prior to re-exposure may be used deliberately for diagnostic purposes.

In other words, the reaction of a person to a given amount of a specific substance may be interpreted meaningfully only when the individual's degree of specific susceptibility and his stage of adaptation to that and related exposures are known. From this standpoint, a sliding scale relationship exists between reactions on the basis of individual susceptibility and those attributable to toxicity, rather than the rigid quantitative distinctions alleged to exist.[167]

It is evident that many past decisions regarding the safety of environmental chemical exposures have not been made on an ecologically sound basis. Otherwise, there would not have been a continued increase in the number of persons becoming susceptible to and made ill from the impingement of various "accepted" chemical environmental exposures.

Since a working knowledge of the stages of adaptation permits either an accidental or an intentional bias in the interpretation of long-term chronic toxicity studies, and since such tests involve the interests of both the manufacturer and the public, the manufacturer who hopes to profit from the sale of an item should not be given sole responsibility for conducting tests as to its safety. Although the public's interest in these matters is said to be protected by governmental regulatory agencies, actual protection varies from being reasonably good in the case of new drugs to exceedingly poor in respect to airborne insecticides, gas fired cooking and heating devices, and janitorial supplies.

If man is to regain his health and productive capacity with any hope of retaining them as he moves farther into the chemical age, much greater emphasis must be placed on the ecologic aspects of human existence. Human ecology, a concept which regards the individual as an integrated biologic unit in constantly changing relationship to his intake and surroundings, must be recognized and explored. Its dynamic approach opposes current relatively static analytical approaches. The latter includes the current medical emphasis on bodily mechanisms of disease and the dominant toxicologic approach of studying the environment by means of an analysis of fractions of it.

More specifically, we need to incorporate the principles of human ecology and specific adaptation into testing techniques to evaluate new chemical agents and to reevaluate certain presently accepted applications.

In addition to these basic changes in techniques, we also need to elevate to a professional status those responsible for testing and approving chemical applications, freeing them more completely from production ties and associated economic pressures. As a society, we do not accept statements of corporations as to the status of their finances; independent professionals, i.e., certified public accountants, render this service. In view of the public stake in the matter of controlling the chemical environment, we should not accept the producer's statement as to the safety of his product; independent professionals, i.e., "certified public testers," should render this service.

If this appraisal of the health hazards associated with the chemical environment[137] and this critique of testing techniques[167] are correct, no time should be lost in establishing the educational facilities, licensure provisions, and ethical standards of so-called "eco-testers."

Summary

1. The chemical environment, including gas, oil, coal, their combustion products and derivatives, is to be regarded as a composite environmental exposure, in view of the tendency for reactions to spread to synthetic materials of common genesis.

2. A sliding scale relationship between reactions on the basis of individual susceptibility and toxicity exists, instead of the rigid quantitative distinctions presently alleged.

3. The impact of lesser chemical exposures known to be toxic in greater concentrations is magnified in susceptible persons and manifests as a wide range of chronic physical and mental illnesses.

4. The resulting health problem is best approached ecologically in keeping with the stages of specific adaptation. Medically, this means greater emphasis on demonstrating environmental incitants of illness. Toxicologically, this entails improved techniques to establish the safety of chemical exposures. Recognition of the independent professional status of the testers is also needed.

Relationships Between Allergy and Immunology

In keeping with the dissociation of immunology and allergy as previously presented was Ethan Allen Brown's comment that allergy is a practice in search of a science and immunology is a science in search of a practice. Dr. Brown also stated that in the past 30 years immunology had made no significant contribution to clinical allergy. (Brown EA. The New Review of Allergy. Review of Allergy 19:300, 1965)

Jerome Glaser, writing on The Menace of Immunology in the Clinical Practice of Allergy, (Review of Allergy 21:1120, 1967) agreed in general with Brown's statement. Glaser also commented on the vague relationships between immunology and clinical allergy. It should be noted that Dr. Glaser had probably trained more pediatric allergists than any other physician.

It is also of note that identification of IgE antibodies as a carrier of reaginic activity (Ishizaka K, Ishizaka T. J Immunology 99:1187, 1967) and the description of RAST which measures it, occurred at about this same time (Johansson SGO. Raised Level of a New Immunoglobulin Class [IGND] in Asthma. Lancet 2:951, 1967)

Relationships with Psychiatrists

As stated earlier, it had been difficult bringing the ecological point of view to psychiatrists and gaining sufficient acceptance for them to try it, and finally changing their practices to emphasize the causative roles of nonpersonal environmental exposures (i.e., foods and chemicals) in individually studied patients. I will now relate my failures and successes in this endeavor. Simply lecturing on the subject is relatively ineffective in gaining this type of acceptance.

Motion pictures of experimentally induced reactions have been tried repeatedly[79,130,149] without convincing acceptance. When this motion picture was shown at the National Institute of Mental Health, the Chairman's comment was: "Obviously, a very rare occurrence."

On several different occasions I have had the following type of experience during such a lecture presentation to psychiatrists in which only the positive nonpersonal aspects of illustrative case reports were emphasized. The comments afterward often went like this: "This is all very interesting, doctor, but don't you think that all etiologic factors should be presented?" Each time my reply was as follows: "Yes, but would you mind taking up and applying the most demonstrable personal-environmental relationship first? When that is done correctly, it is ordinarily not necessary to consider other more speculative interpretations of etiology." The discussion has usually stopped at this point.

I have tried the more direct approach by means of the title "Psychiatry Exteriorized," tracing the transition from introspection to objective observation of environmentally related reactions, including behaviorism, unsuccessfully.[122]

The most effective way is to have psychiatrists visit one's hospital or office in the course of this work, but this can rarely ever be arranged. The way that this point of view was actively transferred to and accepted by some psychiatrists will now be described. For several years I had made presentations on the subject of ecologic mental illness at international meetings of psychiatrists, either showing motion pictures, and/or distributing reprints, or both. At one of these meetings in Europe, a Philadelphia psychiatrist picked up a reprint and mailed it to his friend, Dr. William Philpott, who in turn read it, thought it was interesting and filed it. A year or so later, another reprint reached Dr. Philpott in much the same way. Meanwhile, having heard of this through Solomon Klotz, M.D., Philpott tried it and confirmed it. Enlisting the help of Marshall Mandell, M.D., another clinical ecologist, they set this up in a research hospital for mental illness in North Attleboro, Massachusetts. Philpott then presented this concept to other psychiatrists. He carried through with other diagnostic facilities for this work and eventually wrote a widely selling book on this subject. Apparently more than any other person, Philpott, and to a lesser extent Abram Hoffer, has brought this message to psychiatrists which I, as a non-psychiatrist, could not do as effectively.

My most effective presentations of this subject during this period have been 1961[131], 1966[184], 1970[206], and 1971[224].

Ecolony

The need for a self-contained relatively chemically free environmental community for those having the chemical susceptibility problem to an advanced degree has long been recognized. In 1968, I spent several days trying to find a satisfactory location west of Denver, Colorado. But to minimize pollution from metropolitan Denver entailed an elevation at this time of at least 7,500 feet. This elevation, however, was too cold for most persons interested in building permanent homes in such a community. Although the advantages of warmer locations are obvious, the increased volatility of traffic fumes and other chemical pollutants are troublesome. Absence of the means of livelihood in such a development is also restrictive. To my knowledge, no one has as yet succeeded in setting up such an economically self-supporting establishment of this type in the United States.

Publications

At this time, I prepared a series of three connected manuscripts for publication, totalling 126 double spaced typewritten pages, under the following titles:

ECOLOGICALLY ORIENTED MEDICINE; ITS NEED
I. COMPARISON WITH ANTHROPOCENTRIC MEDICINE (25 pages)
Presented on the Program of Instructional Seminars of the Society for Clinical Ecology, 1965 and 1967.

ECOLOGICALLY ORIENTED MEDICINE;
II. THE ROLES OF SPECIFIC ADAPTATION AND INDIVIDUAL SUSCEPTIBILITY IN CHRONIC ILLNESS (56 PAGES)
Presented on the Program of Instructional Seminars of the Society for Clinical Ecology, 1966 and 1967.

ECOLOGICALLY ORIENTED MEDICINE;
III. STIMULATORY AND WITHDRAWAL MANIFESTATIONS OF SPECIFIC REACTIONS (42 pages)
Presented on the Program of the 22nd Annual Congress of the American College of Allergists, April 1966.

Unfortunately, these manuscripts, which are still in my files, were never published other than in abstract form.[188,189,197] I don't recall the reasons for this failure to publish these manuscripts. Neither do I remember having submitted them for publication or their rejection for publication. However, much of this material was published later under other titles. Other parts are presented in the final chapters of this book.

The Archives for Clinical Ecology — the Official Organ of the Society for Clinical Ecology — was started in 1969, with Johnathan Forman, M.D. as editor, Guy O. Pfeiffer, M.D. as managing editor, and Joseph Interlandi, M.D. for advertising. As stated under the title: "This is a periodical which approaches the problems of the healing arts from the viewpoint of Human Ecology."

Unfortunately, only two volumes were published, the first in 1969, the second in 1971, because of Dr. Forman's terminal illness. Since these issues are relatively unavailable, the table of contents of the first, Volume 1, Number 1, is reproduced herewith:

FOREWORD 1
HUMAN ECOLOGY 3
Human Comfort In Environmental Transition 5
  Peter W. Gygax
Allergy Testing in Otolaryngology 8
  Richard H. Stahl, M.D.
Allergic Factors in Oral Glycerol Therapy. 14
  Irwin E. Gaynon, M.D.
The Clinical Management of Infant Allergy 17
  Tennie Mae Lunceford, M.D.
Possibilities of Non-Allergic Extracts 24
  Charles G. Gabelman, M.D.
Gastro-Intestinal Tissue Eosinophilia In Patients With ... 29
  Clinical Allergy — Lawrence D. Dickey, M.D.
Reactions to Hidden Drugs 39
  Stephen D. Lockey, M.D.
A Review of Sublingual Provocative Testing and Therapy
Of the Sensitive Patient 45
  Guy O. Pfeiffer, M.D.
Human Conservation and Health 51
  Theron G. Randolph, M.D.
Carbon Monoxide Poisoning 64
  Francis Silver, P.E.
The Factors Favorable to Survival in Man's Environment ... 69
  Jonathan Forman, M.D.
The Ecological Basis of Environmental Health 73
  James A. Lee
An Irresponsible Experiment 78
  Ebba Waerland
A Decade of Regulating Food Additives 84
  F. J. McFarland and Alan T. Sphiher, Jr.
Natural Food Poisons 90
  A. D. Campbell, Ph.D.
Petro-Chemical and Synthetic Gases 96
  S. J. Heiman
The Antigenic Uniqueness of Each Species of Hay Fever
Pollen Demonstrated by Provocative Inhalation, ... 104
  Marshall Mandell, M.D.
Ragweed Pollinosis and The Barr Body Structure 113
  Eloise W. Kailin, M.D. and Lois I. Platt, M.D.
Some Aspects of Carbohydrate Metabolism 123
  Eugene W. Higgins, M.D.
LETTERS TO THE EDITOR 132
BOOK REVIEWS 139

    Since my own publication, "Human Conservation and Health," which was presented on the program for the Second Annual Meeting of the Society for Clinical Ecology, is not available other than from the Human Ecology

1965-1970

Research Foundation, I will quote from its introduction and summary:[221]

> Reprinted from the Archives for Clinical Ecology,
> Volume 1, 1969
> **Human Conservation and Health:**
> **An Ecological Orientation in Medicine**
>
> Human conservation, which aims at optimal utilization of human resources, focuses on the development, preservation and rehabilitation of productive capacities through maximal life spans. The health aspects of human conservation include prevention of disease and detection and early correction of illness. Human conservation, in keeping with other conservation programs, emphasizes *ecologic* aspects of medicine.
>
> Medicine's evolution from its traditional anatomical focus to ecologically oriented medical care will be traced before describing the medical and ecologic implications of human conservation.
>
> *Bodily Centered Versus Environmentally Focused Medicine*
>
> The first steps in the development of modern ecologically oriented medicine were the recognition of toxic and occupational aspects of disease occurring as by-products of the industrial revolution. This took the form of associating specific illnesses with given environmental exposures. Although many of the abuses that led to the initially recognized occupational diseases have been corrected, much still remains to be identified and controlled.
>
> The next major step in the development of the environmentally focused medicine was the demonstration of the infectious nature of certain diseases. Application of the infectious concept of disease culminated in present public health procedures on the one hand, and immunization practices and antibiotic therapy on the other.
>
> Another development toward ecologically oriented medicine was the concept of allergic disease. Unfortunately, this has not been applied widely, due to inaccurate diagnostic techniques and difficulties — both individually and in general — in controlling exposures to environmental incitants. Moreover, the initial environmental emphasis associated with the specialty of allergy has also been blunted and partially reversed by the treatment of various allergic states similarly by means of antihistamines, steroids and other drugs — irrespective of their demonstrated causes.
>
> A more recently recognized environmental development, the chemical susceptibility problem, is characterized by cross-reactions to a wide range of man-made synthetic chemicals. In addition to the daily exposures of *all* persons, this individualized reaction pattern encompasses many toxic and occupational factors and is frequently complicated by infection. Relatively unappreciated and underdeveloped at present, this field is characterized by clinically demonstrable responses

to materials known to be toxic in various concentrations. It is concerned primarily with man-made, chemically derived air pollutants (ambient air, occupational air and household air), food and water additives and contaminants, synthetic drugs, hydrocarbon solvents and other combustion products and derivatives of oil, gas and coal.[137]

No large group of practicing physicians is as yet especially interested in this new area. This relative disinterest appears to be attributable to the fact that only recently the stages of specific adaptation and diagnostic techniques based upon them[137,165,166] became available for the demonstration of etiology in this important area. Also, since these concepts and techniques have only begun to be applied, illness inducing and illness perpetuating aspects of the chemical susceptibility problem remain relatively unappreciated.

The final area of environmentally oriented medicine said to be ecologically significant is the alleged effect of other persons and groups on the health and behavior of given individuals. Physicians thinking in terms of psychiatric and sociologic concepts have championed the causative roles of such exogenous personal factors, predicting that these relationships would characterize medicine of the future. (Hinkle LE, Wolff HG. Ecologic Investigations of the Relationship Between Illness, Life Experiences and the Social Environment. Ann Internal Medicine 49:1378, 1958; Rosen G. Community Orientation and the Next Stage of Medical Evolution. Arch. Environmental Health 7:625, 1964) Although these apparent relationships are within the broad meaning of the term, human ecology, their roles in the writer's experience are far more speculative than demonstrable. If one accepts as facts cause and effect relationships that are demonstrable under relatively controlled conditions,[166] such experimentally induced reactions implicate other persons on the basis of propinquity rather than on the basis of a fixed personal interrelation.

The apparent etiologic roles of interpersonal situations on the health and behavior of individuals are not to be confused with the demonstrable roles of specific foods, chemicals and other environmental substances in the production and perpetuation of cerebral and related behavioral syndromes encompassed by the term, ecologic mental illness.[131,184] Differentiated from psychiatrically interpreted mental illness by the ability to demonstrate nonpersonal environment behavior interrelationships by means of techniques to be described,[166] this advancement from descriptive to etiologically demonstrable concepts and procedures is opening new horizons in mental illness.

This focus on the demonstrable role of nonpersonal factors in mental illness and abnormalities of behavior is also being adversely affected by the current vogue of treating mental symptoms by means of sedatives, tranquilizers, mood elevators and other drugs. Since these medications are largely man-made chemical derivatives capable of inducing reactive symptoms in their own right,[167] such therapeutic approaches often add to the burden to which a given patient is attempting to adapt.

In summary, individualized environmentally focused medical programs have been overshadowed by bodily centered medicine which is more readily applied by mass techniques. Current medical approaches, summarized on the left of Figure 25, aim at demonstrating the mechanisms of illness by means of fixed routines. Diagnostic dependence is placed on X rays and other analytical laboratory procedures and tests of organ function. Therapy, not closely integrated with diagnosis, depends principally upon such mass expediences as the treatment of similar mechanisms and symptoms by means of similarly acting drugs.

Since the overall record of bodily centered medicine is not regarded as ecologically sound, obviously more is needed if medicine is to keep abreast of other current conservation movements and is to be integrated with them. Ecologically oriented medicine answers many of these criticisms. This holistic approach which aims to demonstrate the environmental causes of illnesses emphasizes the dynamic long-term interplay between bodily and environmental factors in medicine. Its relative position in respect to current practices, outlined on the right of Figure 25, is best understood and applied through a workable knowledge of specific adaptation.

Figure 25. Bodily centered versus ecologically oriented medical care

	Current Bodily Centered Medicine	Ecologically Oriented Medicine
Aims	Demonstration of Bodily Mechanisms of Illnesses Fragmented*	Demonstration of Environmental Causes of Illnesses Holistic
General Approach	Static Episodic	Dynamic Long-Term
Diagnostic Techniques	X rays and other Analytical Laboratory Tests	Avoidance and Test Re-exposures and other Provocative Tests
Treatment	Drug Therapy of Limited Mechanisms or Symptoms	Avoidance of Incitants or Maintenance of Specific Adaptation

*Anatomically demarcated specialties of medicine

Summary

Human conservation — the optimal utilization of human resources — demands the development of ecologically oriented medicine.

An individual's ecologic status in respect to a given environmental exposure apparently depends upon the stage of specific adaptation existing at the time. Although adaptation usually proceeds smoothly, in the presence of individual susceptibility specific adaptation becomes disruptive and symptom related due to acceleration of the process, augmentation of the impact of environmental agents and enhancement of the responsiveness of the subject. Since the development of specific susceptibility depends, to a great extent, upon repetitive exposures which seem to be increasingly characteristic of modern life, individual susceptibility appears to be one of today's keys to the environmental health problems of tomorrow.

Change of Name
The group of patients who first called themselves the "Chemical Victims" experienced some difficulty in attracting speakers. In 1968 they changed their name to The Human Ecology Study Group.

Albert H. Rowe, M.D. — Continued
The fact remains that Albert H. Rowe was a clinical observer far ahead of his time. His early recognition that food allergy was a factor to be considered in all cases of headache, fatigue, bronchitis, bronchiectasis, emphysema, ulcerative colitis, regional ileitis as well as many cerebral, neural, musculoskeletal and other acute and chronic manifestations, represented relatively new interpretations when these reports were first published. His description of interreactions between infectious agents and allergic manifestation, especially in children, was one of the first accounts of this important area. Also, his report of refractoriness after one acute allergic manifestation minimizing another similar episode immediately was the first reference to this important phenomenon.

But despite Albert Rowe's accomplishments in respect to calling attention to the subject of food allergy, he did not neglect other environmental factors contributing to the allergic state. His contributions in the areas of pollens, fungi, infections and many other exposures, as indicated in the title of his major work, the 1937 book, attests to his comprehensive coverage of the subject of the role of the environment and health in respect to specific foods existing at that time.

Dr. Rowe was awarded the Forman Gold Medal — the highest award bestowed by the Society for Clinical Ecology — in 1970. Unfortunately, he died a few weeks prior to the annual meeting of that organization at which time this award was to be presented. I had the pleasure of giving this award to Mildred Rowe, his wife, in their Piedmont, California home. This gave me the opportunity to inspect once more the remarkably complete allergy library in his home — in my opinion the most complete collection of books dealing with the subject of food allergy assembled in one place.

Albert Rowe's last book, written in conjunction with his son Albert Rowe, Jr., M.D., entitled *Food Allergy; Its Manifestations and Control, and the Elimination Diets — A Compendium,* was not published until 1972.

Favorable Governmental Environmental Protection

Earth Day was celebrated April 1, 1970 as a fitting tribute to the legislative activities of both the Democratic and Republican administrations in respect to their environmental interests and concerns during the previous decade. Both the Environmental Protection Agency (EPA) and the Occupational Safety and Health Administration (OSHA) were created in 1970. William D. Ruckelshaus was appointed the first EPA administrator on December 2, 1970.

In contrast to most other physicians in Chicago who seemed to be apparently oblivious of the significance of Earth Day, I closed my office. My staff and I spent the entire day lecturing at various places in Chicago on the importance of environmental-medical interrelationships and health. (Clean Air News. Volumes 1 and 2, 1967, 1968; Clean Air and Water News. Volumes 1 and 2, 1969, 1970)

Chapter X — 1971-1976

Continued Ecological-Medical Enlightenment
Environmental Protection and
The National Legislative Scene

Analytic Versus Biologic Dietetics
In General

These distinctions, first presented in 1971,[232] were presented in greater detail in 1976.[257] Since analysis and synthesis as applied to nutrition had not previously been differentiated, this will be described herewith.

As medicine in both Europe and America began to emerge as a science in the nineteenth century, especially as a result of microscopy, medical investigation became increasingly analytical. Nowhere has investigation by analysis been carried out more thoroughly and in greater detail than in the field of nutrition and in its practical application, dietetics.

Much useful information has resulted but, unfortunately, this has been associated with considerable obfuscation. Before attempting to differentiate analytic dietetics from biologic dietetics and to describe their historical development, the hazards of investigation by analysis as they apply to dynamic biological phenomena will be reviewed briefly.

Smuts emphasized potential error of the analytical method when it is applied in the field of biology by pointing out that when the isolated elements or factors of a complex biological phenomenon have been studied separately, then combined to reconstitute the original situation, something commonly escapes. (Smuts JC. *Holism and Evolution.* New York: Macmillan, 1926) He went on to say that investigation by analysis makes the artificial situation as reconstructed something quite different from the original situation which was explored. Moreover, investigators are prone to look upon the data resulting from analysis as the natural factors of the situation, whereas actually they are merely the results of analysis. Smuts referred to this confusion between reality and analytical fragments as an error of generalization.

DuNouy also criticized results obtained by the application of the scientific method, based on analysis, by the comment: "The more deeply man analyzes, the farther away he gets from the problem he meant to solve. He loses sight of it and is absolutely incapable of regaining it by means of the phenomena which he studies, although logically he feels that there should be a link between them." (DuNouy L. *Human Destiny*. New York: Longmans, Green, 1947) DuNouy referred to this as the error of passing from one scale of observation to another analytically. He illustrated this common error by the following example which is apropos in respect to this presentation. In setting out to study the behavior of man, one would necessarily turn to anatomy and physiology. This would lead him to biological chemistry and, in turn, to molecules, atoms, electrons, etc. Although scientific investigation by analysis might result in the discovery of new phenomena, it would also lead the scientist farther and farther away from the original problem he set out to study.

With these errors inherent in the methodology of investigation by analysis in mind, let us turn to historical developments in the field of nutrition.

Analytic Dietetics

According to Lusk, (Lusk F. *The Science of Nutrition*. Philadelphia: Saunders, 1906) who pioneered nutritional science in America, the modern science of nutrition was initiated by Lavoisier and Liebig and their respective pupils. Lavoisier's brilliant contribution of the combustion of food by means of oxygen and Liebig's development of modern methods of organic analysis laid the foundation of modern analytical nutrition. Voit and his pupils, Rubner, Atwater, Lusk and others working in Voit's laboratory in Munich, developed the principles and techniques of calorimetry. As a result of this application, nutritionists have been primarily concerned with calories, protein, fat, carbohydrate, and other analytical constituents of an individual's 24-hour food intake.

Most subsequent advances in nutrition have been concerned with an even more detailed analysis of the 24-hour intake. Between 1910 and 1925 the first vitamins were described. Several subdivisions of the original vitamins and a few additional ones have been added since. With time and continued investigation, both quantitative and qualitative aspects of the total daily food intake were emphasized. For instance, adequacy of the protein intake was considered in terms of the relative amounts of the essential amino acids. Similarly, the relative proportion of saturated versus unsaturated fatty acids was said to have clinical significance. Other advances in respect to data emerging from fractional analysis of the 24-hour food intake could be cited.

Continuing at about the same time, information concerning the relative proportions of these constituent fractions obtainable from the analysis of given foods became available. Simply by adding these fractional constituents, the adequacy of the 24-hour food intake could be determined. Although there were some exceptions, such as the need for several essential amino acids to be present concurrently for maximal nutrition, the above statement has been regarded as generally correct.

One by one, adequacy of the 24-hour diet in terms of various minerals and later enzymes has been postulated. More recently, the role of trace elements in nutrition has received increasing attention.

Indeed, the analytical or fractional approach to the total 24-hour food intake, irrespective of the specific foods comprising that intake, has almost completely eclipsed interest in the specific effects of given foods. For instance, food allergy has been relegated to near oblivion. Different ways in which different persons handle specific foods, in view of their individual susceptibilities, and to some extent deficiencies of enzymes and other processes by means of which given foods are broken down, have also been minimized by this current vogue on general intake in nutrition and dietetics.

Biologic Dietetics

Biologic dietetics concern the roles in nutrition of foods in respect to their biological origins and interrelationships, the impact of which is accentuated by the existence of individual susceptibility to specific foods or by deficiencies in the ability of certain persons to metabolize given foods. In contrast to analytic dietetics, in which major interest is concerned with a fractional analysis of the 24-hour intake of food, interest in biologic dietetics is concentrated on food in the form actually ingested. More specifically, this means food in the sense of its biologic identity such as wheat, corn, and other closely related cereal grains; with milk and the closely related beef; or with potato and tomato, red and green pepper, and tobacco — all members of the same biological family. Individual susceptibility not only commonly exists to such biologically identified whole foods or their refined products but also readily spreads to biologically related dietary items. Biologic dietetics is also concerned with chemical additives and contaminants of given foods as well as of the cumulative intake of chemical additives and contaminants of the food and water supplies.[232,257]

Individuals adapt (in the relative absence of troublesome reactive symptoms) or maladapt (manifesting various levels of clinical responses) to the cumulative ingestion of given excitants. This may be a single food and/or those closely related biologically, or multiple foods.This may involve the ingestion of a given chemical additive or contaminant, or, more commonly in this instance, various closely related chemical additives and contaminants. The speed with which this adaptation-maladaptation process occurs and the relative severity of the resulting clinical manifestations occurs in approximate proportion to the degree of individual susceptibility involved. Under these circumstances, multiple cumulative environmental exposures to which various degrees of individual susceptibility and maladaptation exist are often impinging on a given person at any given time.

Biologic dietetics dates back to antiquity. (Brothwell D, Brothwell P. *Food in Antiquity: A Survey of the Diet of Early Peoples.* New York: Prager, 1969; Tannahil R. *Food in History.* New York: Stein & Day, 1973) Early Greek physicians were aware of the fact that such uncommonly eaten foods as shrimp, rhubarb, and buckwheat sometimes made selected patients acutely ill whenever these foods were eaten. Such promptly occurring acute reactions, first referred to as idiosyncracies, were later called allergies. Despite such responses sometimes requiring medical treatment, these are not usually diagnostic problems inasmuch as such causative agents are ordinarily readily apparent. Although this is food allergy as it is usually considered, it actually is a relatively unimportant part of this large subject.

The modern era of biologic dietetics started slightly later than the modern era of analytic dietetics. Despite the overlapping development of the two points of view, biologic dietetics has remained relatively eclipsed by the overwhelming scientific interest in analytic dietetics.

Clinical reactions to the cumulative use of common foods were not recognized until the present century. Hare, (see Appendix, page 330) and later Brown, (Brown TR. Role of Diet in Etiology and Treatment of Migraine and Other Types of Headache. JAMA, 77:1396, 1921) pointed out the role of carbohydrates, especially cereal grains and sugars, on the one hand, and protein foods, especially beef and milk, on the other, in the etiology of migraine headaches. Although benefits were reported by means of diets avoiding these food groups and infractions of such diets were said to be associated with recurrences of headaches, neither author apparently employed direct feeding tests under observation to substantiate and demonstrate their claims.

Schloss was apparently the first to report allergy to a common food confirmed by clinical observation. (Schloss OM. A Case of Allergy to Common Foods. Am J Dis Child 3:341, 1912) Although similar observations were also made by Shannon, (Shannon WR. Neuropathic Manifestations in Infants and Children as a Result of Anaphylactic Reactions to Foods Contained in Their Diet. Am J Dis Child 24:89, 1922) contributions of Duke, Rowe, Rinkel and Coca have already been presented.

Unfortunately, Rowe's elimination diet approach to the specific diagnosis of food allergy is less accurate as a scientific method than desired, as it permits errors of deduction. For instance, major emphasis is placed on clinical improvement resulting from the use of a given diet in the patient's home setting. Improvement might have resulted from some unrelated concurrent change. Also, instances of partial improvement are difficult to evaluate.

Another error in performing individual food ingestion tests, as described by Rinkel (see Appendix, page 334) and in conjunction with Zeller and the writer, was the fact that these early descriptions did not differentiate allergic reactions to foods per se from those due to chemical additives and contaminants of the same food. The necessity to use chemically less contaminated foods for test purposes was not recommended until later.[137] Neither did the original descriptions of the individual food ingestion test stipulate that such testing be carried out under controlled conditions as far as airborne and contactant chemical exposures are concerned.[102,137]

It was also observed that the performance of individual food ingestion tests on an office or outpatient basis in which the patient had full responsibility for avoiding given foods prior to testing was often inaccurate, especially in respect to foods widely distributed and difficult to eliminate. Similar testing in the hospital following preliminary fasting in the program of comprehensive environmental control[166] in an ecologic unit,[155,156] which has been employed since 1956, is relatively more accurate.

The provocative test is sometimes criticized on the basis that some of the evidence constituting the presence of a positive reaction is subjective in nature. Although this criticism is valid, subjective information, repeatedly observed, is not only convincing to a patient but is also valid. Moreover, there usually are lesser objective components associated with most positive provocative tests observable by those trained in the use of these techniques.

Perhaps the greatest changes in respect to the diet during the past 25 years has been the increasing awareness of the fact that food allergy must be differentiated from apparently similar reactions occurring to various chemical additives and contaminants of the diet capable of precipitating acute reactions or inducing or perpetuating chronic ones.[103,137] Demand has been increased by specifically diagnosed ill people who require this type of food to remain well, and also by interest of younger people in chemically less contaminated food on general principles. Supply is decreasing, both relatively and absolutely, as chemical contamination of soils accumulates and agrochemical production techniques continue to be more widely employed.

In summary, the development of biologic dietetics in respect to the means of diagnosis may be epitomized by Rinkel's comment that there is no foolproof method for the diagnosis of food allergy. (See Appendix, page 334) This might be paraphrased positively as follows. Collecting proof in the form of evidence to establish the existence of food allergy is exacting, difficult, time consuming, and individualized — but very rewarding, both to the chronically ill patient and his etiology-seeking physician.

The position is taken that the fields of nutrition and dietetics, especially as dietary information is applied in the daily practice of medicine, is presently monopolized by a relatively narrow and essentially one-sided point of view characterized by a fundamental error in fact gathering. This error, inherent in the analytical approach as applied to dynamic biological interactions, imparts a relatively static orientation to one of nature's most dynamic biological relationships. There is nothing static about the fact that clinical responses of the same person at different times, depending on the stage of specific adaptation at the moment and of different individuals to the same food and depending on the degree of specific individual susceptibility, is usually dissimilar. Indeed, relationships involved are as dynamic as a pot of bubbling porridge. In order to understand this bubbling phenomenon, let us look at it in greater detail. In focusing down on it with a high-powered microscope, one is reminded of volcanic activity. Most commonly, we remove the porridge from the interreactive dynamic context which we started out to study, subjecting it to chemical analysis by determining the protein, fat, carbohydrate, minerals, and vitamins. Not finding the answer, we might investigate the chemistry of such subfractions as proteins and glycerides carrying on to their respective molecular structures. Or, we could subject the pot and the fire under it to their ultimate constituents. Although these answers would provide much new information of essentially a static nature, we would not have any better understanding of the bubbling phenomenon analogous to a living changeable thing.

All diagnostic testing procedures must be focused on the dynamic interrelationship occurring between a given person and the ingestion of a specific foodstuff under precisely controlled circumstances. The latter means in a phase of specific adaptation in the host which favors maximal immediate acute and, therefore, convincing test reactions and under environment conditions which minimize other possible etiologic relationships. A rapidly absorbed physical form of the food in question often provides more striking test results than a more slowly absorbed form of the same food.[70]

Reducing the intake of specific foods but not avoiding them not only accentuates the clinical features of food addiction, but virtually precludes following such a diet by many patients. Far more satisfactory therapeutic results may be obtained by identifying and avoiding specifically incriminated foods.[111] (Matsumura T, Kuroume T, Amada K. Close Relationship Between Lactose Intolerance and Allergy to Milk Protein. J Asthma Res 9:13, 1971)

Although the treatment of alcoholism should be voluntary abstinence of alcoholic beverages, alcoholics responding to their intake of alcoholic beverages in the sense of maintaining relatively adapted responses to one or more specific foods entering the manufacture of alcoholic beverages find it exceedingly difficult to comply with such a program. Identification of specific food addictants and the complete avoidance of their intake in both their edible and potable forms provides for superior therapeutic results in that such a program minimizes addictive urges. When alcoholics cease drinking they tend to substitute the edible form of their specific addictants, especially the more rapidly absorbed physical forms of such foods. For instance, the majority of alcoholics in America are susceptible to corn.[111] Upon abstaining from alcoholic beverages, they usually substitute corn sugar found in candy and ice cream or other commercially available foods containing corn sugar, dextrose, or glucose. Such a program not only perpetuates their specific food addiction at a slightly lower level, but also continues the craving for more effective measures to treat or postpone their corn related withdrawal syndromes. Multiple cereal grain allergy and susceptibility to yeast, grape, potato, and other foods entering the manufacture of alcoholic beverages may be assumed to exist until shown not to be present. These observations point up the significance of specific foods entering the manufacture of alcoholic beverages, shown in Figures 26 and 27.[255] Many alcoholics are also found to be susceptible to coffee. It is well known that the social activities of abstaining alcoholics center around the addictive intake of coffee.[111]

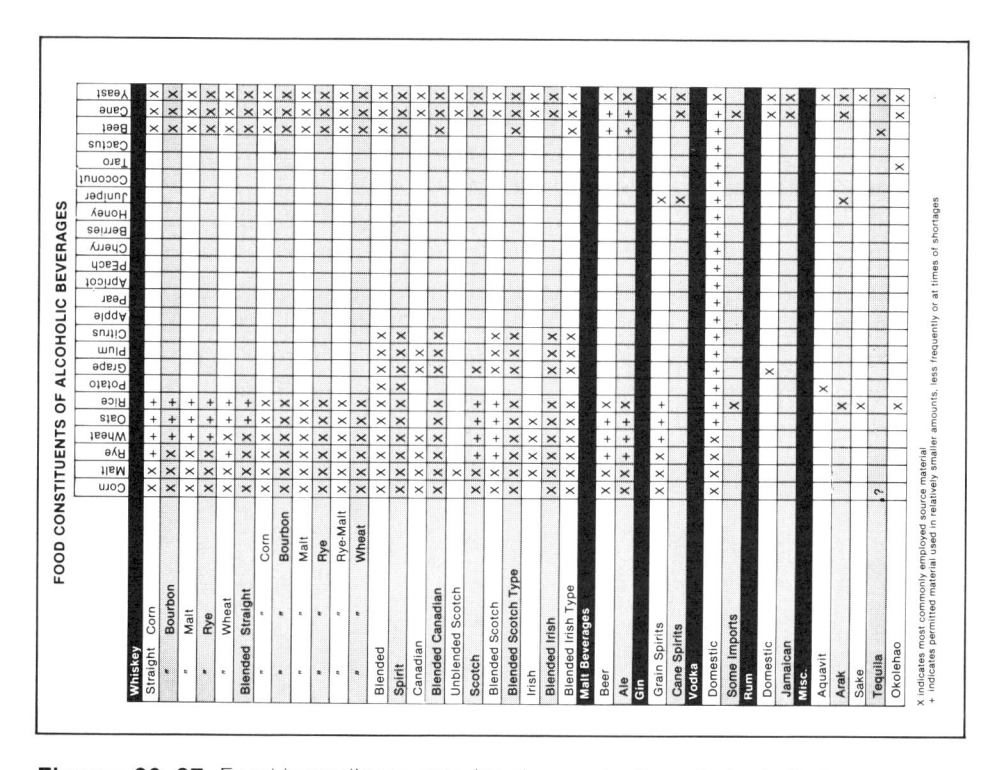

Figures 26, 27. Food ingredients entering the production of alcoholic beverages

The greatest single failing of analytic dietetics is the unfortunate deduction that specific foods in the sense of wheat, corn, milk, egg, etc., are not important in the management of a wide range of physical and mental manifestations. Indeed, rigid adherence to the principles of analytic dietetics in the sense of focusing on the 24-hour food intake of a given patient has virtually eclipsed interest in the subject of food allergy in the common form in which it most frequently occurs. Although there may be some recognition that such uncommonly eaten foods as strawberries, seafood, or certain nuts may be involved in selected cases, there tends to be almost a complete void of interest and knowledge on the part of those adhering to the principles of analytic dietetics that commonly eaten foods to which essentially addictive behavior may be demonstrated actually exists. Indeed, it is more the vogue of traditionally trained physicians to ridicule the existence of food related chronic illness than to admit to such possibilities, to say nothing of attempting to confirm or not confirm such reported relationships by means of acceptable diagnostico-therapeutic techniques.

Many other instances of the application of analytic and biologic dietetics in the practice of medicine could be cited, but this is sufficient to illustrate the fact that there is much more to diet therapy than is included in the traditional analytical approach to this subject which has dominated it for the past 60 years.

Distinctions between analytic and biologic dietetics at the educational level should also be emphasized. Since there are no courses, to the writer's knowledge, on the subject of biologic dietetics and little instruction otherwise, it must be assumed that all physicians and graduate dietitians have been trained almost exclusively in analytic dietetics.

In summary, the application of analytic dietetics simply fails to provide effective medical approaches to many nutritionally related chronic allergic disturbances in which the etiologic roles of specific foods may often be demonstrated. Despite this relative therapeutic failure of analytic dietetics, this chemically oriented analytical approach to nutrition dominates this area to the virtual exclusion of biologic dietetics. Biologic dietetics deserves to escape from its long eclipse and to become an integral part of the training of physicians, dietitians, and others interested in the application of nutrition to medicine.

Figure 28 summarizes the differences between analytic and biologic dietetics.

	ANALYTIC DIETETICS	BIOLOGIC DIETETICS
Definition	Concerned principally with the twenty-four-hour food intake, especially with such analytical fractions as calories, protein, fat, carbohydrate, vitamins, minerals, triglycerides, acidity, alkalinity, roughage, and other fractional constituents.	Concerned principally with the response of a given person to primary foods, such as wheat, corn, milk, egg, potato, beef, beet, yeast and others and/or their chemical additives and contaminants.
Variations	The nutritional roles played by these constituent fractions are modified by deficiencies, excesses, impaired absorption, or altered utilization.	Clinical effects resulting from primary foods are accentuated by individual susceptibility and relatively obscured by adaptation (addiction) to oft-repeated specific intake.
Diagnosis	These constituents of the twenty-four-hour food intake are measured by calorimetry (burning under controlled conditions) or by means of analytical chemical laboratory procedures.	Clinical effects resulting from the impingement of primary foods on an individual's health and behavior are determined by performing feeding tests under correct conditions of testing or by means of other techniques which correlate therewith.
Measurement	Clinical deficiency states are confirmed by analytical laboratory procedures and animal experimentation.	The degree of individual susceptibility is determined by the timing of onset and the degree of severity of test reactions.
Treatment	Based on dietary regimens providing an excess or deficiency of given analytical fractions, the administration of supplements of given materials or program to improve absorption and/or utilization.	Based on avoidance of incriminated foods and rotation and diversification of compatible foods in keeping with a working knowledge of the biological classification of foods and the stages of specific adaptation.
Summary	Analytic dietetics involves concepts of nutrition which are relatively applicable en masse. Most popular presentations of nutrition and dietetics deal exclusively with analytic dietetics.	Biologic dietetics involves relationships which are individualized to person and highly specific to the primary food involved.

As yet, no book has been written on the subject of biologic dietetics.

Figure 28. Analytic versus biologic dietetics

1971-1976

Ecologic Mental Illness Updated
The History of Ecologic Mental Illness

I was asked in 1973 by Claude A. Frazier, M.D., editor of the *Annual Review of Allergy*, to present the History of Ecologic Mental Illness.[251] This report started with an overview which began as follows:

> The most characteristic feature of the respective dark ages of science, medicine and psychiatry has been the tendency to linger, especially in the United States and particularly in the field of psychiatry. As far as the demonstration of causality in mental and behavioral disturbances is concerned, darkness based on introspection as well as speculative and nondemonstrable etiology is still with us. In my opinion, the chief reasons for this retarded development have been: 1) The wide acceptance and continued dominance of psychoanalytical concepts and practices and, more recently, pharmacotherapy in psychiatry, and 2) The delayed recognition and extraordinarily slow acceptance of the etiologic roles of specific foods, simple chemicals and other exogenous nonpersonal exposures demonstrated to be important in mental and behavioral syndromes....
>
> In tracing the history of ecologic mental illness, major emphasis will be placed on the existence of demonstrable and reproducible etiologically related syndromes. The naturally occurring course of these illnesses will be stressed, focusing on the responsible exogenous agents, levels of manifestations as well as the mediating and determining role in these responses played by specific adaptation. Historical points leading up to or distracting from this experimental point of view and observations supporting it will receive major attention....
>
> Savage, an English psychiatrist (Savage GM. *Insanity and Allied Neuroses; Practical and Clinical.* Philadelphia: Henry C. Lea's Son and Co., 1884), was apparently the first to point out the alternation of mental and physical syndromes. He wrote in 1884: "I have met with several cases in which insanity has alternated with spasmodic asthma in patients who have for years been subject to recurrent attacks of asthma, and who have become almost suddenly well, as far as the asthma was concerned, but who at the same time developed insanity, and as long as the insanity was present the asthma was absent."
>
> Beginning in the mid to late 1890s and culminating at about the turn of the century were three independent developments destined to change the course of modern psychiatry. For the most part, each point of view was initiated by persons working alone in widely separated parts of the world.

1. The Objective Approach

Although one of the first attempts to study psychological phenomena by means of an objective, experimental approach was Thorndike's (Thorndike EL. *Animal Intelligence — An Experimental Study of the Associative Processes in Animals.* New York, 1898) observations in America in 1898, Sechenov's (Sechenov IM: Quoted by Wells HK. *Ivan P. Pavlov; Toward a Scientific Psychology and Psychiatry.* New York: Internat. Publishers, 1956) observations and, especially, Pavlov's (Pavlov IP:Trans. by Gantt WH. *Lectures on Conditioned Reflexes; Twenty-five Years of Objective Study of the Higher Nervous Activity (Behavior) of Animals.* London: Lawrence and Wishart, Ltd., Vol. I and II, 1928) contributions, both in Russia, laid the groundwork for the later development of objective psychology and psychiatry. From the turn of the century to his death in 1936, Pavlov clung tenaciously to the concept stated in a lecture at the International Congress of Medicine in Madrid in 1903:

"Only by proceeding along the path of objective investigation can we step by step arrive at the complete analysis of that infinite adaptability in every direction which constitutes life on this earth...."

Pavlov's observations formed the basis of behaviorism first stated by Watson (Watson JB. *Behaviorism.* Chicago: University of Chicago Press, 1957) in his lectures at Columbia University in 1912, in which he commented that man is an animal different from other animals only in the types of behavior he displays. In the introduction to an edition of his book published in 1930, Watson commented that behaviorism had brought out the same type of resistance that appeared when Darwin's *Origin of the Species* was first published. He concluded that human beings do not want to class themselves with other animals. "But through it all," he continued, "without behaviorism being overtly accepted, its influence has been profound during the 18 years of its existence." Commenting further, in comparing journal titles and books written before and after, not only have the subjects studied become behavioristic, but the words of the presentations have become behavioristic.

Until very recent years, Pavlovian concepts of behavior have been championed in this country almost solely by Gantt (Gantt WH. *Experimental Basis for Neurotic Behavior.* New York: Hoeber, 1944; Gantt WH. [Editor]. *Physiological Basis of Psychiatry.* Springfield, Illinois: Thomas, 1958), who not only founded the Pavlovian Laboratory at Johns Hopkins in 1932, but who also works at the laboratory in the Veteran's Hospital at Perry Point, Maryland.

Although objective or behavioristically oriented psychiatry has resulted in much new descriptive and etiologic information, it has failed thus far to provide effective and widely applicable approaches to mental illness.

1971-1976

2. The Introspective Approach

Wells (Wells HK. *Pavlov and Freud: II. Sigmund Freud — A Pavlovian Critique.* New York: International Publishers, 1960) introduces his book, *Sigmund Freud — a Pavlovian Critique,* with the statement: "Pavlov and Freud, each in his own way, were striving to fill one of the last great gaps in human knowledge." He described the gap in question as the lacuna in our knowledge of ourselves.

Wells goes on to describe how Freud's scientific background had been in neurological anatomy where main reliance had been on observation through the microscope, not experimentation. By 1895, physiology of the higher parts of the brain had not been established. Although Freud was interested in discovering the cause of neurosis, his background and training according to Wells did not equip him to do this. He began seeking answers beyond the borderline of experimental science. Coming upon the concept of psychoanalysis in 1896, Freud worked alone, elaborating this in detail until 1902.

Brill from New York translated Freud's work into English and Freud delivered a series of lectures in America in 1909. These lectures were well received, according to Wells, and launched psychoanalysis in the United States. Despite bitter opposition from the first, it took a firm root and has been a dominant factor in American psychiatry since this time. As previously mentioned, this emphasis on the subjective in contrast to emphasizing the objective and experimental has not only dominated American psychiatry, but also seems to have been to a large extent responsible for psychiatry in the United States becoming relatively isolated from other fields of medicine. However, this status has changed somewhat during the past several years with the increasing use of psychotherapeutic drugs in the treatment of mental and behavioral disturbances. In this sense, both psychiatrists and members of the medical profession in general may be criticized for their reliance on drugs in the treatment of medical syndromes in which little effort has been made to demonstrate inter-relationships between nonpersonal environmental exposures and individuals as holistic units of the ecosystem of which they are functional parts.

Another unfortunate consequence of the relatively uncritical acceptance and application of speculative concepts of etiology in mental disease is the extension of subjective approaches and interpretations to other medical syndromes of obscure etiology in the name of psychosomatic medicine (Dorfman W. *Closing the Gap Between Medicine and Psychiatry.* Springfield, Illinois: Thomas, 1966) As will be related, many of these apparent psychosomatic manifestations are those for which etiologic roles of environmental factors have been demonstrated.

In my opinion, subjective psychiatry, especially psychoanalysis, despite its wide application, has been devoid of any demonstrable evidence of etiology and has been relatively ineffective therapeutically.

3. *The Specific Environmental Approach*

The quest for demonstrable evidence of etiology in interrelated physical and mental ills apparently began with the observations of Hare (see Appendix, page 330) working with various chronic illnesses at the Diamantina Hospital in Brisbane, Australia. Without access to medical reference libraries, he simply recorded the observations of the effect of foods on behavior in a 1,000 page, two-volume report. He observed that many chronically ill patients benefitted from either one or two general types of diet: 1) those avoiding starches, sugar, alcohol and carbonaceous materials, and 2) those avoiding or minimizing the intake of proteins....

Shannon (Shannon WR. Neuropathic Manifestations in Infants and Children as a Result of Anaphylactic Reactions to Foods Contained in Their Diet. Amer J Dis Child 24:89, 1922) was apparently the first to observe that nervousness followed the ingestion of specific foods, such as wheat.

Duke also related specific foods and simple chemicals to a wide range of allergic symptoms, including headache and bewilderment resembling delirium. (See Appendix, page 320). Galtman (Galtman AM. The Mechanism of Migraine. J Allergy 7:351, 1936) and Crowe (Crowe WR. Cerebral Allergic Edema. J Allergy 13:173, 1942) were apparently the first to associate edema with specific reactions. Duke and Coca both incriminated specific sugars. (See Appendix, page 315). Coca made the additional observation that all cases of dementia praecox thus far studied exhibited the typical food allergy pulse changes. Coca also noted reactions to such simple chemical substances as aspirin, gasoline and mineral oil.

Since other early contributors of cerebral and neurological manifestations up to the mid-twentieth century have been well documented elsewhere in reviews and surveys, major emphasis in this review will be placed on what has happened since 1950 that bears on the demonstrable roles of exogenous factors in the etiology of mental and behavioral disturbances.

Development of Ecologic Mental Illness

In the course of performing individual food ingestion tests in the late 1940s, acute psychotic episodes were induced not infrequently.[74,79] At first these were thought to have been adventitious, but when the precise dietary intake and test conditions were reproduced, it was established beyond doubt that these feeding tests and subsequent acute mental and behavioral disturbances were causally related. Moreover, when such incriminated foods were removed from the diet of these patients, their chronic mental symptomatology often ceased, only to recur immediately after the reingestion of such dietary items. Although such responses recurred acutely and promptly following intermittent exposures, chronic reactions developed subtly and without apparent relation to meals when these foods were eaten frequently and regularly. These were characteristic features of what Rinkel first described, respectively, as unmasked and masked food allergy,[38,75] later interpreted as food addiction[111] and, respectively, as nonadapted and specifically adapted responses....

Manifestations — Levels of Reaction

The apparent range and interrelationships between physical and mental manifestations resulting from demonstrable reactions to foods, simple chemicals and drugs evolved slowly during the past two decades.[74,106,109,120,122,131,140,184,233,248] There was confusion initially in differentiating stimulatory and withdrawal phases of reaction and interpreting their alternation[89] until it was realized that these constituent levels represented extensions, respectively, of the observations of Kraepelin (Kraepelin E. *Manic-Depressive Insanity and Paranoia*, Trans. by R. Mary Barclay and Ed. by George M. Robertson. Edinburgh: Livingstone, 1921) and Savage. Lesser stimulatory manifestations, see levels (+++, ++ and +) and lesser withdrawal syndromes, see levels (---, -- and -) in Figure 29, are to be considered as lesser extensions toward normalcy (level 0), respectively, of the manic (++++) and depressed (----) phases of Kraepelin's manic-depressive psychosis.

DIRECTIONS:	Start at zero (0) Read up for predominantly Stimulatory Levels Read down for predominantly Withdrawal Levels
++++MANIC WITH OR WITHOUT CONVULSIONS	Distraught, excited, agitated, enraged and panicky. Circuitous or one-track thoughts, muscle twitching and jerking of extremities, convulsive seizures and altered consciousness may develop.
+++HYPOMANIC, TOXIC, ANXIOUS AND EGOCENTRIC	Aggressive, loquacious, clumsy, (ataxic), anxious, fearful and apprehensive; alternating chills and flushing, ravenous hunger, excessive thirst. Giggling or pathological laughter may occur.
++HYPERACTIVE, IRRITABLE, HUNGRY AND THIRSTY	Tense, jittery, hopped up, talkative, argumentative, sensitive, overly responsive, self-centered, hungry and thirsty, flushing, sweating and chilling may occur as well as insomnia, alcoholism and obesity.
+STIMULATED BUT RELATIVELY SYMPTOM-FREE (SUB-CLINICAL)	Active, alert, lively, responsive, and enthusiastic with unimpaired ambition, energy, initiative and wit. Considerate of the views and actions of others. This usually comes to be regarded as "normal" behavior.
0 BEHAVIOR ON AN EVEN KEEL AS IN HOMEOSTASIS	Children expect this from their parents and teachers. Parents and teachers expect this from their children. We all expect this from our associates.
-LOCALIZED ALLERGIC MANIFESTATIONS	Running or stuffy nose, clearing throat, coughing, wheezing (asthma), itching (eczema or hives), gas, diarrhea, constipation (colitis), urgency and frequency of urination and various eye and ear syndromes.
--SYSTEMIC ALLERGIC MANIFESTATIONS	Tired, dopey, somnolent, mildly depressed, edematous with painful syndromes (headache, neckache, backache, neuralgia, myalgia, myositis, arthralgia, arthritis, arteritis, chest pain) and cardiovascular effects*.
---DEPRESSION AND DISTURBED MENTATION	Confused, indecisive, moody, sad, sullen, withdrawn or apathetic. Emotional instability and impaired attention, concentration, comprehension and thought processes (aphasia, mental lapse and blackouts).
----SEVERE DEPRESSION WITH OR WITHOUT ALTERED CONSCIOUSNESS	Nonresponsive, lethargic, stuporous, disoriented, melancholic, incontinent, regressive thinking, paranoid orientations, delusions, hallucinations, sometimes amnesia, and finally comatose.

*Marked pulse changes or skipped beats may occur at any level....

Figure 29. Manifestations — Levels of reaction in ecologic mental illness

1971-1976

Another more graphic way of representing these levels of reaction is shown in Figure 30.

CLINICAL LEVELS OF SPECIFIC REACTIONS

STIMULATORY

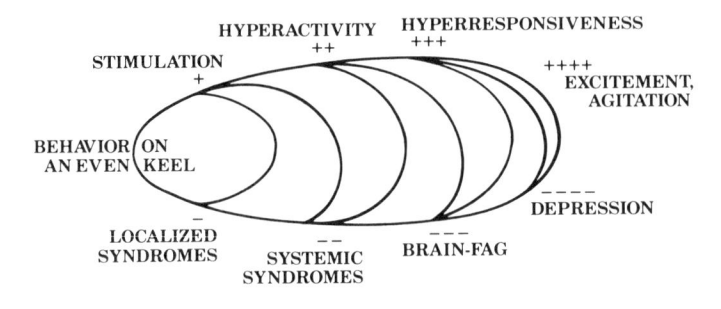

HYPERACTIVITY ++ HYPERRESPONSIVENESS +++
STIMULATION +
++++ EXCITEMENT, AGITATION
BEHAVIOR ON AN EVEN KEEL
DEPRESSION ----
LOCALIZED SYNDROMES -
SYSTEMIC SYNDROMES --
BRAIN-FAG ---

WITHDRAWAL

Start at center left, follow clockwise, induced reactions may reach any stimulatory level before merging with the approximately corresponding withdrawal level before receding

Figure 30. Clinical levels of specific reactions in ecologic mental illness

Changes in Perception and Sexuality

Proprioception tends to be unchanged in level (+), accentuated in (++), aberrant in (+++) and diminished to absent in (++++).[98]

Exteroception is ordinarily unaltered in level (-), accentuated in (--), aberrant (sometimes hallucinatory) in (---) and either aberrant, diminished or absent in (----). Although any exteroceptive function may be involved, the senses of smell, taste, touch, hearing and sight are most commonly affected in this order.

Both sexes tend to manifest normal to slightly heightened sexual desires at (+) and hypersexuality at (++). Although excessive desires may occur at more advanced stimulatory levels, performance of the integrated sexual act tends to be poorly coordinated, if possible, at (+++) and usually impossible at (++++).[128]

Although relatively normal sexual desires and performance may continue in both sexes at (-), fatigue and debility frequently reduces desire at (-) and other manifestations may interfere with performance. Female frigidity and male impotence may characterize (---) and especially (----), but frigidity may occur at any withdrawal level.[128]

Environmental Excitants of Ecologic Mental Illness

The most commonly incriminated materials are specific foods.[344] Those eaten most frequently by a given subject such as coffee, wheat, corn, milk, egg, potato, beet sugar, cane sugar, yeast, orange and others are most commonly involved. Simple chemical environmental exposures (including food additives and contaminants; synthetically derived drugs, such as sedatives, tranquilizers and most psychotherapeutic agents; air pollutants, especially domiciliary sources such as fumes from gas fired and other fossil fuel utilities, sponge rubber bedding, padding and upholstery, synthetically derived textiles and carpets, insecticides, solvents, cosmetics, leaking refrigerants and other hydrocarbons) to which highly susceptible persons may also be essentially addicted, may be demonstrated by means of proper techniques[155,156,166] (Dickey LD. Ecologic Illness: Investigations by Provocative Tests with Foods and Chemicals, 1963-1970. Rocky Mountain Med J 68:23, 1971) (Miller JB. *Food Allergy: Provocative Testing and Injection Therapy.* Springfield, Illinois: Thomas, 1972)[165] Drugs, especially opiates, to which addictive responses develop readily are probably closely related in their overall effects.

As may be noted, most presentations bearing on psychiatric aspects of this work have either been made abroad or at international meetings. If psychiatrists of this country are unfamiliar with these views, this is attributable to the fact that program chairmen of their national organizations and editors of domestic journals in their field have consistently rejected submissions for scientific exhibits, program presentations and publications with the exception of a scientific exhibit in 1956.[108,109] However, these views have been given in yearly meetings of the Section of Allergy of the Nervous System of the American College of Allergists since its founding in 1957. The first meeting set up for the purpose of bringing ecologically oriented views to the attention of psychiatrists was held at Attleboro, Massachusetts in November, 1971. The following individuals were in attendance: M. Mandell [Chairman], T. G. Randolph, E. L. Binkley, F. Waickman, E. W. Kailin, H. Hosen, J. G. Maclennan, L. Rosenzweig, G. C. Pfeiffer, J. J. Miller, H. Dorn, G. Von Helsheimer and W. H. Philpott.

Domiciliary Chemical Air Pollution

Several presentations on this subject[131,140,179] culminated in a publication in 1970.[206] For the most part, these environmental relationships were demonstrated as a result of the diagnostico-therapeutic program of comprehensive environmental control[166] in a hospital ecologic unit[155,156] followed by test re-exposures to given home and work related chemical environmental exposures.

All the manifestations previously described may be evoked under these circumstances. These observations are summarized as follows:

Ecologic mental illness is differentiated from psychiatrically interpreted mental syndromes by the ability to demonstrate cause and effect relationships by means of the application of ecologically orientated techniques.

Domiciliary air pollution is more important than ambient air pollution in the etiology of ecologic mental illness.

The crux of these relationships is individual susceptibility which builds up insidiously as a result of cumulative exposures, enhances the impact of domiciliary airborne substances and accounts for their selective impingement on the health and behavior of occupants of the home.

Full appreciation of domiciliary chemical air pollution awaited the development of: (1) The concept of specific adaptation which helps to explain the interrelationships between a susceptible person and this impingement of his everyday environment, and (2) A full appreciation of what constitutes man's chemical environment.

Ecologic mental illness usually culminates after a long history of multiple complaints which manifest initially as lesser localized physical and/or systematic disturbances. Such chronically ill persons and/or their most immediate associates are commonly maligned by the application of speculative interpretations of causality which are not amenable to proof.

Treatment consists principally of the avoidance of incriminated exposures.

Discussion and Summary

It is of historical interest that the first report on the association of commonly eaten foods as incitants and perpetuants of mental illness occurred in 1950,[74] the same year that psychotropic drugs were brought out. The detailed individualized application of the former versus the en masse applicability and advertising hype associated with the latter provided an enormous competitive advantage for psychotropic drugs. Only after addictive drug responses, tardive dyskinesia and other major complications from the use of psychotropic drugs became apparent were alternative approaches to the management of mental illness even considered.

Ecologic mental illness is differentiated from psychiatrically interpreted mental illness by the ability to demonstrate etiology, in the sense of the impingement of external environmental factors, on the mental and behavioral manifestations of individually studied patients. This is accomplished by observing chronically reacting patients, as probable environmental incitants — especially given foods and chemical exposures — are first avoided simultaneously and later returned for test purposes, in keeping with a working knowledge of the stages of specific adaptation.

Although ecologic mental illness does not exclude the possibility of reactions to other persons and situations, these are far less demonstrable than reactions to specific nonpersonal noninfectious agents. The overall illness related effect of clinical responses to exogenous materials is enhanced in the presence of individual susceptibility.

Both the sequential stages of the development of ecologic mental illness and of induced test reactions are classified in respect to their stimulatory and withdrawal features. These represent lesser extensions, respectively, of the manic and depressed phases of the manic-depressive psychosis as described by Kraepelin. Physical manifestations of chronic reactions (localized and systemic allergic effects) and more advanced mental and behavioral expressions tend to alternate as described by Savage.

As might be expected, the application of this experimentally oriented objective approach to mental and behavioral disturbances has resulted in inductively derived generalizations and techniques to demonstrate them, which are at variance from current approaches in psychiatry. This variation is most marked when compared with the application of deductively derived desiderations, introspective approaches and techniques which characterize psychoanalysis.

Etiology may also be established by demonstrating physiologic mechanisms of disease. To date, attempting to alter such specifically undiagnosed processes therapeutically by means of drugs has been less effective in mental illness than in certain other fields of medicine. In either instance, as well as when drugs are used purely symptomatically, sustained drug therapy for chronic afflictions tends to be associated with adverse reactions and other complications.

The development of ecologic mental illness, as contrasted with other trends in psychiatry and medicine has been traced. Despite their restrictive nature, presently applied programs offer more objective, effective and ecologically sound diagnostico-therapeutic approaches to chronic mental and behavioral disturbances than alternate programs. As medical knowledge of mechanisms increases, it is quite possible that the application of combined ecologic and physiologic approaches may be more effective and desirable than either alone.

Observed Togetherness: An Experimental Demonstrable Approach In Studying Apparent Interpersonal Reactions

The application of ecologic mental illness which demonstrates the occurrence of reactions to specific foods and environmental chemical exposures in given persons has been differentiated from psychiatrically interpreted mental and behavioral syndromes. But instead of eliminating contact with alleged "scapegoats," which often results in isolating an already withdrawn individual and relegates that person to coping with his circuitous thoughts in loneliness, and / or decreasing his cerebration by means of drugs so that one may not object to another person, I proposed a different program. Since spouses, parents and children or even business associates are often involved in these apparent illness-related hassles, the alleged protagonists are observed conjointly as their food intake and other environmental exposures are varied. This is best carried out over a period of several days in a single room, preferably in an ecologic unit of a hospital.[155,156,166] This program was reported in 1973 and 1974.[248,262]

The design of this setup not only provides a physician with the opportunity of observing interactions between the apparent accusor and accusee, but more importantly it enables a physician to observe both persons during the course of experimentally induced reactions. Even more significant is the opportunity that this regimen provides for the alleged protagonists to observe each other both during and between such reactions.

More particularly, this program of observed togetherness has the following advantages: 1) It usually demonstrates cause and effect relationships between the nonpersonal physical environment and the specifically afflicted person as well as upon the allegedly well "protagonist." 2) It takes the accusor off the "hook," so to speak, educating that person in understanding the basic nature of the medical problems of both the accusor and the accusee. 3) The mutually beneficial insight engendered by this program of observed togetherness virtually precludes the common misunderstandings and failure to cooperate in therapeutic programs, based on the avoidance of demonstrated exposures by only one member of such a joint matter.

The following illustration was suggested from the histories provided by a man and his wife. She, an articulate artist, was subject to intermittent unexplained depressions in which she accused her husband of being responsible for them because he was always involved in these interpersonal hassles. He, a truck driver, being unable to defend himself against her intermittent tirades, simply resorted to alcoholic beverages on each of her personally directed attacks on him. Meanwhile, he had become an alcoholic.

Since neither understood the basic nature of their problems, it seemed desirable to hospitalize them in the same room as they were diagnosed specifically. Symptoms in both — including arguments between them — were accentuated initially. Then they got along well until the provocative food ingestion tests. The bourbon drinking alcoholic developed a severe reaction to corn, observed by his wife. Shortly after the artist ate wheat for the first time in five days, she launched into a typical attack on her husband. His comment was classical: "Don't take this out on me. Attack that bowl of porridge in front of you and leave me out of it."

In my experience, the program of observed togetherness is the most effective diagnostico-therapeutic approach to "scapegoatism" and other apparent interpersonal hassles which patients manifesting mental and behavioral syndromes are prone to present. Potential errors inherent in psychiatric approaches and interpretations, as compared with ecologically oriented approaches which demonstrate causation have been discussed in a companion manuscript.[272]

Although we may expect all persons to be sane, reasonable and rational, the fact remains that in the throes of a mental reaction — especially the early phase of such a reaction — these highly desirable features either do not exist or these cerebral functions have been significantly impaired. Under such circumstances, "conclusion jumping" tends to become more the rule than the exception. Case reports were purposely selected in this and a companion presentation to illustrate the common errors of "scapegoatism" either with or without emotionality and psychogenesis. The possible role played by suggestion was also discussed previously.

It should be mentioned that in recent years case reports are sometimes dismissed as being anecdotal and old fashioned, more attention being given to statistical analyses and bodily mechanisms common to multiple subjects. However, in a subject such as this, characterized by high degrees of individuality of manifestations and specificity of the environmental incitants involved in which underlying mechanisms are not yet well understood, selected representative case reporting remains the method of choice.

Fellowship in Clinical Ecology

One of my San Francisco patients, Anna L. Lewes, left a bequest in 1974 for the training of young physicians in the subject of clinical ecology, under my supervision. Recipients of this Fellowship for a period of three months or longer are listed in Appendix B. This training was concerned primarily with assistance in operating the comprehensive environmental care unit.

Hospitalization

Because of the inability to obtain an area of adjacent rooms free of tobacco smoke exposures which could be made into a comprehensive environmental control unit at the Henrotin Hospital, I moved to the American International Hospital at Zion, Illinois in 1975. Mr. Richard Stephenson was very helpful in setting up this diagnostico-therapeutic unit.

The Book — *Clinical Ecology*

The outstanding accomplishment of this period as far as clinical ecology is concerned was the preparation and publication of the book, *Clinical Ecology*, under the able editorship of Lawrence D. Dickey of Fort Collins, Colorado. Since my contributions to this volume, representing about a quarter of it, have been mostly described elsewhere, only the reference numbers to my bibliography in the Appendix of this book will be supplied here: 252, 256, 257, 258, 259, 284, 288, 289, 290, 291 and 137 (pages 25-29). Some of my previously unpublished material was presented here and several other parts were updated.

1971-1976

Fortunately, the book with its 45 contributors is still in print. Anyone with sufficient interest in the subject of clinical ecology to read and/or possess this autobiographic sketch should also read and have access to this outstanding presentation of clinical ecology.

In 1976 I received an Environmental Quality Award from Region V of the United States Environmental Protection Agency in recognition of having written the book, *Human Ecology and Suceptibility to the Chemical Environment*,[137] and development of the hospital based program of Comprehensive Environmental Control in Diagnosis and Therapy.[166]

Environmental Protection and the National Legislative Scene

According to a review in the Environment Index for 1971, some of the bloom went off the environmental rose in 1971 in that a backlash permeated the running environmental debate against the sustained emphasis on environmental quality for the previous several years. This came as an almost constant barrage from representatives of industry, from congressmen, and from government officials.

According to the same source of information, environmentalists gained more than they lost in 1972, especially as the implications of the Occupational Safety and Health Act (OSHA) became more significant. Adding to the impact of the environment on the international scene was the United Nations Conference on the Human Environment in Stockholm, Sweden in 1972. But as the benefits from environmental cleanup grew more visible in 1973, so did the costs.

Compared with a rapid program in pollution control from 1970 to 1973, 1974 was not a good year for environmental interests, according to a review of the national scene in the Environment Index of 1974. Although some cleanup deadlines were extended, no laws were repealed. In a year dominated by the energy "crisis," environmental groups came face to face with economic interests.

In the United States, 1975 was a mixed year for environmental protection. Indeed, all three branches of the Federal government seemed to cool distinctly toward environmental issues.

Also, as reported in the review of environmental issues on the national scene by the Environment Index, environmentalists had a grim year in 1976. Although a major accomplishment was the passage of the Toxic Substances Control legislation after a six year struggle by environmental groups, environmental defeats far outweighed advances. Moreover, funding for environmental enforcement dropped in 1976 for the first time in a decade.

Chapter XI — 1977-1980

Continued Ecological-Medical Enlightenment
Continued Deterioration of the Environment
Decreasing Regulation of Environmental Hazards
Environmental-Industrial Stalemate

The Emergence of Ecology as a New Integrative Discipline

It seems appropriate to start this chapter with reference to Eugene P. Odum's presentation, as above entitled, published in Science 195:1289, 1977, (©1977 AAAS) which starts off as follows:

> It is self-evident that science should not only be reductionist in the sense of seeking to understand phenomena by detailed study of smaller and smaller components, but also synthetic and holistic in the sense of seeking to understand large components as fundamental wholes. A human being, for example, is not only a hierarchal system composed of organs, cells, enzyme systems, and genes as subsystems, but is also a component of supraindividual hierarchal systems such as populations, cultural systems, and ecosystems. Science and technology during the past half-century have been so preoccupied with reductionism that supraindividual systems have suffered benign neglect. We are abysmally ignorant of the ecosystems of which we are dependent parts. As a result, today we have only half a science of man. It is perhaps this situation, as much as any other, that contributes to the current public dissatisfaction with the scientist who has become so specialized that he is unable to respond to the large-scale problems that now require attention....
>
> An important consequence of hierarchal organization is that as components, or subsets, are combined to produce larger functional wholes, new properties emerge that were not present or not evident at the next level below.... The old folk wisdom about "the forest being more than just a collection of trees" is indeed the first working principle for ecology.... This is not to say that we abandon reductionist science, since a great deal of good for mankind has resulted from this approach, and some of our current short-range problems can perhaps

be solved by this approach alone. Rather, the time has come to give equal time, and equal research and development funding to the higher levels of biological organization in the hierarchal sequence. It is in the properties of the large-scale, integrated systems that hold solutions to most of the long-range problems of society....

Finally, there is yet another divided world, the scientific and the politico-legal spheres of action, where holistic thinking might help. In a recent editorial in Science, Gerald Edelman expresses pessimism that these two disciplines will ever intersect, and states that we are left with "two extreme ideological positions — scientism and anti-scientism". (Edelman GM. Science 192:99, 1976) As long as students and practitioners of both disciplines insist on fragmenting their subjects, rigidly adhering to their own way of thinking and calling each other derogatory names, adversary interaction will continue to predominate. I am much more optimistic about the integration of these spheres because I have found that a meeting of minds in study panels and public commissions begins with the general acceptance of the idea that large-scale problems and issues might have common denominators that could be assessed along with the more narrowly defined scientific, political, or legal aspects. If hierarchal theory is indeed applicable, then the way to deal with large-scale complexity is to search for overriding simplicity. Sometimes, it appears, this turns out to be old-fashioned common sense. As noted, the dichotomy inherent in short and long time spans imposes a major stumbling block in acting on common sense judgment.

In summary, going beyond reductionism to holism is now mandated if science and society are to mesh for mutual benefit. To achieve a truly holistic or ecosystematic approach, not only ecology, but other disciplines in the natural, social, and political sciences as well must emerge to new hitherto unrecognized and unresearched levels of thinking and action.

Odum further pointed out that popularization of the subject ecology is having a beneficial effect of focusing on man as a part of, rather than apart from, his natural surroundings — subject matter to be carried through in the introduction of the following chapter.

Odum's comment that science should not only be reductionist but should also be more involved with synthesis and holism is also especially applicable to modern medicine. Indeed, it was awareness of this deficiency in the field of allergy in particular that prompted the founding of the Society for Clinical Ecology in 1965. As basic ecological concepts and techniques developed, their application materially extended the horizons of this field, in that the earlier immunologically restricted mechanisms were modified in favor of multiple mechanisms and a wider range of clinical manifestations became apparent. As this extension of clinical ecology developed — incorporating and applying the concepts as later expressed by Odum — the medical approaches of clinical ecology began to impinge on many long-standing medical principles and practices.

Coincident with these developments in allergy in particular and medicine generally, the paucity of interest and knowledge of the medical profession in the roles played by the nonpersonal human environment in health and illness became increasingly apparent to ecologically focused physicians. Indeed, some of us involved in the application of clinical ecology for the past several years have gradually come to the following conclusion: Because of its consistent reductionist and environmentally alienated points of view for the past half-century, the medical profession is to a large extent responsible for the degree of environmental deterioration that has been allowed to develop and proliferate in this country.

Although the time is late, it is mandatory for the future of our children and grandchildren that these relatively inept environmental attitudes and practices which characterize modern medicine be corrected. The place to start this reorientation is with the medical curriculum and postgraduate medical training programs.

Candidiasis

One of the most important things learned during this period was the significance of candidiasis as a common complication of patients subject to ecologic illnesses, especially in women. This combination is so common that it should be suspected in all women until it is demonstrated not to coexist with food allergy. Recognizing the existence of candidiasis and treating this in addition to treatment of environmentally related illnesses is apt to result in far superior management than when either is diagnosed and treated singly.

Orion Truss pointed out the multiple localized and, especially, the systemic manifestations of yeast infections in 1977 (Truss CO. Tissue Injury Induced by Candida Albicans; Mental and Neurological Manifestations. J Orthomolecular Psychiatry 7:17, 1978.) This point was later extended by Truss, *The Missing Diagnosis,* Birmingham, Alabama, 1983; and Crook WG. *The Yeast Connection — A Medical Breakthrough,* Jackson, Tennessee: Professional Books, 1983.

Dietary management of systemic yeast infections, consisting of the avoidance of rapidly absorbed sugars and minimizing the intake of fruits and starches often combines well with the avoidance of cereal grains on initial diagnostic programs for food allergy.

Publications

About 1976 I was asked by Lippincott-Crowell to write a popular book on mental illness with a do-it-yourself slant. This firm was subsequently taken over by Harper and Row. Being impressed by the rapidly increasing frequency and relative severity of the chemical susceptibility problem, and particularly its applicability to mental illness, I decided against trying to write a do-it-yourself type of book because of the complexity of that problem. But later I agreed to attempt this, providing interrelationship between physical and mental illnesses would be emphasized.

I had reached the decision at about this time that the program of writing about clinical ecology for the medical profession was relatively unproductive and that in the future I would address most of my writing efforts to the public. This was the first serious effort in this direction. Fortunately, I was able to get Ralph W. Moss, Ph.D., an experienced writer for the public to help me. I dictated the gist of this book on a tape recorder and he did most of the actual writing.

I was not pleased with the title of *An Alternative Approach to Allergies* initially, but this has turned out to be a good title, although the sub-title, "The New Field of Clinical Ecology Unravels the Environmental Causes of Mental and Physical Ills," is more descriptively correct.

This book has sold well and is still in print. It seems to be useful to many clinical ecologists in instructing their patients. During this period (1977-1980) there have been over 20 other books written for the public. Indeed, almost all popular books on the subject of allergy have been written either by clinical ecologists or have favored the concepts and techniques of clinical ecology over those employed by traditional allergists. I had written the forewords or chapters of some of them. These books are listed in my bibliography, published in 1980, of which copies are available upon request.

Continued Deterioration of the Environment, Decreasing Regulation of Environmental Hazards, and the Environmental-Industrial Stalemate

President Carter, elected with strong support of environmentalists, brought many of them into his administration. But this move galvanized corporate and industrial opposition to environmentally protective programs, as costs as well as the Federal deficit also mounted.

According to a review of national environmental activities for 1978, the nation completed a cycle of federal environmental legislation which began with the National Environmental Policy Act in 1970. Although new legislation enacted in 1978 extended the deadlines of earlier measures it left the ultimate environmental goals of the early 1970s intact. But unfortunately, both for environmentalists and industry, the regulatory machinery became almost hopelessly clogged in 1978.

However, the Federal Pesticide Act of 1978, signed into law by President Carter on October 1st, a sweeping revision of the Federal Insecticide, Fungicide and Rodenticide Act of 1972, represented progress.

According to the Environment Index of 1979, national and world energy events in 1979 raised serious doubts about the success of the environmental movement in the United States. For this and other reasons, including several environmental disasters, the question was raised: Has the environmental movement run out of steam as the seventies come to a close? But environmental laws passed in 1976 and 1977 finally began to gain momentum.

All in all, these various pros and cons characterizing the Carter administration's environmental record may best be summarized as an environmental stand-off between the Carter administration's environmental proponents and industry's increasingly critical anti-environmental stance. The increasing Federal deficit also exerted a quieting effect on activity in this area.

In short, the seventies as well as the Carter administration ended in essentially an environmental stalemate.

Chapter XII — 1981-1987

Continued Ecological-Medical Enlightenment
Continued Deterioration of the Environment
Increasing Resistance From Academic Medicine
Decreasing Regulation of Environmental Hazards
Increasing Politicization of Environmental Issues
Increasing Governmental Control of Medicine

Emergence of the Specialty of Clinical Ecology

The previous chapter, starting with the emergence of ecology as a new integrative discipline, failed to comment about its application to medicine. This chapter will begin with the emergence of clinical ecology.[384]

Clinical ecology integrates the medical interests of allergy, nutrition and toxicology in terms of adaptation to specific environmental exposures. Unfortunately, however, most physicians in these fields refused to have their medical activities interrelated.

Another reason for the emergence of clinical ecology is the fact that the ABCDs of modern medicine (its analytical, bodily centered and drug-related approach) minimizes and often neglects the roles played by specific environmental factors operating in highly susceptible patients, as shown in Figure 31, in which the point of view of clinical ecology is represented by its ABCDEFs.[386]

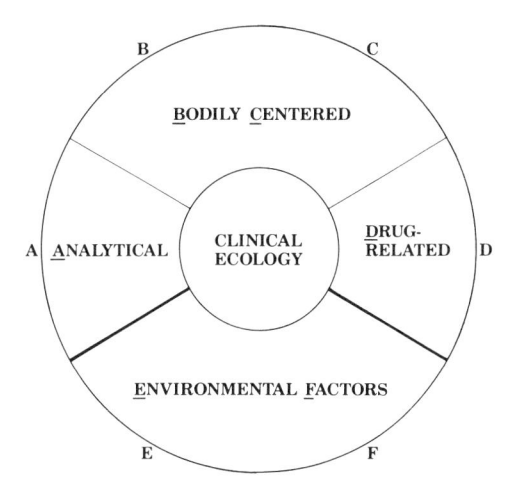

Figure 31. ABCDEFs of clinical ecology.

Indeed, clinical ecology is the only medical discipline on the clinical horizon which is primarily concerned in integrating these bodily and highly personal attributes, on the one hand, and highly individualized specific environmental exposures, on the other hand, which in their interrelationships are impinging on the health, illnesses and behavior of given patients. Under these circumstances it is the only dynamic and holistic specialty of medicine, in contrast to most of the other medical specialties, which are basically analytical and anatomically demarcated. Clinical ecology, also being relatively disinterested in drug suppression of symptoms, is the only medical specialty whose treatment routines do not involve major dependence on drug therapy. Lastly, clinical ecology, in contrast to most other medical specialties which are characterized by a relative discontinuity between diagnosis and treatment, is characterized by a diagnostico-therapeutic continuum.[386]

Larry Dickey, editor of Clinical Ecology — Archives for Human Ecology in Health and Disease, also notes that clinical ecology is an ideal medical specialty in that it is comprised of a group of physicians who devote their activities to evaluating and perfecting procedures and techniques in a specific field of medicine and eventually making these perfected procedures available to all interested practitioners. He further stated, in the fall 1984 issue of Clinical Ecology, that this involves a learning process not only for physicians, but also for patients who must be largely responsible for the success of their own care without undue dependency on the physician.

In keeping with this concept and in the absence of the development of a medical specialty of environmental medicine, the Society for Clinical Ecology changed its name recently to The American Academy of Environmental Medicine — the Discipline of Clinical Ecology. Maintenance of the concept of ecology in the name of this specialty is important in that it indicates that we are not only concerned with the human environment, but especially in demonstrating interrelationships between a patient whose intake and surroundings are impinging on that individual's health and behavior because of his or her specific susceptibilities.

A Prospective Study of Rheumatoid Arthritis of the Hands

Since the effects of fasting on rheumatoid arthritis are poorly understood, a prospective study of rheumatoid arthritis patients utilizing quantifiable measurement techniques was carried out in the comprehensive environmental control units of Chicago, Dallas, and Chadburn, North Carolina.

Forty-three patients with definite or classical rheumatoid arthritis from three hospital centers underwent a water fast lasting six to seven days under controlled environmental conditions in a specialized ward setting. No major medical complications were noted. Seven parameters of arthritic activity (tenderness articular indices, swelling articular indices, grip strength, dolorimeter pain index, arthrocircameter PIP joint circumference, functional activity index questionnaire and ESR) all significantly improved during the fast. These observations were reported in 1984.[388]

A follow through study of rheumatoid arthritis of the hands was carried out by the same physicians in 29 previously fasted patients in the same environmentally controlled units of these hospitals. The summary of this investigation is as follows:

Twenty-seven patients with definite or classical rheumatoid arthritis were water fasted approximately seven days. A previous article reporting this initial study showed that most of the arthritis patients improved significantly by the end of the fast.[388] They subsequently were subjected to single sequential primary organic food challenges. Corn, wheat and animal protein provoked reactions more frequently and intensely than fruits and vegetables. Twenty-two patients experienced significant loss of grip strength, an increase of hand arthrocircameter PIP joint circumference, and an increase in hand dolorimeter tenderness index when compared to the nonreactive foods. Arthritic evaluations compared at admission and discharge showed significant improvement. The patients' functional activity index was significantly improved, staff measurements of grip strength, dolorimeter tenderness index and hand arthrocircameter PIP joint circumferences were all significantly improved. The direct physician examination of the joints showed significant improvement of both the tenderness and swelling indices. The sedimentation rate was not significantly altered. This study shows that selected foods enhance the clinical manifestations associated with active rheumatoid arthritis.[389]

Most rheumatoid patients responded to specific foods with acute inflammatory reactions, as compared with nonreactive foods, although selected foods were associated with more frequent reactions than others. The arthritic response of each patient was highly individualized. In general, vegetables and fruits were less reactive than animal proteins and cereal grains, especially wheat and corn. From this experience, any general diet would not be suitable for all arthritics.

Importance of the preliminary period of fasting and the concurrent avoidance of many other probable environmental excitants of chronic illness prior to and during the course of given oral food challenges is emphasized for the following reasons: 1) This reduces the general level of arthritic activity which enables both physician and patient to observe the symptom response following food challenge more accurately. 2) The regimen of avoiding such frequently eaten foods as common meats and cereal grains until one has recovered from their cumulative effects enhances the clinical effects associated with such challenges. Despite such acute nonadapted test responses sometimes occurring promptly, reactions are more apt to be delayed in onset between 4 and 12 hours.

Responses to a wide range of chemical additives and contaminants of the diet were studied in 24 patients who had not reacted to the same foods in their organic form, but no single additive nor class of additives was implicated.

Hospitalization

Although the hospital in Zion, Illinois was satisfactory in many ways, it simply was too distant from my base of operation (home and office) in Chicago. Consequently, when the Henrotin Hospital in Chicago expressed interest in setting up a comprehensive environmental control unit, I accepted their offer in July 1983. They spent approximately $130,000 in applying the necessary central air-conditioning and ventilatory arrangements with a vent for each room. Ilene Buchholz, my head nurse at Zion, trained the hospital personnel in this work. Despite the location of the hospital — a mile from the center of Chicago — being less satisfactory than the essentially rural location of the Zion unit, it worked out reasonably well from the technical standpoint.

The chief problem was resistance of health insurance companies to pay claims of patients hospitalized for this type of diagnostico-therapeutic medical service. First refused by Blue Cross several years earlier, they resumed payments for a few years after a patient sued them successfully, only to refuse to pay again in 1982 and since. One by one, Medicare, Governmental health insurance programs based on Medicare policies, and private insurance carriers followed suit. The time was reached in early 1986 when an insufficient number of patients with insurance coverage to support the unit made its closure mandatory. This is a prime example of forces outside of medicine, responsible for the payment of medical services, interfering with

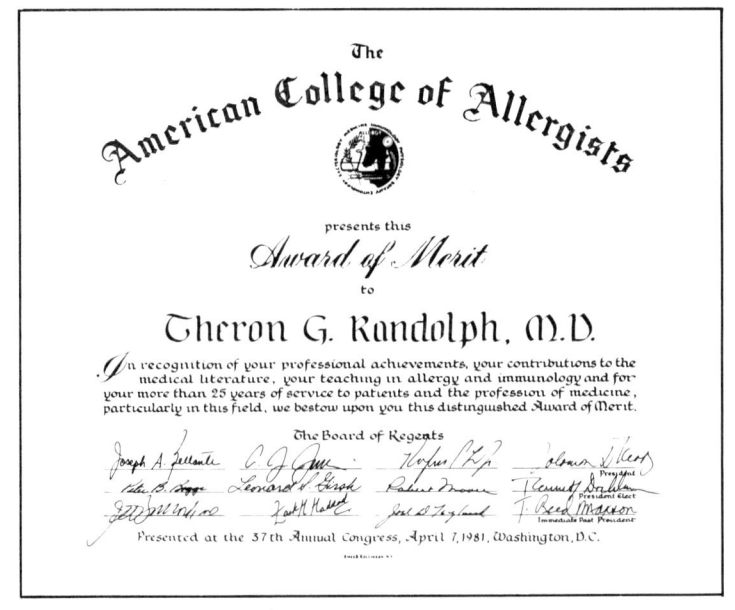

Figure 32. American College of Allergists Award of Merit

the practice of medicine. The sad fact remains that there is a significant number of patients — approximately 15 percent of those seeking my professional services as a clinical ecologist — who are simply too ill to be diagnosed and treated on an outpatient basis. This fact did not seem to enter the decision making of many insurance companies.

Although I have retained my Henrotin Hospital appointment, I rarely use this hospital, inasmuch as my most ill patients usually have the chemical susceptibility problem which cannot be handled adequately on a general medical floor.

Figure 32 shows an Award of Merit which was presented to Theron G. Randolph, M.D. by the American College of Allergists at their 37th Annual Congress in Washington, D.C., April 7, 1981.

Publications

The book, *An Alternative Approach to Allergies,* was republished with Ralph W. Moss, Ph.D. in paperback under the same title by Bantam Press in 1981. It was also published in England under the title of *Allergies, Your Hidden Enemy,* Wellingborough, North Hamptonshire, 1981 and by Thorsons Publishers, Limited, Wellingborough, North Hamptonshire, 1984. This book was published in Germany under the title, *Alternative Konsepte,* Karlosuhe, Verlag, CF. Muller, 1984.

The book, *Human Ecology and Susceptibility to the Chemical Environment,*[137] is still in press. Approximately 20 additional popular books on the subject of clinical ecology were also written between 1981 and 1985.

The China Trip

In late May and early June 1985, several of us were invited to mainland China to present the concepts and techniques of clinical ecology to medical students in six Chinese medical schools (Shanghai, Beijing, Xian, Chongqing, Wuhan and Guangzhou [Canton]). This program was set up by Dr. Shitai Ye, professor and head of the Department of Allergy, Peking Union Medical School, Chinese Academy of Medical Sciences, Beijing, China. Dr. Ye also accompanied us (Doctors William Rea, Lawrence Dickey, Del Stigler, John Boyles, Sherry Rogers and Theron Randolph), serving as interpreter when necessary.

Vera Rea has written an excellent account of this three week trip which was published in Clinical Ecology News (Clinical Ecology, Vol. III, 1986). In brief, the trip included short visits in Shanghai, Beijing, Xian, a boat trip of two days on the Yangtze river from Chongqing to Yi Chang. There also were visits in Wuhan and Guangzhou and, finally, two days in Hong Kong before returning.

One of the most amazing finds was the excellence of the medical libraries in all of the medical schools visited. When I found complete sets in English of medical journals checked in the Peking Union Medical Library (all allergy journals and the Journal of Immunology), I made a point to check for the same journals in the other medical libraries visited, and found the same completeness of bibliographic references. I asked of one librarian how this was accomplished during the cultural revolution, when allegedly many historical records were destroyed. I was told that Chou En-lai had told them to lock the medical libraries and to keep them locked until told otherwise!

It seems that China, a country which has been through many political twists and turns, has been able to maintain a sustained interest in aspects of medical history.

Continued Deterioration of the Environment

The overall trend of environmental deterioration in America since approximately the first World War, continues. Although this rate of increase apparently came under some degree of control during the score of years between 1961 and 1981, relaxation of enforcement of environmentally protective regulations during the present Reagan-Bush administration has apparently been characterized by accelerated environmental deterioration.

Also, failure of most physicians to recognize impingement of the human environment on the health of their patients and consequent failure to minimize or avoid the impact of given environmental exposure on their health and productivity has added to the overall medical burden of continuing environmental deterioration.

Increasing Resistance From Academic Medicine

The term academic medicine is here used synonymously with the so-called medical establishment, which incorporates the official positions emanating from medical schools, major postgraduate training programs in medicine as well as the American Medical Association and many specialty medical societies.

As the physicians who later referred to themselves as clinical ecologists developed environmentally focused concepts and techniques, many of which were not primarily dependent on immunologic mechanisms, they distanced themselves from traditional allergists despite both groups being concerned with environmentally related reactions on the basis of individual susceptibility. Initially both groups were also primarily interested in localized allergic reactions. However, divergence increased as clinical ecologists emphasized the demonstrable roles of foods and environmental chemicals in the etiology of numerous systemic manifestations. As long as these dissidents remained small in number they were dismissed by the medical establishment as "gadflies." But after allergists limited memberships in national allergy societies to board certified internists and pediatricians, differences between those employing ecologically focused techniques and those employing traditional techniques increased progressively. Also, as ecologically focused physicians were denied the opportunity to serve as instructors in instructional courses sponsored by national allergy societies, early clinical ecologists accepted invitations to teach ecologically focused medical techniques to otolaryngologists and other physicians interested in allergy. This divergence between clinical ecologists and traditional allergists was also accentuated by the fact that traditional allergists held most of the teaching positions in allergy departments of medical schools and graduate training programs. During the past decade criticism of clinical ecology and of those physicians employing ecologically focused techniques has emanated largely from the American Academy of Allergy and Immunology (AAAI).

I am writing as a fellow of AAAI as well as all other national allergy societies for the past several decades. But despite this long standing association, I have not been represented by the American Academy of Allergy and Immunology since a presentation on food allergy on the program of the Annual Meeting of this society in the late 1940s was rejected for publication by the editorial board of the Academy's official publication, the Journal of Allergy. Fortunately, this subject, dealing with contamination of packaged food by the cornstarch surfaces of packaging materials, was published elsewhere.[40] It was largely because of such intransigent attitudes of national allergy societies, especially AAAI in regard to food allergy and other nonimmunologic aspects of this field at that time that several of us founded the Society for Clinical Ecology in 1965.[175]

Recognizing that criticism of clinical ecology had been increasing during the past decade, many of us holding joint memberships attempted to minimize this budding controversy by not calling attention to it. But as this criticism has become increasingly strident (American Academy of Allergy and Immunology: Unproved procedures for diagnosis and treatment allergic and immunologic diseases, J Allergy Clin Immunol 78:277, 1987) it necessitates a reply.

To summarize this criticism preliminarily, it alleges that the techniques used by clinical ecologists are unscientific and unproved because they are not based on randomized clinical trials and double-blind placebo based observations. In my opinion, this criticism is unrealistic and not in keeping with the basic dynamic nature of the medical phenomena that clinical ecologists are studying. This deductively derived and reductionist perspective is incompatible with our inductively based and holistic point of view. Indeed, there seems to be little possibility that traditionally trained allergists will be receptive to our clinically oriented approach and findings based upon the demonstrability of multi-factorial adaptive interrelationships observed in individually studied patients, as long as our critics persist in defending their excessively analytical specific etiology point of view at the expense of our holistic adaptive interpretations.

Defense of current concepts and techniques of clinical ecology (medical

1981-1987

care based on demonstrable environmental-individual interrelationships) is best presented by tracing medical developments in respect to these overlapping sequential periods. These three periods were recently reviewed by Melvyn R. Werbach (*Third Line Medicine — Modern Treatment for Persistent Symptoms.* New York, & London: Arkana, 1986). These periods will be described briefly, emphasizing the turmoil that occurs when a previously widely accepted medical doctrine is in the process of being challenged and then superseded by a current alternative medical interpretation.

The Hippocratic Ecologic Period of Medical Practice

Although the beginnings of this period of Grecian medicine are lost in antiquity, it is known to have persisted to at least some degree until modern times. The basic profoundness of medical practices during this age of Hippocratic ecologic medicine still amazes open minded medical historians. Werbach calls attention to the fact that Hippocrates initiated ecologically focused medical care based on a harmonious balance between various components of man's nature. These important components involved an individual's environmental exposures, ways of life and other interrelationships including those between body and mind. These generalizations, constituting the basis of Hippocratic ecologic medicine, were all gained inductively as a result of careful clinical observations.

The Specific Etiology Doctrine of Medical Practice

This view was initiated by the acceptance and application of the germ theory of disease. It was named the specific etiology doctrine of medicine by Rene Dubos (*Mirage of Health,* New York: Harper & Row, 1971). This view postulates that each disease is characterized by a unique primary cause and a distinctive pathology. Unfortunately, many of the fundamentally sound clinical approaches of the Hippocratic ecologic age of medicine, which emphasized interrelationships between individuals and their intake and surroundings reflecting in their health, were discarded or distorted during this modern period of medical practice. This currently accepted medical doctrine will be discussed in respect to its transition from the former Hippocratic era, its acceptance and distinctive features, as well as its over-application and complications.

As pointed out earlier, the transition in medical thinking to accommodate the germ theory of disease had occurred gradually in Europe over a period of several decades. But this transition was retarded in the United States due to our preoccupation with the build-up, actuality and aftermath of our civil war. With acceptance of Koch's postulates by medical academia in 1880, our medical schools suddenly began to teach the germ theory of disease as fact. But as is well known from medical history, many practicing physicians resisted its application for several decades.

Gradually, however, the specific etiology interpretation of medical practice not only became widely applied, but, aided by the development of antibiotics and other technical advances, it has since been over-applied in the name of science. This over-application coincided with the decline of interest in general practice and excessive medical specialization. Werbach emphasizes the fact that present day medicine is still dominated by specialization. He refers to specialization as the glue that cements modern medicine into the specific etiology doctrine. Indeed, the greater the degree of specialization in medicine, the less likely that either the Hippocratic ecologic model of medical care or currently proposed ecologically focused medicine can be appreciated.

A major complication of this over-acceptance of the doctrine of specific etiology and the rigid thinking that accompanies it — all defended in the name of science — is that many alternative interpretations have been dubbed as "unscientific." This course of events has occurred in the controversy between traditional allergy and clinical allergy, the development and present status of which will now be described.

Inductively Derived Versus Deductively Derived Concepts in Allergy

Although differences in how the concepts of clinical ecology and allergy developed have been described earlier, these differences will be reviewed briefly because of their bearing on this controversy.

The concept of allergy in the sense of altered reactivity occurring with time was made on the basis of detailed observations in both animals and in humans at about the turn of the century. It is important to emphasize that these detailed observations led to the generalization of allergy as altered reactivity occurring with time. Two important developments occurred in this field starting about 1920 which were destined to change its course. While the majority of early allergists were preoccupied with the localized manifestations of this new phenomenon, others — the predecessors of clinical ecology — described systemic syndromes of headache, fatigue, arthritis and mental disturbances as allergic manifestations. But before these more general effects from susceptibility to physical agents, foods and environmental chemicals had become widely accepted, allergy was redefined in terms of its apparent immunologic mechanisms. The deduction that allergy was an immunologic phenomenon may have been influenced by the immunologic interpretation of infectious diseases. It was most unfortunate that this immunologic reinterpretation of allergy, which was in keeping with the doctrine of specific etiology, was accepted before this new field had been fully described, especially in respect to the roles played by specific foods and chemicals. Since these environmental agents, which were especially responsible for generalized clinical effects, did not involve demonstrable immunologic mechanisms as a rule, neither these systemic manifestations nor the environmental agents largely responsible for them were acceptable to those who came to be called traditional allergists who were now dominating this new field.

Coca's repudiation of this immunologically restricted redefinition of allergy and his continued emphasis on clinical evidence supporting his view of food allergy has been noted earlier. Although a renowned immunologist, Coca's primary interest was not in extending ideational affinities, but in finding practical solutions to commonly occurring work-a-day medical problems. Perpetuating this pragmatic interest not only sustained his unpopular concern about food allergy, but also led later (1945) to his initial description of individual susceptibility to environmental hydrocarbon exposures. Similar critical views of immunology were also expressed by Brown EA (The new review of allergy and (incidentally) of applied immunology. Review of Allergy 19:300, 1966, Editorial) and by Glaser J (The menace of immunology in the clinical practice of allergy. Review of Allergy 21:1120-1131, 1961) as well as by me.[240] Brown also contributed later to the chemical susceptibility problem (Brown EA, Colombo NJ. Persistent cough and bronchospasm due to exposure to fumes from range oil, Ann Allergy 7:756-760, 1949).

Although this immunologically restricted interpretation of allergy was supported by the discovery of IgE which mediates atopic allergy (Ishizaka L, Ishizaki T, Hornbrook MM. Physiochemical properties of reaginic antibody. V. Correlation of reaginic activity with reaginic antibody. J Immunol 97:840-853, 1966) and RAST which measures it (Johansson HGO. Raised levels of new immunoglobulin class (ND) in asthma. Lancet 2:951-953, 1966), great gaps remain between allergy as defined immunologically and allergy as practiced empirically and pragmatically by clinical ecologists. More recently and especially during the past decade, however, immunological interpretations of allergy have been expanding, as reported by Fauci AS (The Revolution in the approach to allergic and immunologic diseases. Ann Allergy 65:632, 1985, Editorial).

In contrast, concepts and techniques characterizing clinical ecology evolved inductively from detailed environmentally focused clinical observations which were confirmed as they were checked clinically and led to generalizations. It is significant that these inductively derived generalizations did not originate in committees nor academic medical institutions. Rather, they originated from the clinical experiences of physicians practicing in the absence of medical school affiliations. Under these circumstances, it is not surprising that most traditional allergists as well as medical academia in general object to these inductively derived clinical approaches employed by ecologically focused physicians. As mentioned, criticism has been directed especially to the techniques employed by clinical ecologists in reaching their conclusions in that they failed to employ randomized clinical trials and double-blind provocative test procedures. I agree with Doris Rapp that this criticism has been overdrawn. Her views on this subject and her review of publications in this area are published in Appendix C of this book.

It should also be mentioned that differences between deductive and inductive investigational programs have been described in detail by John Stuart Mill (*Philosophy of Scientific Method,* New York: Hafner Publishing Co. Inc., 1950)

Ecologic Versus Traditional Medical Care in Allergy

More specifically, traditional allergists demand that clinical ecologists study their patients by means of randomized clinical trials in which patients are assigned to test and control groups by means of random selection. Results should be evaluated as objectively as possible and expressed in quantitative terms. And if evaluation depends on a patient's or a physician's judgment the trial should be conducted double-blind, so that neither patient nor evaluating physician should know which person received a test or a control procedure.

But dividing chronically ill patients responding dynamically to various environmental exposures capable of eliciting variable degrees of individual susceptibility into "comparable representative" groups and exposing them blindly for statistical analysis of results is an impossibility because of: a) The characteristic multiplicity of their physical and/or mental manifestations and the fact that at any given time these diverse syndromes may alternate. b) The wide range of specific environmental exposures involved (pollens, molds, dusts, insect and animal emanations, food and environmental chemicals) which must be considered in respect to both their intermittent and cumulative contacts. c) The extreme degree of individual susceptibility known to exist between patients in respect to their environmental exposures, depending on their recency, intermittency or cumulative specific contexts. d) Marked individual differences amongst patients in respect to their ability to adapt to the impingement of such exogenous exposures. e) Marked variations in regard to the advancement and complications of the numerous manifestations presented by patients and whether or not at the time of testing they are exhibiting stimulatory or withdrawal type responses.

For these combined reasons, the field known descriptively as allergy does not qualify for study by means of randomized clinical trials and double-blind provocative techniques. Indeed, it is a virtual impossibility to categorically assign all patients seeking medical attention for their chronic illnesses into such alleged "representative" groups. Assume for the moment that one might expect to find a group of patients meeting these requirements as a result of advertising for them and paying them for their time. There would be no assurance that a significant portion of such individuals would not have truncated or distorted their histories to meet the desired requirements.

Of all of the demands of our adversaries, there is one highly significant and primary area of our activities about which our critics remain conspicuously silent. This is the diagnostico-therapeutic program of comprehensive environmental control — the precise circumstances surrounding our actual provocative testing, upon which the conclusions in this book and most of my other publications have been largely based. It is mandatory that while a patient is being observed in the process of provocative testing that the individual is not being exposed simultaneously to other environmental exposures capable of eliciting confusing test responses. The one technique designed to minimize this occurrence or such a complication is the program of comprehensive environmental control in a hospital unit.[155,156,166]

It is to be reemphasized that most of the clinical observations reported in this book about which traditional allergists are quibbling, have been established according to this comprehensive environmentally controlled technique. Between 1950 and 1985 at least 3,000 advanced chronically ill patients voluntarily seeking medical assistance for their chronic illnesses have been hospitalized and observed under these conditions. These provocative tests were usually performed blindly but double blindly only rarely in routine work.

Neither, to my knowledge, have our critics attempted to confirm or not to confirm our findings by becoming proficient in the use of our techniques and actually employing them in a significant group of chronically ill patients seeking medical assistance for their illnesses.

I was recently extremely pleased to find Dr. Melvyn R. Werbach's remarkably well documented book, *Third Line Medicine — Modern Treatment for Persistent Symptoms* (New York and London: Arkana, 1986) as it so completely confirms and extends many of the points made in this chapter. Although I had completed this chapter before meeting Dr. Werbach, fortunately I was able to include his views in this discussion.

He too emphasizes the hazards associated with scientific experiments, such as randomized double-blind controlled studies which suffer from several flaws. According to him, these include the fact that only certain kinds of phenomena can be investigated by means of double-blind studies and that many treatments cannot be evaluated in this way because of the impossibility of creating a truly blind placebo. In my experience, this is especially apt to be true in instances of allergic-type responses to given foods and environmental chemical exposures. Werbach also deplores the fact that those defending the doctrine of specific etiology tend to worship randomized double-blind studies because they appear to offer simple conclusions about what they consider to be simple issues. In my opinion, this technique is especially demanded by traditional allergists who seem to be oversold on the application of limited immunologic interpretations of that subject to the relative exclusion of alternative interpretations.

Application of Clinical Ecology Practices to Medicine Generally

Although clinical ecology started by this challenge of the limited application of traditional allergy, it soon became apparent that much wider implications and applications were at stake. Indeed, the doctrine of specific etiology itself, which has dominated the practice of medicine for the past century, has come under increasing criticism as it became apparent that the concepts and techniques of clinical ecology possessed fundamentally more useful approaches to many other medical problems.

In his book, Werbach traces medical practices from early Greek times to the present. He points out how Hippocrates, through careful clinical observations, originated ecologically oriented medicine, consisting of a harmonious balance between components of man's nature — his environmental exposures, his way of life, and especially, his body-mind interrelationships.

But with the acceptance and application of the germ theory of disease, medicine became increasingly dominated by highly analytical techniques and increasing medical specialization. Werbach quotes Renee Dubos (*Mirage of Health,* New York: Harper & Row, 1971) who referred to this movement in modern medicine by stating that specialism is the glue which cements modern medicine into the specific etiology doctrine. Moreover, he says that specialization is continuing to ravage the ecologic model — the greater the degree of specialization, the less this model can be appreciated. Werbach also points out that medical research has repeatedly confirmed the existence of multi-factorial origins of many illnesses — especially many chronic syndromes. He agrees both with Louis Pasteur and Roger J. Williams on the need to emphasize the roles of individual susceptibility and other personal factors in considering the origins of many different illnesses. He concludes that the failure of the application of science to medicine was less the fault of science than of the way many medical practitioners misused science. He also opines that although science has helped to clarify what is occurring in medicine, it has provided physicians in practice with only pieces of knowledge which, when accepted naively, have led to distortions in their appreciation of the nature of many illnesses as well as of the means of best moving their patients back toward health. Werbach and I agree that despite the stranglehold which the doctrine of specific etiology has had on the practice of medicine, the past decade has witnessed a gradually occurring revival of ecologically focused medical concepts and practices.

Werbach concludes in Chapter 8 of his book, *Third Line Medicine,* that the weight of scientific evidence has disproven the doctrine of specific etiology which has dominated medicine for the past century. Retrospectively, he regards this reductionist doctrine as an unsound fad which developed on the basis of Pasteur's germ theory of disease. He then describes how new physicians are returning to our basic roots in Hippocratic medicine. In so doing, he defends the holistic concept in biology as developed by Jan Christian Smuts in his book *Holism and Evolution,* published in 1926.

In addition to laying the groundwork for the modern subject of clinical ecology, Werbach also emphasizes the split between immunologically focused and ecologically focused allergists, as well as several other alternative medical programs.

1981-1987

Reductionism Ad Absurdum

James A. Lee points out that in the study of man in health and disease there has been less interest shown in a better understanding of the human organism's total response to the total environment than in the biochemical activities of its components. (The ecological basis of environmental health, Arch Clinical Ecology 1:73, 1969) This reductionist approach, while highly productive, has not served to greatly increase our understanding of the interplay between the whole human organism and its environment. Lee agrees with Rene Dubos: "...the time has come to give the study of the responses that the living organism makes to its environment the same dignity and support which is being given at present to the study of the component parts of the organism... exclusive emphasis on the reductionist approach will otherwise lead biology and medicine into blind alleys." (Environmental Biology, Bio Science 14:11, 1964).

Lee goes on to cite several examples. An ecological approach finds in air pollution the necessary and sufficient conditions to produce illness. In contrast, the reductionist oriented biomedical researcher seeks to understand the action of pollutants at the cellular level, through metabolic pathways, enzyme systems, and other higher levels of cellular complexity, but such findings fail to answer the question, does air pollution cause disease. Lee states that if we are seeking methods of prevention and control, a holistic understanding is required.

Dubos (same reference) also speaks of the "diseases of adaptation," saying that to a very large extent disorders of the body and the mind are but the expression of inadequate responses to environmental influences.

Lee points out that to the medical ecologist, the health status of man is determined by the interplay of the ecological universes; these include the internal environment of man, and the external environment to which he relates. He continues by stating that as it is with the individual, it is with the community, in that the health and well-being of both are the outcome of a continual interplay between these two universes and the struggle to adapt, and he emphasizes the need for applying the basic biologic concepts of human ecology and adaptation in medicine and public health. With the realization that it is the total functionally integrated human organism that responds to environmental stimuli, the need for a clinically oriented concept of adaptation is apparent. Lee points out further, with which I am in full accord, that this is especially true in the case of chronic illnesses resulting from the long-term interplay between environmental and bodily factors. More specifically, many environmental insults are received by the human organism in the form of frequent low level exposures resulting in a broad spectrum of physical and related mental syndromes, wherein pathological changes are either minimal or seemingly absent.

I agree with Lee's plea for studies centering around the observation and recording of cause and effect relationships between given environmental stimuli and particular illnesses, including the development of techniques to demonstrate such relationships, as well as with his comment that the role of specific adaptation in environmentally induced illness needs to be elucidated and the ecologic implications of human individuality explored. Indeed, this is exactly what I have been trying to do for the past several decades.

Other Supporting Evidence
Paradigms and Problem Solving in Medicine

Eta S. Berner (Paradigms and Problem-Solving: A Literature Review. J Med Education 59:No.8, 1984) states that medical problem solving is in a state of crisis because data fail to support some of the basic assumptions on this topic. This writer points out how research in medical problem solving has followed the pattern that Kuhn (Kuhn T. *The Structure of Scientific Revolutions* [2nd Edition] Chicago, IL: U of Chicago Press, 1970) documented for the physical sciences. Berner claims that research in medical problem solving is at the stage where new assumptions are necessary if the field is to progress further. After reviewing Kuhn's thesis, Berner then discusses its applications to the research on medical problem solving that has occurred over the past 20 years. He concludes with suggestions for a new research agenda — a new paradigm emphasizing multiple types of problems and varied solution strategies.

Berner went on to say that the body of knowledge about medical problem solving has developed in the way Kuhn described in the context of the physical sciences. But he states that since 1974 the nature of the research on medical problem solving has changed, pointing out that problems are not homogeneous, as previously assumed.

This presentation also points up potential errors in the simplistic view that medical investigations of complex medical problems require approaches other than merely the application of techniques found useful in evaluation of the hypothetico-deductive traditional scientific method.

The Tomato Effect —
Rejection of Highly Efficacious Therapies

Rejection by medical academia appears to be another example of the so-called tomato effect as described by James S. Goodwin, M.D. and Jean M. Goodwin, M.D. (The Tomato Effect: Rejection of Highly Efficacious Therapies. JAMA 251:2387, 1984). They reported confusion between the placebo effect and the tomato effect. Although the tomato was imported from Peru to both Europe and North America in the mid 1500s and was quickly accepted in Europe, it was ignored or actively shunned in North America. Indeed, according to these authors, tomatoes were not even cultivated in North America until the 1800s and commercial cultivation of tomatoes was rare until the twentieth century.

The reason tomatoes were not accepted until relatively recently in North

America was that they were regarded as poisonous, as were the leaves and fruit of several other members of the nightshade family — a botanical term commonly preceded by the word "deadly." Despite the Italians and French eating tomatoes in increasing quantities without seeming harm, they did not encourage colonial Americans to try them until 1820 when an American ate a tomato in public in New Jersey and survived. Only then did the people of this country start to consume tomatoes.

According to the Goodwins, the tomato effect in medicine occurs when an efficacious treatment for a certain disease is ignored or rejected because it does not "make sense" in the light of accepted theories of disease mechanism or drug action. In analogous position, there have been many therapies in the history of medicine that while later shown to be highly effective were at one time rejected because they did not meet preconceived notions or make sense. These authors contend that the tomato effect is in its own way as influential in shaping modern therapeutics as the placebo effect. They assert that while the placebo effect has contributed to the enthusiastic and widespread acceptance of therapies later shown to be useless, or harmful, the tomato effect has stimulated the rejection or nonrecognition of highly efficacious therapies. They pointed out the history of colchicine, gold and aspirin therapy of rheumatoid arthritis as examples of how the tomato effect delayed their therapeutic acceptance.

It would also seem that the rigidity of medical academia's rejection of clinical ecology might also be, at least in part, a tomato effect, in view of the widespread acceptance by medical academia of the narrow immunologic mechanism of allergy. These authors also pointed out that there are only three issues that matter in picking a therapy: Does it help? How toxic is it? How much does it cost? In this evaluation, ecologic management of environmental-personal demonstrable interactions in highly specifically susceptible persons is relatively more helpful than alternative programs; is not toxic; and in the long term tends to be less costly.

In Defense of Detailed Anecdotal Case Studies and Ecological Medical Approaches

Anecdotal case reports in large quantity — in my case 20,000 private patients studied by basically similar techniques during the past 40 years —cannot be ignored. These patients all had received traditional health care from professionals with their health problems unresolved. Results from ecologically focused medical care — whether no change, better or worse — are acceptable to mainline science in the sense of providing a working hypothesis until more detailed understanding is forthcoming.

The clinical ecology diagnostico-therapeutic approach when pursued vigorously — especially by means of comprehensive environmental control and outpatient modifications — provides relief from chronic physical and/or mental illnesses not provided by other techniques.

Tests of Time and Competition in the Application of Clinical Ecology and Traditional Allergy

Demonstration of the clinical effects of specific foods and given environmental chemicals in exciting and perpetuating chronic illnesses have been employed, respectively, for the past 60 and 30 years. The number of physicians employing ecologically oriented techniques has increased rapidly during the past two decades to the point that vis-a-vis traditional allergists they are now only slightly shy of a majority. Whereas traditional allergists are concerned about losing significant portions of their patients (Marketing For a Changing Market. AAAI, "News & notes," Summer 1986) and some are resorting to marketing techniques to minimize continued losses, clinical ecologists remain relatively much more busy. Indeed, a significant reason for the sharpness of the criticism of clinical ecology by traditional allergists appears to be motivated economically.

Moreover, the reading public is receiving the message of the relative superiority of the ecologic approach to many chronic illnesses over traditional allergic approaches and other methodologies depending on excessive drug therapy. Indeed, most new books for the public during the past decade concerned with environmental-allergic relationships have championed the ecologic orientation.

Decreasing Environmental Regulation, and Increasing Politicization of Environmental Issues

Lewis Regenstein, in *How to Survive in America the Poisoned,* (Revised Ed. ©1982 Acropolis Books Ltd., 2400 - 17th St. NW, Washington, D.C. 20009, $9.95 paperback) tells the story of the major environmental and health crises of our times: the pervasive presence in our society of chemicals that may be killing and disabling millions of Americans. This book quotes the U.S. Environmental Protection Agency (EPA) as calling this toxic chemical contamination "the most grievous error in judgment we as a nation have ever made — one of the most serious problems our nation has ever faced."

"Within a few days after taking office," according to Regenstein, "President Reagan and his cabinet began a concerted effort to cancel, postpone or weaken dozens of regulations that protect the public from toxic chemicals. The administration has moved quickly to dismantle many of the hard-won environmental gains of the last few decades."

Indeed, one of the first things President Reagan did on taking office was to cancel President Carter's executive order, issued a month earlier, tightening the exportation of pesticides and other extremely hazardous chemicals banned for sale in the U.S. In the same month of February 1981, President Reagan signed an executive order, called "Regulations Reviewed" by the Washington Post, curtailing the powers of the EPA (Environmental Protection Agency) and OSHA (Occupational Safety and Health Administration) restricting the ability of governmental departments and agencies to issue regulations. Shortly following, Vice-President George Bush, head of the administration's Task Force on Regulatory Relief, announced further restrictions easing Clean Air Act pollution controls.

In March 1981, the Washington Post reported that Vice-President Bush had targeted 27 regulations for review and change — including some of the government's most important environmental and job safety rules.

In April 1981 President Reagan proposed eliminating 35 air quality and safety regulations dealing with the automobile industry.

In August 1981 Vice-President Bush announced additional proposed changes in 30 other regulations as well as several of those under the Toxic Substances Control Act.

Moreover, as Lewis Regenstein continued to report in his above cited book, industry-oriented people with long records of opposition to environmental protection were appointed to head key departments and agencies dealing with toxic chemicals and pollution. President Reagan's Budget Director, David Stockman, was given virtual veto power over spending by the EPA.

But even more importantly, budgets for several environmental protection regulatory agencies were cut by millions of dollars. Especially hard hit was funding for the Consumer Products Safety Commission (CPSC). Indeed, President Reagan recommended in May 1981 that the CPSC — one of the few consumer watch dogs — be eliminated.

Other scandalous actions of the Reagan administration need only to be mentioned to be remembered. These included the appointment of Anne M. Gorsuch as head of the EPA. Others from the ranks of the regulated chemical, oil, coal, steel and paper industries were appointed to regulatory subsidiary positions.

This is but the beginning of a six year Reagan-Bush environmental disaster which is still unfolding. The best brief review of the start of this disaster that I have seen published is Regenstein's book cited above. He concluded by saying: "The solution is for more of the overwhelming majority of people who support conservation to get active and organized. The public must demand that the politicians either support environmental protection or be voted out of office."

In short, prior to this current administration, both Democratic and Republican administrations had been active in protecting the public from environmental hazards. Most unfortunately, these issues in which as consumers we are all concerned have now been politicized for the first time.

Increasing Governmental Control of Medicine

Despite sustained efforts to relax and suspend many regulatory activities, the Reagan administration's imposition of diagnostic related groups (DRGs) on the medical profession in an effort to curb hospitalization costs is an extreme example of precipitous regulation of a complex problem.

Although only recently fully applied as originally conceived, it is too early to be certain of its effectiveness in curbing costs but there is already significant evidence that it is interfering with the quality of medical care.

Chapter XIII

Environmental Aspects of Clinical Ecology Updated

The General Course of Chronic Reactions

Clinical ecology is characterized by dynamic interactions between a given person and that individual's environment in the sense of his or her intake and surroundings capable of eliciting individual susceptibility. It is well known that the response of a person to a specific exposure may change with time, in that he or she may become increasingly susceptible, thereby enhancing the impact of a given exposure(s) and magnifying their manifestations. This often occurs even though the degree of environmental exposure remains unchanged.

But if the exogenous exposure in question also increases in degree with time, overall clinical effects may be accentuated additionally. This change may result in an increasing severity of reactions at any given stimulatory or withdrawal level or the development of a more advanced level of chronic reaction, providing that the dose of the exogenous excitant in question is sustained for an appreciable period of time.

Exogeny Versus Endogeny

It is also necessary to consider the impingement of selected environmental factors on health and behavior in respect to their inherent toxicity and their ability to be stored in bodily tissues. This is especially important in regard to such environmental chemicals as pesticides, solvents and others which are readily stored in the fat deposits of the body. At least in respect to these environmental exposures, we must be concerned with both exogenous and endogenously released substances.

So much for sustained exogenous and endogenous exposures manifesting in chronic levels of reaction. Variable and sometimes alternating stimulatory and withdrawal levels will be updated later.

Update of Exogenous Chemical Exposures
In General
It is especially important that the chemical problem be updated currently, inasmuch as its appreciated significance has changed significantly in recent years.
Incidence
Previously Appreciated Exogenous Chemical Exposures
As far as can be determined on the basis of clinical evidence, there has been a significant increase during recent decades in the previously described environmental chemical exposures capable of inducing and perpetuating the chemical susceptibility problem.[137] This has apparently resulted from the inexorable advancement of the industrial revolution on a worldwide basis. At least experience in my own private practice of clinical allergy and ecology support this claim. For instance, 30 years ago (1956) complaints and manifestations of new patients seeking medical attention were concerned with demonstrable reactions to environmental agents in the order from the most to the least encountered as follows: 1) Biological particles (such as pollens, mold spores, dusts, insects and animal danders); 2) Specific foods; 3) Environmental chemical exposures, as a distinct third. But in 1986, the major reasons for patients seeking my medical assistance are in the following order: 1) Individual susceptibility to exogenously encountered chemicals; 2) Specific foods; and 3) Biological particles. But as previously noted, it is difficult to separate reactions from foods per se from chemical additives and contaminants of commercially available foods. Finally, these statistics are skewed to at least some degree by my relatively greater interest and publications in respect to foods and chemicals.

Previously Not Fully Appreciated Exogenous Chemical Exposures
Formaldehyde
Formaldehyde is easily the most important exogenous environmental chemical exposure to have been recognized since 1975. Although aldehydes had been mentioned earlier as possible troublesome environmental chemical exposures,[129,137] Morris is to be credited in calling this hazard to the attention of clinical ecologists. (Morris DL. Recognition and Treatment of Formaldehyde Sensitivity, Clinical Ecology 1:27, 1982) Small also contributed to this knowledge. (Small B. *Chemical Susceptibility and Urea-formaldehyde Foam Insulation.* Longueuil, P.Q., Canada: Deco Books, 1982)

Environmental formaldehyde exposures are too numerous and important to be described here; their most significant exposures are as follows: urea-formaldehyde foam insulation, plastic interiors of mobile homes and other closed spaces, tobacco smoke, formaldehyde treated cottons, synthetically derived fabrics and floor coverings, substitute agglomerated construction materials including particle board, chip board and indoor plywood and various other occupational exposures.

Sulfur dioxide, Metabisulfites and Sulfites

Sources of sulfur and sulfur dioxide contamination of fruits, vegetables and certain other specific foods were demonstrated as common causes of reactions in individuals known to have been highly susceptible to other exogenous chemicals but not to chemically less contaminated sources of the same foods in the early 1950s[88,96] and early 1960s.[133,137] More specifically, French fried potatoes and potato chips are commonly treated with sulfur dioxide immediately after they are peeled as an anti-browning agent. Such dried fruit as apples, pears, peaches, nectarines, apricots, raisins and prunes, as well as melon, candied citrus peel and marmalade may also be bleached with sulfur dioxide-containing solutions. Also, commercially prepared fresh apples, pears, peaches, asparagus and some other vegetables used in salads are commonly treated with sulfur dioxide-containing solutions.

Shelled corn is soaked in sulfur dioxide-containing solutions to separate its hull, starch and germ parts. This means that such manufactured corn products as cornstarch, corn flour, corn sugar and corn dextrins are sulfur contaminated to some degree. Cane sugar and beet sugar may also be contaminated by sulfur dioxide in the process of their manufacture.

Note that these reactions to sulfur and sulfur dioxide were observed in the 1950s,[88,96] and were published in more detail in the 1960s,[133,137] approximately two decades before they were apparently published independently, as recently reviewed by Simon (Simon R. Sulfite Sensitivity. Ann Allergy 52:281, April, 1986). Although this review and articles it quoted largely concerned acute reactions to sulfites and metabisulfites, Simon admitted that these products usually break down to sulfur dioxide. Some of the cases reviewed also experienced intolerance to ambient levels of air pollution. The common occurrence of this more general emphasis[137] was not emphasized.

Demonstration

The hospital program of comprehensive environmental control in a hospital unit followed by observation of the effects of test re-exposures through the accustomed portal of entry remains the most accurate diagnostic measure to demonstrate the existence of the chemical susceptibility problem.

From the standpoint of medical indications and the accuracy of diagnostic techniques, this is still the most accurate way of diagnosing the chemical susceptibility problem — both qualitatively and quantitatively. Although this is indicated in about 15 percent of chronically ill patients seeking my professional help for management of their apparent chemical susceptibility, this is presently not available in more than one percent of patients because of economic reasons. These reasons include costs of hospitalization for a minimum of two weeks, which Medicare, Blue Cross and many private insurance carriers refuse to cover — irrespective of the relative severity of the patients' illness.

This means that clinical ecologists are forced to diagnose and treat these health insurance "refusniks" under most unsatisfactory conditions, as many have their chronic symptoms perpetuated by their occupational and/or their home chemical exposures. Another problem lies with physicians in general. For instance, the Physicians Review Committee of the Chicago Society of Medicine reviewed the record of an advanced chronic case of rheumatoid arthritis hospitalized on my medical service which had been reported to this Committee by the patient's health insurance company in late 1984. The chairman of this Committee wrote the insurance company on December 10, 1984 as follows: "The Physicians Review Committee has reviewed the above captioned case and believes that the services provided by Dr. Theron Randolph are not usual and customary in this area. As a consequence, the committee is not in a position to determine whether or not the fees associated with those services are usual and customary.

"In reviewing Dr. Randolph's services in the past, although he has been published, the society found that there is not a great deal of scientific basis on which to judge the appropriateness of his care or its effectiveness."

Apparently, as a direct result of these assertions both the insurance company and the patient have refused to pay for medical services and the patient has sued me additionally, despite improving in the course of the hospitalization. She had been found to have reacted acutely with a sharp recurrence of her joint symptoms following a provocative exposure to utility gas. Upon first returning to her completely gas-equipped home her arthritis also recurred acutely. Upon reporting this to me by telephone, I recommended that this family move to an all-electric home, which they did. Unfortunately, however, this new home was located close to a sewage sludge dump and her arthritis was again accentuated acutely. It has been learned recently that this patient has since become symptom-free as a result of following specific instructions.

Both the comprehensive environmental care methodology in question had been reported[155,156,166] as well as its application in the management of rheumatoid arthritis,[291,372,373] although more detailed accounts of our hospital management of rheumatoid arthritis had not been published until the summer and fall of 1984.[388,389] This is an example of how many physicians base their claims on the "scientific merits, appropriateness and effectiveness" of a technique with which they have not had experience, more upon currently accepted practices than with the merits of a new approach. It might also be mentioned that in previous hearings before this Committee when I had defended this hospital program in other chronically ill patients not responding to outpatient management, I had been vigorously defended by two of the 13 members of this Committee. The Henrotin Hospital where these observations had been made had also denied the experimental nature of this program and had attested to its scientific merits. Elsewhere in this book I have defended the basic scientific validity of this inductively derived diagnostico-therapeutic approach.

The fact remains that many patients are either too ill to come to a medical office or become more ill as a result of transportation exposures in attempting to do so. Although the majority of private patients may be diagnosed reasonably well in an office or outpatient basis, subsequent management is far less satisfactory. For economic reasons, only about one-half of private patients are able to make the necessary changes in their homes to accommodate the demands of this illness. Even more troublesome is the fact that only approximately 10 percent of advanced cases are able to accommodate to their working conditions. The most troublesome illness-related home exposures for the chemical susceptible patient are pesticides, gas burning and other hydrocarbon burning utilities, formaldehyde, perfume and tobacco smoke exposures indoors and lawn chemical exposures outdoors. The most troublesome occupational exposures are tobacco smoke, perfumes, formaldehyde emanating from particle board partitions, carpets and glue adhesives as well as fluorescent lights, in selected persons.

An update of the chemical susceptibility problem should include reference to the phenolic food compounds pioneered by Robert Gardner, professor of Animal Science, Brigham Young University. Although I attended the first symposium on this subject, entitled, "The Immunotoxicity of Foodborne Phenolics and Airborne Pollutants," I have not employed these techniques in my own practice.

Dr. Gardner theorized in 1979 that there might be a common denominator in allergic reactions to foods and pollens in respect to possible sensitivity to some aromatic compounds found in all plant foods and pollens. Acquiring some of these pure aromatic compounds, he made dilutions, and performed sublingual tests with them. He found positive clinical reactions to various phenolic compounds and neutralizing doses. He used these neutralizing doses successfully in the treatment of his own allergies and those of other patients. (Ber A. Neutralization of Phenolic (Aromatic) Food Compounds in a Holistic General Practice. J Orthomolecular Psychiatry 12:283, 1983)

The action of these naturally occurring aromatic food compounds, being of small molecular weight and probably not allergenic, is assumed to be on the basis of haptens. (McGovern JJ, Gardner RW, Brenneman LD. The Role of Naturally Occurring Haptens in Allergy. Ann Allergy 47:123, 1981)

Update of Endogenous Chemical Exposures
In General
Pesticides have long been regarded as exceedingly toxic and are absorbed and stored in body tissue. This is especially true for chlorinated hydrocarbons, as represented by DDT and related products. I testified in the Senatorial Pesticide Hearings in 1963 that pesticides constituted the crux of the chemical susceptibility problem,[150] a point of view reiterated in 1966.[173]
Incidence
Nevertheless, the frequency and importance of endogenously stored pesticides and other chemicals awaited the development of sophisticated analytical technique to document their absorption and storage.

Demonstration

Laseter and associates, in a study of 16 different synthetic chlorinated hydrocarbon pesticides and their common metabolites by means of high-resolution gas chromatography and high-resolution gas chromatography-mass spectrometry, found 16 different synthetic chlorinated hydrocarbon pesticides and common metabolites in randomly selected environmentally sensitive patients (Laseter JL, DeLeon IR, Rea WJ and Batter JR. Chlorinated Hydrocarbon Pesticides in Environmentally Sensitive Patients. Clinical Ecology 2:3, 1983) They found that 99 percent of 200 patients screened had residues at or above the 0.05 parts-per-billion level in their sera. From this concentration in the serum, it was postulated that a significantly higher concentration was present in other body tissues, especially in fat and other lipid material.

In the 1970s Hubbard developed a detoxification regimen to address bio-accumulations of drugs, food additives and environmental contaminants which, he concluded, became lodged in the adipose tissue. The technique was widely utilized by drug rehabilitation facilities, but it was not until 1981 that physicians and researchers examined its effectiveness in reducing body burdens of environmentally persistent compounds. The method was reported by D.W. Schnare, G. Denk, M. Schields and S. Brunton in an article entitled, "Evaluation of a Detoxification Regimen for Fat Stored Xenobiotics," (Foundation for Advancements in Science and Education, 4801 Wilshire Blvd., Suite 215, Los Angeles, CA 90010 in Medical Hypotheses 9:265, 1982). They found this detoxification regimen safe, useful and virtually without complications if properly administered. It consists of seven precisely integrated components: a) Physical exercise immediately prior to sauna exposure; b) Forced sweating in a well-ventilated sauna for 2½ to 5 hours daily; c) Nutritional supplements consisting of niacin in proportion to other vitamins and minerals; d) Water, salt and potassium sufficient to avert dehydration or salt depletion; e) Intake of polyunsaturated oil based on individual tolerance; f) Calcium and magnesium supplementation; g) Regular daily schedule with balanced meals and adequate sleep in the absence of medications and alcohol while following their accustomed diet.

Subsequent studies of the program found it effective in reducing body burdens of certain organohalides. (Schnare DW, Ben M, Shields MG. Body Burden Reductions of PCBs, PBBs and Chlorinated Pesticides in Human Subjects. Ambio 13 (5-6) 378-380, 1984; Schnare DW, Robinson P. Reduction of HCB and PCB Human Body Burdens. Proceedings of the International Symposium on Hexachlorobenzene, International Agency for Research on Cancer, World Health Organization, Scientific Publications Series, Vol. 77, Oxford University Press, In press.)

It is apparent from this information that pesticides and certain other environmental chemicals to which individuals are exposed over a period of time, even in relatively low dosage, are capable of accumulating in body tissues. It is especially apparent that physicians interested in environmental medicine must be concerned both with exogenous and endogenous chemical exposures and clinical interrelationships between these two sources of re-exposure. These interrelations are currently under intensive investigation in a pilot study of 400 patients.

Relationships Between Exogenous and Endogenous Chemical Exposures

There are obviously important relationships between exogenous and endogenous chemical exposures which need to be studied and defined. Although these two investigative programs cannot be conducted simultaneously, they do have a common need — namely, a safe place to be housed without additional chemical exposures while each problem is being investigated on an outpatient basis.

I propose a "Get Well Hotel" for this purpose which is to be located on the windward periphery of a metropolitan area not more than 50 miles from a major airport and reasonably near diagnostic facilities for evaluating the chemical susceptibility problems, as well as the detoxification process.

This "Get Well Hotel" minimizing further environmental chemical exposures should meet as many of the following requirements as possible — the most important being listed at the top of the construction, furnishings and operational listings. Each exposure to be avoided has been incriminated in many chronically ill, chemical susceptible patients. For the most part, materials recommended for use have been tolerated by the same patients.

— USE —

CONSTRUCTION	FURNISHINGS	OPERATIONS
Stone	Electric utilities	Baking soda
Glass	Incandescent lights	Unscented soap and water
Brick	Steel or wood	
Cement	furniture with natural	'Bon Ami'
Ceramic tile	fabric or leather	Steel wool
Enamel steel walls	upholstery	Vacuum cleaners
Lath & plaster walls	Non-vinalized	Brooms, brushes
Wood floors	wallpaper	and cotton rags
Mix paints with baking soda		

— AVOID —

CONSTRUCTION	FURNISHINGS	OPERATIONS
Gas, oil and coal	Man-made products	Pesticides, herbicides
Formaldehyde	Treated cottons	and fungicides
Particle board	Polyester, etc.	Pesticide-containing
Chip board	Synthetic textiles	wallpaper paste
Indoor plywood	Glues	Smoking
Pesticide Treatment	Solvents	Perfumes and scents
Solvents	Rubber	Dust repellants
Rubber	Plastics	Solvents
Plastic	Fluorescent lights	Soil retardants
Sealants	Vinyl surfaced	Detergents
Asbestos	wallpaper	Disinfectants
		Deodorants
		Flame resistance
		Static controls
		Furniture polish
		Glass cleaners
		Air fresheners
		Household cleansers
		and bleaches
		Photographic chemicals
		Art supplies
		Aerosol cannisters

One does not necessarily remain free from chronic symptoms simply by checking into a "Get Well Hotel" and/or moving into a "Stay Well Home" because of multiple personal exposures known to perpetuate environmentally related illnesses. The most important of these are commonly eaten foods to which one may be susceptible unknowingly and often addicted. Maintenance doses of drugs, odorous cosmetics, and synthetically derived clothing may also perpetuate chronic manifestations.

Neither are exposures to dusts, pollens, molds and other environmental allergens, to which one may be highly sensitive, controlled by these restrictions. The "Get Well Hotel" is a temporary measure to minimize occupational, home and transportation chemical exposures while undergoing specific diagnostic measures or while being detoxified for the past endogenous chemical deposits which may also be perpetuating chronic manifestations. The "Stay Well Home" is primarily a therapeutic program for the purpose of minimizing recurrent acute reactions to environmental chemical exposures. It may also be a prophylactic measure.

Update of the Food Problem
In General
During the past 43 years, I have studied the food allergy problem within a quarter of a mile of the same location in the north side of Chicago in the course of seeing approximately 20,000 private chronically ill patients studied by means of basically similar diagnostico-therapeutic techniques. From mid-1956 to mid-1986 approximately one-third of these patients have been hospitalized under the program of comprehensive diagnostico-therapeutic technique in several different hospitals.

Incidence
The great majority of these patients were found to have one or more common foods to which they were reacting. It was relatively rare to find any of these chronically ill patients who did not have evidence of allergic-type reactions to one or more commonly eaten foods. About one percent reacted adversely to most commonly eaten foods. Although there has been considerable publicity about alleged allergy to the so-called twentieth century, an encompassing allergy to virtually all foods, this does occur but it is exceedingly rare.

Labeling, Geography and Utilization of Sugars and Alcohols
Since a ruling of the Department of Agriculture in 1922, labeling of specific sugars has not been mandatory. Although cane sugar tends to be labeled voluntarily, beet sugar is never labeled. Corn sugar, representing up to 40 percent of sweetening agents in the American diet, is not labeled consistently.

Because of these facts, the geography of sugar production and distribution become important. Despite corn sugar being distributed universally, this is far less true for cane and for beet sugar which are produced and more commonly used regionally. Cane sugar dominates the marketplace east of the Appalachian Mountains, up navigable rivers to the heads of navigation, and in southern states east of the Rocky Mountains. Although cane sugar and beet sugar are used about equally in states bordering the Pacific coast, beet sugar production and utilization dominates other western states, including western Texas, and all northern states except for areas immediately adjacent to such navigable rivers as the Missouri, Mississippi and Ohio. Nationally distributed manufactured foods are usually sweetened with the sugar prevalent in the area of their processing. For example, peaches canned in Georgia tend to be sweetened with cane sugar; cherries canned in Wisconsin and Michigan are ordinarily sweetened with beet sugar.

Although the labeling of specific sugars has long been desirable, this has become relatively impossible in recent years due to the fact that mixtures of corn, beet and cane sugar in liquid form are now piped into the moving food processing line. Moreover, such mixtures now tend to be transported not as dry bulk as formerly, but by tank boats, tank trucks and tank cars.

Although the percentage of alcohol in alcoholic beverages is carefully regulated because of the tax angle, the food constituents entering the manufacture of alcoholic beverages have never been labeled adequately. However, this subject is being reevaluated currently.

Demonstration

The most common feature of techniques to demonstrate the existence of a reaction to a commonly eaten food consists of the regimes of preliminary avoidance, followed by observation of the acute effects of specific re-exposure. Provocative re-exposures may be through the accustomed portal of entry, as in the program of comprehensive environmental control[155,156,166] or through a more rapidly absorbed or unaccustomed portal of entry, as with intradermal and sublingual exposures. In all instances, it is mandatory that maximally controlled environmental controls be maintained in order to minimize inadvertant or accidental concommittent environmental exposures. These control measures are essentially those previously outlined in detail for the proposed "Get Well Hotel." Intradermal tests, when performed one at a time for the purpose of inducing an acute test reaction, also provide evidence of wheal production. It is generally recognized, however, that any type of skin testing with food extracts are not reliable indices of sensitivity when performed multiply simultaneously and reading only the degree of the whealing response. Intradermal skin tests perform the same function and additionally may provide evidence of wheal formation, although skin testings with foods are not reliable indices of sensitivity.

Update of the Inhalant Allergy Problem

In General

Pollens, molds, dusts, dust mites and other insects and animal danders remain major environmental causes of illness and often coexist with individual susceptibility to environmental chemical exposures and foods.

Incidence

Although the incidence of clinical reactions to these inhaled biological particles apparently remains about the same, their relative importance is somewhat less than formerly because of the rapidly increasing clinical significance of the chemical problem and food allergy — both being major complications of the continued application of the industrial revolution. Incidence is also lessened by the prevalence of air-conditioning.

Demonstration

The intradermal skin test performed with serial dilutions of potent extracts are both qualitatively and quantitatively accurate in evaluating inhalant allergy, especially as performed with 1-5 serial dilutions as recommended by Rinkel (see Appendix page 334) and [61]. Similar information is found by performing RAST tests.

Chapter XIV

Adapted (Stimulatory) and Maladapted (Withdrawal) Responses Updated

Stimulatory and Withdrawal Ecological Responses Updated

Introduction

At the height of the clinical activity leading up to the concept of clinical ecology in 1952, my friend, George A. Zindler, M.D. of Battle Creek, Michigan who had followed through and confirmed many previous clinical impressions, asked me what I was trying to do. Whatever it was, would I please quiet down for a while, as I was getting him into trouble with his medical friends.

Quite truthfully, I did not know what I was attempting to do in an overall sense, as I had not yet arrived at that point. Until then, I was simply trying to report environmentally focused clinical observations which were occurring more rapidly than I was able to describe them. Indeed, at this time, these findings were only beginning to come together into an alternative concept of environmentally focused medical care integrating physical and mental illnesses. Realizing that this was too early to present such a preliminary point of view, I simply followed the advice of my friend for a year during which time I collected my thoughts, fully aware of the importance of what I was doing and that I was on the edge of a clinically oriented medical breakthrough, quite in contrast to the mass of analytical bodily centered and excessively drug related approach emanating from medical academia. The general direction to be followed in rounding out this concept, with the development of techniques to bring this about, became increasingly apparent. As this program progressed, miscellaneous clinical observations began to come together into a workable environmentally focused approach.

Very important in this development was a more satisfactory classification of the manifestation of clinical ecology. What appears to be the culmination of this development as far as such a classification is concerned only evolved recently in the course of preparing a chapter on food addiction for an English psychiatry book. Many relatively disconnected clinical observations finally coalesced into a classification which will be outlined in this chapter — a relationship which had started 40 years earlier as a relatively confused mismash. Although many of the observations to be cited have already been described, something may be gained by hooking them together chronologically. Fortunately most of these preliminary concepts and data have been presented in published abstracts or program announcements. Indeed, dating these developments is the chief reason for extending my bibliographic references to include these preliminary accounts, in contrast to the other bibliographies listed in the appendix.

Early Clinical Observations

As previously noted, I entered this field after the localized manifestations of allergy (rhinitis, bronchitis, asthma, eczema, hives, colitis as well as urgency and frequency of urination) had already been described. At that time, the systemic manifestations of allergy, especially headache and fatigue, had only recently become appreciated. I added to the description of headache,[1,18,21] fatigue,[20,23,25,30,32,39] myalgia and arthralgia[41,42,48,67,83] as well as arthritis.[75,118,241,291]

But other more sustained allergic manifestations, in the sense of the original descriptive definition of allergy, did not quite fit this localized-generalized classification of syndromes. These included hyperactivity in children[34] obesity,[35,57,87,111] alcoholism,[70,72,73,81,84,89,111] and, especially, mental and behavioral disturbances.[74,79,89] These cerebral responses, first demonstrated in 1950, became better understood when they were interpreted in keeping with Rinkel's description of masked and unmasked food allergy.[75]

The keys to this classification were interpretations of the common form of food allergy as food addiction[90,92,95,111] as well as Rinkel's masking phenomenon in food allergy in terms of specific adaptation in the presence of individual susceptibility.[113,115,117,119,121] This reorientation was patterned after Selye's observations. (Selye H. The general adaptation syndrome and the diseases of adaptation, J Allerg 17:231,289,358, 1946) In accordance with this concept, masked food allergy has been referred to as adapted stimulatory responses in the relative absence of symptoms and maladapted withdrawal responses characterized by delayed manifestations. The important point to emphasize here is that both stimulatory and withdrawal levels of reactions are perpetuated by cumulative food exposures which, for the most part, are not suspected. This is food allergy as it most commonly exists, preferably called food addiction.

But if such oft-repeated exposures to a given food are avoided until the most recent feeding of the food in question has cleared the gastrointestinal tract, a reexposure then precipitates a nonadapted acute test reaction. The relative immediacy and severity of this test response indicates the degree of individual susceptibility existing to that food at this time. Acute reactions of this type, constituting food allergy as it is ordinarily considered, have been known for over 2,000 years. This type of food response from such uncommonly eaten foods as shrimp, cashews, and buckwheat, is ordinarily not a diagnostic problem as patients usually recognize the existence of these reactions and avoid eating these foods. On the contrary, cumulative reactions to wheat, corn, milk, egg and other foods eaten once in three days or more frequently are rarely ever suspected. Because patients understand these terms much better, I refer to the former as food allergy and the latter as food addiction. The commonly occurring combination is called food allergy/addiction.

Although the overall concept of interrelated physical and mental manifestations from the selective impingement of nonpersonal environmental agents on specifically susceptible persons was first appreciated in reactions to given foods, similar responses to environmental chemicals were also recognized in 1952[88] and 1954.[99,103] By 1956, the stimulatory and withdrawal levels of the manic-depressive psychosis (Kraepelin E. Lectures on Clinical Psychiatry, Authorized translation from the German by Johnstone T. New York, Hafner Pub. Co. 1958) had been extended to include lesser allergic responses to foods and exogenous chemicals. A scientific exhibit revealing relationships between the stimulatory (adapted) and withdrawal (maladapted) interrelated levels of allergic and psychiatric syndromes was displayed at the Annual Meetings of the American Psychiatric Association and the American Medical Association, both in Chicago in 1956.[108-110] A brochure distributed on these occasions is shown in Figures 17 and 18, on pages 111 and 112. Although this scientific exhibit was entitled Specific Adaptive Illness, this overall nonpersonal environmental focus in psychiatry has been referred to as ecologic mental illness since the late 1950s[122,124,131] and this subject in general as clinical ecology.[117,125]

The first two times that I attempted to present the concept of ecologic mental illness and the techniques to demonstrate its existence to psychiatrists in the United States in the late 1950s, I stepped back from the podium expecting to answer questions, but none were forthcoming. When this subject along with a motion picture of the first photographed case[79] was presented to the staff of the National Institute of Mental Health in Washington, D.C. in the early 1960s, the comment was made that this was interesting but obviously very rare. My response was to the effect that this might be regarded as rare, if one rarely, if ever, looked for it by means of proper techniques to demonstrate its existence. It was of frequent occurrence in my experience. I did not dare to state more precisely as to how common I expected that it might be, as I had already been discredited. This as well as other presentations of ecologic mental illness in the United States apparently made little impression at the time, although I am still receiving occasional comments about this work presented then. The history of psychiatry and ecologic mental illness — extending from introspection to extrospection — was reviewed in 1973.[251]

Always encountering resistance and disbelief whenever this interpretation of ecologic mental illness was presented in this country, despite illustrative case reports and motion pictures — the few times when I had the opportunity to do so — I held back for a few years in making further attempts. Meanwhile, details of this point of view and supporting techniques were worked out and published.[155,156,157,166] By that time, however, this environmentally focused point of view had become so detailed and comprehensive that it was impossible to present it briefly in the time allotted by program chairmen for typical analytical reports. The only other alternative seemed to be to present this subject abroad, inasmuch as up until 1970 not a single psychiatrist in the United States had evinced sufficient interest in this subject as to attempt to confirm it. Regarding this experience, it is suggested that the reader check my bibliography for items written between 1959 and 1980. (See appendix pages 353-357)

Finally, a reprint distributed at a London psychiatric conference was picked up by a friend and sent to Dr. William H. Philpott. He read with interest but relative disbelief and filed it. Dr. Solomon Klotz, an allergist in Orlando, Florida called Dr. Philpott's attention in 1970 to statements which we had made in our book *Food Allergy*[75] about mental and behavioral reactions to foods in 1951. To his great credit, Dr. Philpott dug out this 1966 reprint, read it again and acted upon it in conjuction with Dr. Marshall Mandell of Norwalk, Connecticut. They confirmed it. Dr. Philpott has not only written extensively on this subject since then but, as a psychiatrist, has also been far more successful in interesting other psychiatrists in this concept and how to demonstrate it than I had been as an internist and allergist. He recorded his experience in a book written in 1980. (Philpott WH. Kalita DF. *Brain Allergies, the Psycho-Nutrient Connection.* New Canaan, Connecticut, Keats Pub Co. 1980)

Dr. Philpott's initial impression on reading the report of the application of this point of view in psychiatric patients in 1966 is worth noting. "My response in respect to the content of the paper was: incredible, impossible! None of my patients were reacting to foods and chemicals. Instead, it seemed to be clear to me at the time, that their symptoms were, by and large, related to more immediate life circumstances, or sometimes due to deficiencies that I could demonstrate by laboratory examination."

After confirming my observations, Dr. Philpott commented as follows: "According to Randolph — and my experience bears this out — more than half of the so-called psychosomatic reactions are in reality undiagnosed allergic reactions... It is my conviction that diagnoses such as schizophrenia, manic-depressive and other psychotic, neurotic and other psychosomatic labels are relatively meaningless and tend only to aggravate the illness." Dr. Philpott went on to say in a preliminary summary: "I believe it is time to sort out such people by allergy induction testing and give them a chance... Inductive allergy testing rather than deductive reasoning, I believe, is the method that will allow psychiatrists to function not as philosophers but as true scientists."

He went on to elaborate: "My own practice as a psychiatrist has shown that for 250 consecutive unselected emotionally disturbed patients, there is convincing evidence that the majority developed major symptoms on exposure to their commonly consumed foods and frequently encountered chemicals. The highest percentage of symptom formation occurred in those diagnosed as psychotic..." He said additionally: "This group also developed physical reactions to an even greater extent. Headache, dizziness, unsteadiness, inability to read or write, neuralgia, myalgia, arthralgia, tension, hyperactivity, weakness, sleepiness, insomnia, tachycardia, gastritis, diarrhea, colitis, constipation, hypertension, hypotension, itching, hives, psoriasis, and many, many other such responses were observed."

The fact that many psychiatric patients also manifest physical symptoms suggesting a physical (hence somatoform) disorder for which there are no demonstrable organic findings or known physiological mechanisms, introduces the chapter on Somatoform Disorders. (American Psychiatric Association, Diagnostic and Statistical Manual of Mental Disorders DSM III. Washington, D.C. American Psychiatric Association, 1980) Diagnostic criteria for this disorder include: "Gastrointestinal symptoms: Abdominal pain, nausea, vomiting spells, bloating (gassy), intolerance (e.g. get sick) of a variety of foods, diarrhea," has been recognized by psychiatrists officially. In short, there is nothing new about the interrelationships between mental and physical manifestations. Most unfortunately, workable concepts to demonstrate these relationships have not been forthcoming from the field of psychiatry prior to Dr. Philpott's confirming observations.

Despite a few ecologically oriented psychiatrists applying this point of view in their domestic practices during the past decade, to my knowledge there has been no other comprehensive report emanating from them. Neither to my knowledge is this concept of mental illness and/or alcoholism taught in any academic medical set up, and certainly no provisions have been established for handling these medical problems from this standpoint.

Adapted (Stimulatory) and Maladapted (Withdrawal) Levels
In General — Classification of Reactions

The present classification of interrelated physical and mental manifestations in respect to their intermittent and more sustained stimulatory and withdrawal levels of reaction in specifically susceptible persons is presented in Figure 33. Although these relationships were first worked out in food allergy/addiction, they apply equally well to the chemical susceptibility problem and sometimes to other cumulative environmental exposures to which selected individuals may be demonstrated to be highly susceptible.

It is suggested that this chart be viewed "globally" so to speak, as one would think "globally" in looking at a flat map. In view of this analogy, I will employ map-globe designations in referring to part of this overall classification of ecologic manifestations. Start with the "equator" (Behavior on an Even Keel, as in Homeostasis). Move toward the north "pole" for increasing stimulatory levels and toward the south "pole" for increasing withdrawal levels. Focus on the westward or left side of this chart for intermittent stimulatory-withdrawal responses and eastward or the right side of the chart for more protracted and complicated stimulatory-withdrawal responses.

In view of these cartographic appellations, this chart divides itself into four quadrants: southwest (lower left), northwest (upper left), northeast (upper right) and southeast (lower right). After a brief introduction comparing the north and south sides of the left side of Figure 33 (the upper left and the lower left portions) these areas will be taken up in that order.*

Despite the common practice of referring to the chief complaint as the onset of the present illness, this is inappropriate in clincial ecology for several reasons. First of all, by the time that new patients finally seek medical advice from a clinical ecologist, there tends to be a multiplicity of complaints.[256] Merely to claim that the most recent one is the onset of the present illness is inappropriate. Secondly, ecological disturbances usually start subtly with lesser stimulatory responses, such as levels (+) or (++) which are *not* registered as complaints. Such specifically adapted persons know how to keep well merely by maintaining their immediately stimulatory post-exposure effects, thereby avoiding their more delayed withdrawal hangovers.

ENVIRONMENTAL — PERSONAL INTERRELATIONSHIPS

	INTERMITTENT RESPONSES	LEVELS	SUSTAINED RESPONSES
SPECIFICALLY ADAPTED STIMULATORY LEVELS	MANIA (agitation, excitement, blackouts, with or without convulsions)	++++	DRUG ADDICTION (both natural and synthetic)
	HYPOMANIA (hyperresponsiveness, anxiety, panic reactions, mental lapses)	+++	ALCOHOLISM (addictive, drinking)
	HYPERACTIVITY (restless legs, insomnia, aggressive forceful behavior)	++	OBESITY (addictive eating)
	STIMULATION (active, self centered with suppressed symptoms)	+	ABSENT COMPLAINTS (the desired way to feel)
	BEHAVIOR ON AN EVEN KEEL, AS IN HOMEOSTASIS	0	BEHAVIOR ON AN EVEN KEEL
SPECIFICALLY MALADAPTED WITHDRAWAL LEVELS	LOCALIZED PHYSICAL ECOLOGIC MANIFESTATIONS (rhinitis, bronchitis, asthma, dermatitis, gastrointestinal, genitourinary syndromes)	-	IMPAIRED SENSES OF TASTE AND SMELL MENIERE'S SYNDROME
	SYSTEMIC PHYSICAL ECOLOGIC MANIFESTATIONS (fatigue, headache, myalgia, arthralgia, arthritis, edema, tachycardia, arrythmia)	- -	SMALL VESSEL VASCULITIS, HYPERTENSION, COLLAGEN DISEASES
	BRAIN FAG — MODERATELY ADVANCED CEREBRAL SYNDROMES (mood changes, irritability, impaired thinking, reading ability & memory)	- - -	MENTAL CONFUSION AND OBFUSCATION, MOROSE INEBRIATION
	DEPRESSION — ADVANCED CEREBRAL & BEHAVIORAL SYNDROMES (confabulation, hallucinosis, obsessions, delusions & temporary amnesia)	- - - -	DEMENTIA, STUPOR, COMA, CATATONIA, RESIDUAL AMNESIA

Figure 33. Intermittent and sustained stimulatory and withdrawal levels of physical and mental reactions

*The reader may be surprised by a physician writing in cartographic terms, but this physician tends to think that way, having had a collection of historical maps of the Great Lakes — St. Lawrence geographic area in his ecology library where he prepares manuscripts. I am also aware that this global analogy may prompt some eager critic with the aim of discrediting this presentation to refer to it as "globaloney" — a term coined by Clare Boothe Luce in 1943 in referring to our surge of interest in world affairs at that time.

Intermittent and Sustained Stimulatory and Withdrawal Levels of Reaction

When first seen by psychiatrists, most food addicted/allergic patients are presenting either advanced (----) or (---) maladapted levels of reaction as shown in the lower left portion of Figure 33.

But when histories of developmental events are obtained, lesser stimulatory adapted levels of reaction, (+), (++) or more advanced adapted levels, (+++) or (++++), as shown on the upper left portion of Figure 33, are seen to precede the development of these maladapted and more consistently symptom-related levels (-), (--), (---) or (----) withdrawal responses. The most common courses of developmental events consist of the following sequences: (+) (-) or (+) (++) (--) or (---); (+) (++) (+++) (---) or (----).

However, when a person presenting the manifestations of maladaption withdrawals is exposed intermittently to a relatively greater than accustomed specific environmental exposure to which individual susceptibility exists, one of the following courses of test reactions may occur before the reacting person reverts to the starting point: (----) (++++) (----); (---) (++++) (----) (---); (--) (+++) (---) (--); (-) (++) (--) (-).

In addition, the overall course of unipolar physical-mental interrelated reactions tends to move from left to right in both stimulatory and withdrawal levels. Eventually and unless specifically diagnosed and treated, advanced sustained stimulatory responses (upper right) tend to move toward advanced sustained withdrawal reactions (lower right).

Upon recovery, coincident with identification of etiological environmental factors and their avoidance or minimization of their impingement, any given stimulatory levels tends to revert to the approximately corresponding withdrawal level, from which patients capable of responding tend to ascend through lesser withdrawal levels to behavior on an even keel (0).

Unipolar Maladapted Reactions — Onset of the Present Illness

The real onset of the present illness, as far as both patients and their physicians are concerned, occurs only when such previously food addicted victims are no longer able to maintain a relatively effective stimulatory symptom-suppressed existence. And, for the first time, as far as they and their doctors are concerned, they seek medical assistance, complaining of stuffy nose, coughing, wheezing, itching, diarrhea or other localized manifestations characteristic of withdrawal level (-) or other more advanced withdrawals. This course of events is also confusing to both patients and their physicians as there has been no recent change in their apparent environmental exposures. The change responsible for this patient ceasing to be stimulated and becoming withdrawn is the individual's failure to adapt to the same exposures to which he or she had long been accustomed. This is a very important point for physicians to grasp, as most of them are accustomed to inquiring of recently made ill patients as to what changes there had been in their environmental exposures to account for their "recently deteriorating health". This is one of the chief reasons that clinical ecologists, thinking in terms of adaptation, are not on the same wave length as are most of their medical contemporaries.

It is the "real onset of the present illness," as far as patients and their physicians are concerned, but it is the "unreal onset of the present illness" as far as clinical ecologists are concerned in their discussions with their contemporaries, thinking in terms of specific adaptation. This is a third reason why I never speak of the onset of the present illness in talking with other doctors about this type of environmentally related problem. Provided given foods or food combinations are not suspected of being beneficial or that eating in general is not regarded by a patient as being helpful, addicted persons soon begin to experience morning accentuated symptoms as shown in withdrawal level (−). Slightly later, as increased individual susceptibility develops to one or more foods or regularly consumed food mixtures, withdrawal effects tend to occur later in the afternoons and/or before lunch, especially if regularly scheduled meals are delayed.[75] These findings in food addiction have also been observed with other cummulative-intermittent environmental exposures occurring in susceptible persons.

The most commonly occurring course for food addicted victims quite unaware of what is happening to them, is to descend gradually into and through successive withdrawal levels of Figure 33. Or they may come to the attention of the medical profession — family practitioners, allergists, clinical ecologists or other experts in handling physical illness initially. Later they tend to reach the attention of psychologists and psychiatrists as brain fag (−−−) and/or depression (−−−−) come to dominate the clinical picture. Another too common course of events is for these patients, failing to respond to traditional medical management, to be referred by traditional non-psychiatric physicians to psychologists, psychosomatic specialists or psychiatrists for permanent care. More will be written later about these levels in discussing bipolar responses.

Unipolar Adapted Reactions

In contrast to being totally unaware of the stimulatory role of given foods in respect to food-related physical and mental illnesses are those with slightly greater awareness of dietary relationships. For instance, awareness that one food, such as coffee or tea, is beneficial; despite nonawareness of this potential in other foods, this item may be used effectively to postpone the drift to withdrawals, maintain stimulatory level (+) or even propel the addict to level (++). Depending largely on the degree of individual specific susceptibility involved, as well as the inherent ability of individuals to be stimulated, food addicted persons may or may not advance to higher stimulatory levels of reaction. The number of food addictants recognized and the relative ease that their intake may be manipulated also seem to be major contributing factors. Although one living at level (++) may hang up an enviable sales record, such persons often tend to burn out, so to speak, as they progress to

Stimulatory and Withdrawal Ecologic Responses 251

level (+++), often antagonizing their best customers. More will be said later about hypomanic and manic levels (+++) and (++++).

Unipolar Sustained Adapted Reactions — The Addiction Pyramid

Others may follow the more sustained stimulatory course listed on the upper right side of Figure 33. Easily the most common complication of this trend is obesity which is estimated to involve approximately 40 per cent of Americans and at increasingly younger ages. This level may persist for years without advancing or at any time may revert to the withdrawal stages. Others, when threatened by inevitable withdrawal syndromes, despite manipulating their intake of edibles, resort to alcoholic beverages to remain stimulated. Alcoholic drinks manufactured from foods to which individual susceptibility exists, are especially effective in this respect. Some — ever seeking more effective stimulants — may resort to the use of drugs in an effect to prolong their stimulatory manifestations. These may be either physician or self-prescribed and may include either synthetically derived or those of natural origin such as heroin or cocaine. A more detailed description of the climb up the addiction pyramid follows.

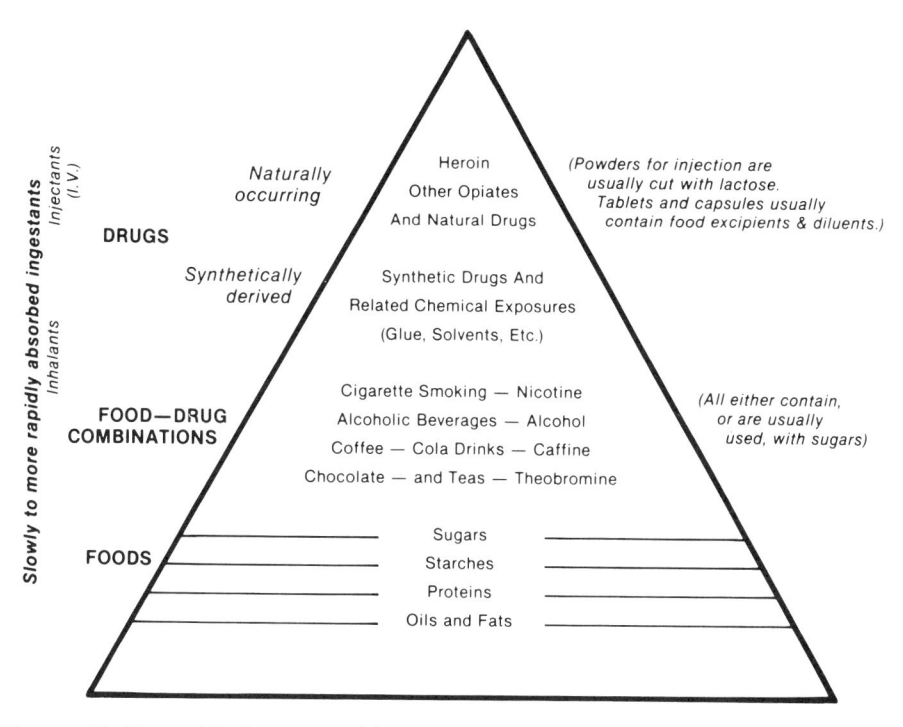

Figure 34. The addiction pyramid

As shown in Figure 34, addiction to heroin is the peak of this addiction pryamid and the acme of the addictive phenomena in general. When considering changing the name of food allergy to food addiction in the early 1950s, I read the 25 best descriptions of opiate addiction written between 1875 and

1925. The decision to make this change was made as a result of three considerations: 1) The relatively poor acceptance of the term, food allergy, when dealing with a person's favorite dietary items to which victims were essentially addicted; 2) Absence in the English language of a term connoting allergy to, or even addiction to, an unidentified substance; and 3) The clinical aspects of addiction to heroin. For instance, all the descriptive features of heroin addiction had been observed in food addiction except for death following an excessive dose after a temporary period of abstinence. In short, food addiction is simply a low key phenomenon of which opiate addiction is the ultimate. Perhaps cocaine addiction is now the ultimate.

Once addiction manifests, the overall course of developments is from lesser to greater addictions, as addicted persons ever seek relatively immediate postexposure stimulatory effects from more potent addictants or from combinations of them. This rule may be stated in the converse. Once one starts to react to given environmental exposures to which susceptibility is developing and learns that such recurrences may be postponed or minimized by specific re-exposures, such a person tends to resort to such an agent(s) or routines including it (or them) as often as necessary to avoid these recurrences. It is important to emphasize that the specific material(s) involved in these subtle relationships may or may not be recognized for this immediately beneficial or protective effect. The most commonly encountered course of developing addictive behavior is from slowly absorbed to increasingly more rapidly absorbed materials.

The degree of individual susceptibility to such items is of paramount importance in the development of addictive phenomena. In the absence of specific susceptibility, a person may be exposed to a given substance at any rate in the absence of becoming specifically addicted. The more rapidly individual susceptibility develops, the shorter the incubation period before addiction manifests. There are also marked differences in the potentiality for developing addiction possessed by different environmental substances. Drugs (either natural or synthetically derived), food-drug combinations and foods seem to meet these requirements especially well — the more rapidly absorbed the greater their addictogenic qualifications.

Of these materials exposure to foods is obligatory, whereas exposure to food-drug combinations and drugs is optional. It is not surprising, therefore, that food addiction forms the base of the addiction pyramid. Oils and fats, being absorbed slowly, are less apt to be addictants than proteins and starches. But starches of corn, wheat and potato origin are exceedingly common food addictants, although their specific identity is not apt to be recognized by the patients themselves. Their excessive intake both in amount per feeding and the frequency of eating increases calories above needs and often leads to obesity. This may persist progressively for many years without the addiction process developing further. More commonly, however, the person addicted to the starch form of a food soon resorts to more immediately effective sugars and/or alcoholic beverages and/or drugs.

Crystalline Sugar

Corn, beet or cane sugars are easily the most potent edible fractions of a food as far as their addictive potentialities are concerned. As previously stated, specific sugars not only carry the allergens of the foods from which they are derived, but their relatively rapid absorption enhances these properties.[288] Moreover, such sugars soon usurp addictive aspects of other less rapidly absorbed forms of the same food.

The problem of controlling the intake of given sugars is complicated by the fact that commercially manufactured processed foods usually contain multiple sugars. Food manufacturers are currently using solutions of mixed sugars rather than single sugars in bulk form. For example, solutions of corn derived sugar as well as sucrose and/or invert sugar (derived from either beet or cane or both) are piped to the moving food processing line. The desired mixture is selected according to the degree of sweetness, taste and other features.

There are also important geographic factors involved with sugars. Although corn sugar is used uniformly throughout the country, cane sugar is dominant in Atlantic and Gulf states and to the heads of navigation on inland waterways. Beet sugar consumption dominates the intake of sugar in states north of the Mason-Dixon line and all northern and western states, except that cane and beet sugar are used about equally in the continental Pacific states. Cane sugar is dominant in Hawaii.

In summary, the majority of food addicts take off up the addiction pyramid from food addiction, especially from reactions to one or more sugars.

Food-Drug Combinations

Various courses may be followed up the addiction pyramid through the food-drug combinations, as addicted victims seek more and more effective stimulatory responses. The most common course is the addictive use of chocolate, cola drinks (colas and chocolate are closely related), coffee and/or tea. These items either contain one or more sugars or are customarily consumed therewith. They also contain various combinations of theobromine and caffeine.

Others go more directly to cigarette smoking — which is also essentially a food-drug combination. The blended cigarette which came on the market during World War I differed from its predecessors in containing added sugars. Any kind of sugar may be used and the precise contents of cigarettes are closely held trade secrets. Tobacco also is biologically related to other members of the nightshade family, namely potato, tomato, and eggplant as well as green and red pepper and chili. Multiple susceptibility to several of these foods is common in persons reacting to tobacco smoke. Of course, nicotine in tobacco smoke may be an additional problem in its own right. There is no doubt that addiction to tobacco in this overall sense is a valid addictive response. Note the extreme difficulty experienced by the person addicted to cigarette use to break this habit.

Alcoholic Beverages

Another common route through the food-drug combinations and up the addiction pyramid is the use of alcoholic drinks. In this connection, one is reminded of Ogden Nash's comment: "Candy is dandy, but likker is quicker." One must differentiate between social drinking and alcoholism, with the full realization that this distinction is often blurred by the transition between.

Social drinkers using alcoholic beverages intermittently who experience either an acute inebriation type of response or a more delayed headache or other withdrawal response from a single drink of an alcoholic beverage are usually reacting acutely to some food(s) ingredient(s) of this drink. A similar involvement of all alcoholic beverages usually suggests the presence of individual susceptibility to yeast. In the absence of an intolerance to yeast, a physician knowledgeable of foods entering the manufacture of alcoholic beverages often obtains considerable valuable diagnostic information from observations of discerning social drinkers. Such reactions, repeatedly, to a single drink of beer or bourbon in the absence of a concomitant edible food intake usually suggests a reaction to corn and/or wheat-barley-malt. Similarly, an immediate reaction to wine under these circumstances suggests intolerance to grape, or to rum (unblended with grape sherry) suggests susceptibility to cane.[255]

But if such corn susceptible persons, for instance, continue to drink corn derived alcoholic beverages frequently and regularly, sooner or later they tend to become specifically addicted. This condition resulting from the sustained cumulative use of such a corn-derived alcoholic beverage by a person susceptible to corn tends to manifest as alcoholism. Similar relationships hold for other foods commonly employed in the manufacture of alcoholic beverages, such as wheat-barley-malt, grape, potato and yeast. Alcoholics who have been studied from the standpoint of demonstrable reactions to specific foods are usually susceptible to more than one constituent food entering their alcoholic beverage(s) of choice.

There also tends to be a close relationship between given sugars and alcoholic beverages, in that many spiritous beverages contain added sugars originally or are served as mixed drinks in which some type of sugar has been used. Moreover, drinking and smoking commonly occur in the same person, as may often be observed in the smoking sections of airplanes and restaurants. Indeed, the use of multiple food-drug combinations are of common occurrence in addicted persons.

Drugs

Marihuana is probably the most commonly self-prescribed biological drug that tends to bring its users into contact with other persons using more potent drugs, including those addicted to heroin. It is not uncommon for marihuana users to experiment with cocaine, heroin or other potent drugs or combinations of drugs.

Synthetically derived drugs and related materials are both commonly available and frequently employed as addictants. Glue sniffing is a simple example of this type of self-inflicted response. Sometimes a therapeutically prescribed drug will lead to the self-prescribed addictive use of other drugs. For instance, a recent patient with a background of reactions to multiple foods became excessively obese in high school. In college she was prescribed amphetamines in the form of diet pills. Their use led her into the company of other drug users and she drifted into the use of LSD and other synthetically derived addictants before she became addicted to heroin. After recovering from heroin dependence through methadone therapy, she became an alcoholic. Joining Alcoholics Anonymous helped with this. Currently she is living in a half-way house for abstaining alcoholics but experiencing reactions to several different foods. As illustrated in this brief case report, one may go up and back down the addiction pyramid, provided they are fortunate enough to survive the hazards of advanced addictive illness.

In summary, the concept that food addiction, especially the addictive use of sugars, is the basis from which more advanced addictions take off is appropriate. It might also be said that even the more advanced patterns of addiction are never completely unrelated to food addiction because of the obligatory nature of food consumption by all addicts. Most drugs in the form of tablets and capsules also contain cornstarch and other major food addictants.[62] Even heroin administered intravenously is usually cut with lactose (milk sugar) if it is available.

Addiction to Corn — An Example

Parents are often amazed and astonished when their teenage daughter is suddenly found to be a drug addict without realizing that their family lifestyle, eating habits and other addictive routines may have been preparing for this eventuality for several years. To illustrate how this may occur let us review common practices in respect to a common food — corn — currently the leading cause of food addiction, from infancy on.

Provided the infant is not nursed, he or she is usually fed a cow's milk formula sweetened with corn sugar (dextrose). Young children are commonly rewarded for good behavior by all-day suckers or other corn sugar-containing candy. Virtually all hard candy is made from corn as it does not become sticky (hygroscopic) on standing and packages well for the same reason. Chocolate candy comes a little later — a food-drug combination also containing a great proportion of corn sugar, as this crystallizes to give chocolate candy a velvety texture. Cane and beet sugar, on the contrary, crystallize in course granular crystals. Other treats prized by children such as soft drinks, ice cream and many other commercially available desserts also usually contain corn sugar. Indeed, corn-derived sweeteners constitute more than 40 percent of all sweetening agents in the U.S.A.

In addition to other sources of corn in the diet, binge eating of junk food commonly manufactured from corn and other cereal grains became the desired fare for food addicted pre-teen and other juveniles. Extreme instances of such a sustained self-perpetuated practice, sometimes called bulimia, tends to result in obesity. Other food-drug combinations, such as the use of coffee, tea, as well as alcoholic beverages and cigarette smoking, either containing corn or served with corn or other rapidly absorbed sugars tend to be the next step.

Although all of these preliminary practices may be socially acceptable and without disapprobation, one further step up the addiction pyramid to using marihuana or some more potent addictant and association with other recognized addicts tends to be associated with intense disapproval.

So much for the overall courses of unipolar or unidirectional developments of food addiction, ranging from so called blind food addiction to voluntarily propelled addictive responses. The important point to remember about the maintenance of any stimulatory level is that these victims of food addiction have no complaints and rarely ever seek medical care voluntarily.

Bipolar Intermittent Adapted-Maladapted Reactions

Chronically ill food addicted persons living at any given withdrawal level may become temporarily stimulated if exposed to a relatively larger or more rapidly absorbed dose of one or more of their addictants than customarily. The takeoff of these acute bipolar responses may be from any withdrawal level. Although these may involve approximately the same withdrawal and stimulatory levels, more ovoid courses occur more commonly. Examples will be cited which may be followed on the left of Figure 33.

A bipolar reaction taking off from level (-) often goes to (++) temporarily before reverting to (--) and then back to (-). Similarly, a bipolar acute reaction starting from (--), may go to (+++) temporarily, then (---) before returning to (--). Those taking off from (---) often go to (++++), then (----) before returning to their brain fag starting place (---). Finally, the course of the well known manic-depressive response may occur with the reacting individual remaining in the (++++) phase for variable periods before severe depressive symptoms again ensue (----).

Since these severe bipolar responses are often traumatizing affairs, even originally blind food addicted patients often learn by experience either to avoid or minimize them by being careful of how they eat and/or drink — especially both simultaneously. For example, such stimulatory-withdrawal bipolar sequences may be precipitated as a result of doubling or tripling the accustomed amount of a given addictant or even by the standard dose taken with an alcoholic beverage. For instance, one susceptible to corn and usually reacting to lesser amounts of it with a chronic stuffy nose (-) tends to become hyperactive and garrulous after one bourbon drink (++), only to develop a severe headache a few hours later or the following morning (--) before reverting to the accustomed chronic nasal congestion (-). The fact that this acute reaction is not the effect of alcohol as a drug may be demonstrated in specifically food addicted persons who are avoiding all of their addictants, as they ordinarily tolerate an isoalcoholic quantity of rum or corn-free brandy in the absence of a residual headache. Even more disturbing to both the addicted person and his or her associates is the ordinarily tired person manifesting relative constant headaches and tachycardia (--) who becomes hypomanic temporarily (+++) and attempts to promote some "out of character" scheme which, in turn, is followed by such a degree of brain fag (---) that the victim is unable to comprehend their objections until he or she has reverted to the accustomed level of physical symptoms (--). Also unknowingly addicted persons who have been on advanced "round trips" previously, often become exceedingly anxious and sometimes panic stricken upon becoming hypomanic (+++), for fear that this reaction may go on to level (++++) in which one may lose control or develop a seizure followed by a postictal depression (----) before reverting to the more accustomed state of being brain-fagged (---).

Unipolar Sustained Maladapted Reactions

The area listed on the lower right of Figure 33 lists more chronic and less reversible complications of both the sustained stimulatory levels of the upper right part of Figure 33 as well as the more chronic and less reversible complications of the withdrawals listed in the lower left portion of this figure. These complications as shown on the lower right portion of this figure are also sometimes characterized by structural changes.

There are limits in both time and relative effectiveness in attempts to remain relatively stimulated occuring at various (+) to (+++) levels. As noted previously, level (+) reactions, often followed by localized responses (-), may be complicated with time and become less relatively reversible. For instance, rhinitis may become associated with Meniere's syndrome consisting of the triad of deafness, dizziness and tinnitus, in which intermittent deafness becomes persistent and irreversible. Or asthma, also a manifestation of level (-) may become complicated by bronchiectasis or other structural changes and be less reversible than initially. Other examples of complicated localized reactions could be cited. Although obesity (++) may coexist with or be followed by any of the intermittent physical manifestations (--), either obesity alone or in combination with any of the physical syndromes listed at level (--) may also be complicated by vasculitis or hypertension with time and chronicity. Collagen disease is a notoriously persistent complication of uncontrolled persistent allergic-type responses. In my clinical experience, most alcoholics (+++) only agree to abstain from drinking voluntarily when each drink, instead of being associated with the desired stimulatory effect, is followed by so-called morose inebriation (---). Brain-fag, a lesser degree of depression, characterized by mood changes, irritability, etc., may also eventuate in relatively irreversible mental confusion and obfuscation with chronicity and presumed structural changes. It is more difficult to generalize on the long term complications of drug addiction. Many drug addicted persons in the hazardous pursuit of attempting to remain relatively stimulated, die as a result of accidental overdosages before reaching the stage of developing sustained or more structurally altered complications. However, advanced stages of depression (----) may drift into sustained dementia with stupor, coma, catatonia and/or residual amnesia with chronicity of these manifestations and concurrent structural changes in the brain (----).

Recovery

When patients recover from any stimulatory level listed in the upper half of Figure 33 as a result of having their food, drug and/or exogenous chemical addictants identified, avoided and/or minimized in their impact by specific therapy, any given stimulatory level is superceded initially by the approximately corresponding withdrawal level. Continued improvement occurs by ascent through lesser withdrawal levels to behavior on an even keel (0).

Discussion

Both physical and mental responses have long been known to be closely related and in need of better integration, but the question has remained as to how this should be done. One of the problems has been the predominately objective aspects of physical manifestations of illness and the largely subjective nature of so-called mental and emotional illnesses.

In my experience, however, these alleged objective and subjective aspects, respectively, of physical and mental illnesses are more apparent than real, in that there are objective and subjective aspects of both physical and mental illnesses. The fact that localized and systemic physical manifestations merge into brain-fag and depression with time and advancement of this ecological process minimizes much of the either/or dichotomy that has previously existed in medicine between physical and mental syndromes. Both are to be regarded as demonstrable manifestations of clinical ecology.

In this connection, I have just completed a series of lectures on the exteriorization of psychology, psychosomatic medicine and psychiatry, tracing the history of this overall field from introspection to extrospection, based on the classification of stimulatory/withdrawal levels of combined physical/mental responses as classified in Figure 33.

I am in complete agreement with Melvyn R. Werbach, M.D., who is board certified in psychiatry and neurology, who in his book (*Third Line Medicine—Modern Treatment for Persistent Symptoms,* New York & London, Arkana, 1986) discusses body-mind relationships in detail. He claims that one of the most serious deficits of present day medicine, characterized by the specific etiology doctrine, is the failure to integrate into treatment a combined biological and psychological approach. Robert Ornstein and David Sobel (*The Healing Brain,* New York, Simon & Schuster, 1987) also refer to the brain as the major organ of adaptation to both physical and psychological stimuli, integrating the external and the internal world.

In summary, the classification of physical and cerebral responses provided in Figure 33, not only serves to integrate these manifestations, but also provides medicine with the means of demonstrating these individual-environmental interrelationships.

Summary*

Ecologically Focused Medical Care

*References will be deleted in this summary since no new materials will be cited.

Introduction

Ecologically focused medical care was designed initially for the management of selected chronically ill patients not responding satisfactorily to conventional medical care offered by traditionally trained allergists. But as these therapeutic failures were studied in greater detail, it soon became apparent that many of them had not been adequately appraised diagnostically. In addition to their localized manifestations of rhinitis, asthma, eczema, colitis and others, many also presented more generalized systemic syndromes. Previously unsuspected common foods were often demonstrated to be responsible for both their localized and systemic manifestations.

As techniques developed to demonstrate these wider manifestations of clinical allergy, it became apparent that these physical syndromes often alternated with more advanced cerebral and behavioral disturbances. Also, in addition to inhaled particles (such as pollens, dusts, danders and others) as well as commonly eaten foods, environmental chemical exposures were demonstrable as major causative factors in many of these chronic heretofore unexplained illnesses. Techniques then emerged in which these personal-environmental interrelationships were demonstrable under relatively controlled conditions. The program of comprehensive environmental control in a hospital setting and the regimen of finding treatment doses by means of performing intradermal or sublingual provocative tests were especially helpful. Since many of these concepts and techniques did not depend on apparent immunologic mechanisms, it soon became evident that allergy as defined immunologically was too limited to account for these illness-related demonstrable roles of chemicals and many foods to which these poorly understood chronically ill persons were found to be extremely susceptible. In the hope of avoiding confusion with traditionally trained allergists, this diagnostico-therapeutic program including both immunologic and non-immunologic mechanisms, was simply referred to as clinical ecology.

Meanwhile modern medicine drifted into an untenable position vis-a-vis the human environment — man's intake and surroundings. In response to pressures from both within and from outside of the medical profession, medical academia had become progressively bodily centered, analytical, relatively static, mass-applicable and increasingly dependent on drug therapy. Most unfortunately, this relative environmental alienation of medicine occurred at the same time that the human environment was also becoming increasingly repetitious, noxious and frequently toxic. Concurrently with academia's progressive environmental disenchantment, an increasing number of people were becoming highly susceptible not only to commonly eaten foods, but also to chemical additives and contaminants of both the food and water supplies as well as to other environmental chemicals. This combination has been demonstrated to be impinging progressively on the health and welfare of an increasing number of humans. These environmental relationships, relatively unrecognized by the medical profession steeped in the specific etiology doctrine emanating from medical academia, are manifesting in apparently increasing levels of physical and mental illnesses. Moreover, medicine's relatively untenable position in respect to the human environment has also been accentuated by poorly advised Federal regulatory efforts. These poorly considered regulations are not only impinging on the medical profession by accentuating bodily, analytical and drug related mass-applicable medical care, but laxity at various levels in other governmental regulations has utterly failed to stem environmental pollution and contamination. The overall result of these trends has been increasing levels of chronic illness. In selected instances characterized by extremes of individual susceptibility to this so-called chemical environment, this encroachment of the environment on health is manifesting as impaired physical and/or mental health characterized by diminished productivity and, later, is often associated with functional disability.

In view of these relatively unrecognized illness-related present day environmental exposures and medicine's continued preoccupation with its limited specific etiology doctrine, the basic holistic nature of clinical ecology was found to have much wider medical applications than with illnesses previously designated as being of allergic origin. Indeed, clinical ecologists have come to take on many of the failures of the doctrine of specific etiology characterized additionally by excessive specialization which has dominated medical care for the past 100 years. My own case is an example of this present trend countering specialization in medicine. Although trained and certified in internal medicine and subcertified in allergy and immunology, during the past several decades I am practicing not as a medical specialist, as are most of my collegues, but I am interested in all sustained chronic physical and mental illness not responding to conventional medical care, in which the history suggests the operation of environmental factors.

Definitions, Limitations and Overall Course

Adaptation is defined as the ability of an organism to adjust to gradually changing but sustained environmental circumstances of its existence. Environmental stimuli eliciting such progressive dynamic modifications must be either constant or sufficiently repetitive to exert cumulative unidirectional effects. But the overall courses must neither be so abrupt in onset nor so intense in degree as to result in either immediate and/or irreversible injury to the host.

Physiologists first described adaptation in terms of its general features and apparent mechanisms as observed in groups of laboratory animals exposed experimentally to single environmental substances. Selye's observations were representative in that there was little emphasis on either the individuality, specificity or totality of multiple exogenous agents impinging on a given organism simultaneously. Despite the observations of Adolph who emphasized both individuality and specificity, adaptation remained more of an academic than a clinically useful and widely applicable medical concept. Indeed, this term only rarely enters the working vocabulary of most physicians.

In contrast, this update of personal adaptation in humans places emphasis on interrelationships between a person and his or her intake and surroundings capable of inducing and maintaining individual susceptibility and chronic illnesses. Specific adaptation in this clinical sense is far more holistic than adaptation described by most physiologists. It is also concerned with the responses of whole persons to the hammering impact of their environment to which they are highly susceptible. Although adaptation in this holistic, individualized and specific sense is not disinterested in the mechanisms involved, mechanisms are not its chief current concern. Neither is individualized specific adaptation primarily involved with an analysis of the body into its constituent organs and processes, nor with an analysis of the environmental exposures into their ultimate chemical makeup. Indeed, excessive analysis of constituent interreacting factors tends to obscure overall interrelationships between them. Computerization of such details does likewise.

Although this concept of adaptation has many other medical applications, the main concern of adaptation in clinical ecology is with foods and chemicals — including both exogenous and endogenous sources of chemicals.

In contrast to the relatively stereotyped features of modern mass applicable medicine, dynamic concepts of clinical ecology emphasize: 1) synthesis and holism in respect to both the host and the host's environmental factors; and 2) the demonstration of causality between these wholes in terms of adaptation.

Specific Adaptation and Individual Susceptibility in General

The dynamic interplay operating between environmental exposures and reacting individuals presenting acute and chronic diseases as well as intermittent and sustained manifestations is best expressed in terms of specific adaptation. A working knowledge of the stages of adaptation to given exogenous nonpersonal exposures as enhanced by individual susceptibility serves several useful medical functions:

1) It aids in understanding the development of ecologic disturbances, being especially valuable in the interpretation of medical histories.

2) It serves as a blueprint for establishing specific diagnosis, in that its application changes chronic illness of obscure causation to acute illness of demonstrably apparent etiology. In so doing, it also demonstrates the degree of individual susceptibility currently existing to such incriminated materials.

3) It provides the basis of ecologically sound medical treatment which most commonly may be carried out in the absence of drug therapy. In short, clinical ecology based on the application of specific adaptation involves holistic relationships between a person and his or her environmental existence. It also provides a constructive alternative to anthropocentric medicine (traditional analytical, bodily centered, drug related, relatively mass applicable, and largely environmentally alienated current medical practices).

4) Finally, evidence accumulated from the ecologic investigation of many individuals pinpoints repetitious lesser dose exposures and/or noxious environmental exposures potentially hazardous for all. Such medically derived factual information may be expected to constitute the basis for prophylactic decisions and programs in the field of public health.

In short, a working knowledge of the stages and techniques of specific adaptation in the presence of individual susceptibility becomes a key to a better understanding of the development, diagnosis, treatment and prophylaxis of many interrelated physical and mental illnesses. Moreover, this application is not only furnishing new avenues to old syndromes, but is also breaking down the artificial dichotomy between many physical and mental illnesses.

Developments in Allergy
Allergy as Originally Considered

Allergy, originally described at the beginning of the century as altered reactivity, was concerned with interrelationships between a given person and his or her intake and surroundings to which specific susceptibility existed. Early concepts and generalizations — based on detailed observations in both laboratory animals and humans — emerged inductively. By this is meant the process of reasoning from parts to wholes — i.e., from details to generalizations. This clinically based inductive approach characterized the subject of allergy for its first quarter-century. Indeed, early allergists were the original environmental experts of medicine. Their diagnosis depended on the demonstration of environmental-personal interrelationships. Avoidance of incriminated environmental exposures and specific therapy constituted the mainstays of treatment. But starting at the quarter-century point, several events both inside and outside of this new specialty changed the course of allergy in particular and of medicine in general.

Allergy as Defined Immunologically

European allergists induced their American contemporaries to redefine allergy in terms of antigens and antibodies about 1925. In contrast to the original descriptive and inductively derived definition of the subject, this deductively derived limited definition of allergy emphasized bodily centered mechanisms at the relative expense of specific environmental exposures: By deduction is meant the process of reasoning from wholes to parts, i.e., in this instance, from a stated generalization to clinical details. This new definition of allergy was later largely restricted to IgE mediated responses following the identification of the IgE form of immunoglobulin which mediates atopic allergy and the description of RAST which measures it.

Most unfortunately, however, this limited interpretation of allergy by so-called traditional allergists has been used by them to deny the existence as allergic phenomena of many clinical manifestations not meeting these restrictive criteria.

Developments in Medicine Generally

Although the hazards of excessive analysis have long been known in science, especially in studying interrelationships between biologic wholes, these disadvantages have largely been overlooked in medicine.

Unfortunately, many generally positive environmental developments before 1981, which had been supported by recent administrations, have been largely replaced by less favorable governmental trends since 1981. As far as the interests of clinical ecologists are concerned, environmental concerns have been blunted and politicized by the present administration on the one hand, and on the other hand, in their apparent misguided effort to hold down medical inflation, prepayment plans for hospitalization have anchored anatomically demarcated medical diagnoses at the relative expense of etiologic diagnoses. Medical interest in the significance of environmental factors in medicine seem to be suffering from both policies.

Analytical Approaches in Medicine

Coincident with the explosion of medical knowledge in the basic sciences and with technical progress starting with microscopy and advanced by spin-offs from the space age, medical diagnostic procedures have become increasingly analytical. Statistical methodology, computerization and data retrieval systems also favor analysis and fragmentation at the expense of synthesis and holism.

Entire fields of medicine have also suffered from excessive analysis. For instance, analytic dietetics, emphasizing constituent subdivisions of the 24-hour diet in respect to proteins, fats, carbohydrates, calories, minerals, vitamins, etc., has dominated the field of nutrition since early in the century. This relatively stereotyped emphasis contrasts with the more dynamic response of individuals to common foods in the form consumed in terms of biologic dietetics. Also, consideration of ambient air pollution in respect to such chemically identified analytic constituents as sulfur dioxide, nitrogen dioxide, carbon monoxide and others detracts from an adequate appraisal of the totality of air pollutants encountered by a given person.

Bodily Centered Considerations in Medicine

Starting about a century ago with acceptance and application of the germ theory of disease by medical academia in this country, modern medicine began to depart from its earlier ecological considerations of the Hippocratic era to emphasize origins of illnesses in terms of a unique primary specific etiology for each disease based on its pathology and apparent bodily mechanisms. Most unfortunately, earlier ecologically focused holistic views emphasizing environmental-individual interrelationships were discarded progressively with increasing specialization in medicine. This trend was hastened by the formation of specialty boards starting in 1917. With specialization largely demarcated on the basis of anatomy, super-specialization in respect to limited anatomical parts plus the attrition of general practitioners with time, resulted in an acute paucity of environmentally concerned primary care physicians.

In keeping with the full acceptance of the specific etiology doctrine of modern medicine along with the application of the tenets of modern science, the demonstration of etiology in disease has been increasingly concerned with apparent bodily mechanisms at the relative expense of the impingement of environmental exposures on afflicted individuals. Despite formation of the American Board of Family Practice and founding of the Society for Clinical Ecology in the 1960s, there are still far too few holistically oriented and environmentally focused medical practitioners. Those existing are almost always found outside of medical academia. This emphasis on bodily centered factors in medicine not only became the rule, but anatomically demarcated diagnoses have also been locked into this system by computerization and by Federal interference with the practice of medicine by stipulating DRG regulations. These trends, resulting in a bodily centered emphasis in medicine at the relative expense of any significant environmental focus also dominates both data retrieval systems as well as third party payments.

This one-sided bodily centered emphasis in medicine has also been accentuated by technological advances and industrial expansion during and following the two World Wars. Vested interests at the relative expense of the human environment continue relatively uncontrolled despite the formation of several Federal regulatory agencies between 1960 and 1980. Moreover, such acts and agencies as EPA (Environmental Protection Agency), OSHA

(Occupational Safety and Health Act of 1970) and TSCA (Toxic Substances Control Act of 1976) and others are dominated at various levels by super analysts and chemically trained toxicologists rather than by biologists and physicians. Indeed, the relative few biologists and physicians found in these organizations tend to suffer not only from their minority representation, but also from their apparent relative unfamiliarity with the medical significance of specific adaptation.

Instead of a unique bodily centered primary etiology for each disease entity, various disease manifestations are characterized by several demonstrable *etiologies* which are associated with readily confirmable interrelationships between specific environmentally related individual susceptibility and readily identified environmental exposures. Necessary ecologically oriented medical concepts and techniques to accomplish an environmentally focused reorientation of medical care in keeping with the environmental realities of the present age and in keeping with specified adaptation are available and waiting to be applied more widely. The stumbling block to this application is the relative absence of physicians in medical academia experienced in this type of work at the level of the practice of medicine.

As a starter, this ecologically focused medical point of view needs to be introduced into the teaching routines of medical academia, including medical school, graduate instructional courses and instructional programs of specialty medical societies. More specifically, we need to have a half dozen medical schools — one in each major geographic section of this country — which are sufficiently environmentally oriented to counter the stereotyped bodily centered medical training already in existence to foster ecologically focused medical care. Preferably, these institutions should be parts of the private sector of medical education to be less accountable to governmental and industrial funding and economic pressures.

It is curious and also symptomatic of present day bodily centered medicine that the concept of adaptation — central to understanding interrelationships between a given individual and that person's intake and surroundings as these reflect in health, have not been applied to any significant degree in the practice of medicine.

Drug-Related Symptomatic Therapy

The increased availability of drug therapy during the past half-century has not only altered most medical practices but has also led to many major complications. The excessive use of antibiotics is frequently responsible for precipitating candidiasis. Overuse of antihistamines and other synthetically derived drugs in the management of allergies and other illness has often been responsible for inducing the chemical susceptibility problem. Steroids have characteristically been overused and abused in the management of allergic-type responses. If instances of asthma and arthritis are diagnosed and treated specifically from the standpoint of their environmental incitants and perpetuants, steroids are rarely necessary. Although I had the unique opportunity of prescribing the first steroids for allergic disturbances in 1949, I have not started new cases of either asthma or arthritis on steroid therapy since 1953. Dominance of psychotropic drugs in psychiatry has materially interferred with the application of the techniques of ecologic mental illness, as both were reported initially in 1950.

The Constructive Alternative of Clinical Ecology

It is important to note that allergy to common foods, which initiated clinical ecology, had been described at about the time that allergy was redefined immunologically. But despite later contributions by Coca, Rowe, Rinkel, me and others, this common form of food allergy — later called food addiction, has neither been widely accepted nor applied. Similarly, the existence of the chemical susceptibility problem, which started about mid-century and was based on clinical observations, has also been poorly appreciated.

Many ecologically oriented new techniques also started about 1925. These included elimination diets (Rowe), the individual food ingestion test (Rinkel), the rotary diversified diet (Rinkel), the titration technique for determining dosage in inhalant allergy (Rinkel), the therapeutic use of alkali bicarbonate salts in keeping with the acid-anoxia-endocrine-enzymatic interpretation of allergy (Randolph), testing techniques using 1-5 serial dilutions (Rinkel), the provocative-neutralization technique (Lee, Rinkel), comprehensive environmental control, and observed togetherness — techniques useful in ecologic mental illness (Randolph).

Synthesis and Holism Versus Analysis and Summation

In clinical ecology, a patient is regarded as an intact biologic unit of his or her personal ecosystem. Each person is unique, both in respect to his or her genetic makeup and ecosystems. But in addition to these relatively stereotyped parameters, it is the dynamic equilibrium existing between a given person and that patient's intake and surroundings which is most individualized. This dynamic equilibrium, which is best expressed in terms of adaptation to given exogenous materials capable of eliciting individual susceptibility, is subject to derangements which often manifest as illness. Although this adaptation process which may lead to maladaptation and either acute or chronic sickness has been described in detail, its stages are summarized in Figure 35.

It is important to emphasize that the personal and environmental factors involved in these dynamic relationships of specific adaptation are wholes, not the sums of constituent parts. Indeed, this subject deals with synthesis — not excessive analysis, which tends to destroy or distort interrelationships between wholes.

ADAPTATION AS OBSERVED BY CLINICIANS

I	II	III
PREADAPTIVE (Nonadapted)	**ADDICTED**	**POSTADAPTIVE** (Nonadapted)

	IIa **ADAPTED**	IIb **MALADAPTED**	

TEST REACTION (Nonadapted)	**ONSET OF THE PRESENT ILLNESS**	**STATE OF EXHAUSTION** (Nonadapted)

SIGNIFICANCE OF SPECIFIC ADAPTATION

Specific adaptation, which extends the concept of general adaptation by emphasizing the specificity of adaptation in the presence of individual susceptibility, is involved in the dynamic interelationships between Bodily Centered and Environmental Factors which characterize clinical ecology.

A working knowledge of the stages of specific adaptation is essential in interpreting medical histories and various other diagnostic and therapeutic measures in clinical ecology.

Figure 35. Adaptation as observed by clinicians

Clinical Ecology Versus the Doctrine of Specific Etiology

The aims and approaches of clinical ecology differ fundamentally from those of conventional allergy and those of most other modern medical fields. The primary aim in clinical ecology is the demonstration of cause and effect relationships between given environmental exposures and individuals undergoing medical investigation. In contrast, traditional allergists and most other physicians are less interested in demonstrating environmental-personal interrelationships and more concerned with nonspecific bodily centered mechanisms. In view of these differing aims, it is not surprising that fundamental distinctions exist in the diagnostico-therapeutic approaches employed by these groups of medical practitioners.

The most troublesome environmental exposures in both clinical ecology and conventional allergy are those which readily induce individual susceptibility. Although physicians in both areas are interested in pollens, dusts, mold spores, animal danders and insect emanations, clinical ecologists are also especially concerned with common foods, food-drug combinations, drugs and a wide range of environmental chemical exposures. The latter includes synthetically derived drugs, ambient and domiciliary air pollutants as well as food and water additives and contaminants. Specific exposures to biological particulate allergens are more apt to be seasonal, variable and intermittent and to manifest in localized syndromes. On the other hand, commonly eaten foods and environmental chemicals — to which multiple cross-reactions usually exist — tend to present cumulatively as more advanced systemic and cerebral effects. These are referred to collectively as chronic polysymptomatic syndromes. Indeed, it was the repeated impact of the food and chemical environment on specifically susceptible chronically ill persons and manifesting as addictive phenomena that retarded the recognition of these relationships for several years.

Environmental exposures are also classified as to whether their effects are primarily on the basis of individual susceptibility or toxicity. Despite both clinical ecology and conventional allergy being characterized by individual susceptibility, clinical ecology also overlaps toxicology. Although most environmental chemical exposures are known to be toxic in greater concentrations, lesser amounts are generally regarded as safe. But in the presence of individual susceptibility, these lesser doses, especially if impinging cumulatively, frequently induce and maintain advanced physical and/or mental syndromes. In dealing with these combined relationships, it should be remembered that clinical ecology involving exogenous exposures is characterized by low dosage which is enhanced in the presence of high degrees of individual susceptibility, whereas toxicology is associated with high dosage and assumed normal host susceptibility. But in the presence of individual susceptibility, the potency of these lesser exposures (known to be toxic in greater concentrations) is markedly increased. Under these circumstances, such lesser sustained exposures as the fumes from gas kitchen ranges, for instance, become major unsuspected causes of chronic ills. In short, the combustion products and derivatives of gas, oil and coal — combining susceptibility and toxicity — become major demonstrable factors in the etiology of chronic ecologic illnesses.

In addition to these exogenous environmental chemical exposures, lesser doses of given environmental chemicals — especially pesticides, solvents and other lipid soluble agents — accumulate in bodily fats. The intermittent release of these endogenous chemicals also is able to induce acute or chronic reactions in highly susceptible persons. Preliminary investigations indicate that a mix of exogenous and endogenous chemical exposures is increasingly common.

Although clinical ecologists are not disinterested in the bodily mechanisms of ecologic disturbances, these remain theoretical at present. Despite some foods being associated with the production of IgE antibodies, the nearly endemic form of allergy to common foods and reactions to chemical environmental exposures are not measured adequately by presently available immunological techniques, although various immunologic abnormalities have been noted. In summary, environmentally focused etiologic considerations of clinical ecologists differ from the more bodily centered and immunologically based interpretations of conventional allergists. These more dynamic etiologic relationships of ecologists are best presented in terms of specific adaptation, the stages of which are summarized in Figure 35.

Adaptation Interrelationships

Most persons adapt to their intake and surroundings without apparent ill effects but in the presence of individual susceptibility, environmental materials manifest lesser or advanced syndromes depending largely on the degree of susceptibility which has developed and the capability of the host to respond. Although individual susceptibility may be induced as a result of a massive dose, its gradual onset following repeated intermittent or cumulative exposures is a much more common occurrence. But once individual susceptibility has developed to one or more substances, manifestations are either acute or chronic — depending largely on whether specific environmental contacts remain intermittent or cumulative.

The overall developmental clinical picture of specifically adapted responses to environmental exposures, to which individual susceptibility exists, consists of successive stimulatory levels of reaction followed by successive withdrawal responses. These levels have been described as they present in Stage I — Nonadapted responses from intermittent environmental exposures, and in Stage II — Addicted responses. The latter, consisting of IIa Adapted and IIb Maladapted subdivisions, results from cumulative environmental exposure. Stage III — Nonadapted responses, characterized by apparent exhaustion of the defense mechanisms involved, is more apt to be approached than actually reached. (See Figure 35)

Application of the concepts and techniques of specific adaptation provides a continuum between diagnosis and treatment, instead of the commonly occurring discontinuity between present day diagnostic and therapeutic measures. Finally, in contrast to the relatively depersonalized and mass applicable features of modern production-line medicine, ecologically focused medical care is both highly individualized to a person and highly specific to that person's intake and surroundings which are capable of inducing individual susceptibility. Thus, clinical ecology is relatively more acceptable to patients and provides a more medically and ecologically sound program than many conventional alternatives.

Manifestations

The basis for classifying allergic manifestations emerged approximately 40 years ago as a result of Rinkel's observations of unmasked and masked food allergy. Unmasked responses, characterized by an immediate and obviously acute reaction after eating a given food to which one is sensitive, have been known for over 2,000 years. Indeed, this is food allergy as it is usually considered. But a masked response to the same food consumed once in three days or more frequently is characterized by an entirely different two phase response, consisting of an immediate stimulatory or symptom suppressive effect and a more delayed symptom-related reaction. Most importantly, the former may be maintained for long periods simply by eating such a dietary item (either blindly or deliberately) as often as necessary to maintain this symptom-relieving effect. Otherwise, chronic illness of unknown cause could be perpetuated. The reader may be startled to know that this is food allergy as it usually exists. Also, significantly, the latter undiagnosed reaction reverts to the former obviously related food reaction, simply by avoiding such a commonly eaten food to which one is reacting cumulatively for four days or slightly longer before eating an isolated dose of the same food.

In keeping with the development of specific adaptation in the presence of individual susceptibility, unmasked responses are referred to as nonadapted acute symptom-related reactions. Masked food allergy, synonymously called specifically adapted food allergy or susceptibility presents either as symptom-suppressed responses present as various (+) to (++++) stimulatory levels, as shown on the left upper side of Figure 33, page 248, whereas the symptom-related withdrawal levels of (-) to (----) are shown on the left lower side of the same figure. Constituent clinical features of these lesser allergic/ecologic adapted stimulatory responses, (+) stimulatory and (++) hyperactivity, take off from (0) Behavior on an Even Keel, as in Homeostasis; the (-) localized physical withdrawal reactions and the (--) systemic physical withdrawals are shown directly below level (0). More advanced (+++) hypomanic and (++++) manic responses are shown at top left of Figure 33; more advanced (---) brainfag and (----) depression levels are shown at the lower left of this same figure.

In contrast to these intermittent stimulatory and withdrawal levels as shown on the left side of Figure 33, more sustained stimulatory and withdrawal responses are presented on the right side of this figure. Those sustained stimulatory responses from foods manifest as (+) characterized by absent complaints, (++) by obesity, (+++) by alcoholism, and (++++) by drug addiction. It should be noted that all stimulatory responses of Figure 33 (those listed on the upper half of this figure), are *not* associated with patients' complaints, as far as afflicted individuals are concerned. In short, these patients who are able to maintain relatively stimulatory responses do not seek medical attention of their own volition, although their friends or family members sometimes take them to see physicians. Although these stimulatory levels may represent the "onset of the present illness" as far as a clinical ecologist is concerned who is familiar with specific adaptation, this is not true either for the involved patient or that person's physician. Only the onset of withdrawal symptoms, which the patient is unable to alter by his own initiatives represent the "onset of the present illness", as far as the patient and this person's physicians are concerned.

More sustained withdrawal responses listed on the right lower portion of Figure 33 represent both extensions of the sustained stimulatory responses shown in the upper right portion of Figure 33, as well as less reversible complications of withdrawal levels shown in the left lower portion of the same figure. Many of these responses at levels (-) to (----) may be relatively irreversible.

Although the adaptive interrelationships between physical and mental syndromes, adapted and maladapted responses to the same food consumed under different circumstances, as well as intermittent and more sustained clinical manifestations began to emerge 30 years ago and were first exhibited in 1956, the full interrelationships between these factors in respect to specific foods, environmental chemicals and certain other environmental agents associated with enhanced individual susceptibility did not become evident until preparing this manuscript. Consequently, this chart has not been published previously, although it has been submitted for publication.

Diagnostic Applications
History

When the medical history is taken emphasizing interreactions between an individual and that person's intake and surroundings capable of eliciting individual susceptibility, it assumes its rightful position as the single most important diagnostic measure in clinical ecology. And since the horizons of this subject have not been demarcated as yet, ecologically focused medical histories promise to elucidate much wider medical areas than those presently recognized as being ecological.

Inasmuch as the medical history is the most important part of the medical investigation as well as the initial contact between a new patient and his or her doctor, it should be recorded by the most knowledgeable person of a medical team — not the least knowledgeable underling. Ecologically focused physicians gain insight into their patients' medical problems by recording their histories chronologically, emphasizing all environmental exposures possessing the potentiality of inducing individual susceptibility and manifesting either acute or chronic illness. And as far as chronic illness is concerned, the historian must always remember that the most alert and astute patient is often unaware of being specifically and unknowingly addicted. This paradoxical nature of allergy histories cannot be overemphasized, on the one hand. And on the other hand are patients who, although aware of the identity of specific exposures maintaining temporary relatively beneficial postexposure immediate effects, sometimes refuse to disclose such relationships even though they are seeking medical help.

Thus, there are several reasons for truncated and inaccurate medical histories. Lack of knowledge of clinical ecology on the part of physicians is the most important, followed by spending inadequate time and/or failure to record what the patient offers. I obviated these problems during the first forty years of my practice by taking my own histories on a typewriter. But patients, experienced in how physicians commonly respond, are often adept in truncating and simplifying their histories, otherwise they are apt to be referred to the psychiatry department. I have handled this potential problem by taking "poker faced" histories. In my experience, this combination of typewritten, poker faced medical histories has been more responsible for any medical contributions I may have made to this field than any other measures. In short, the medical history provides the direction of the medical investigation.

In my experience, and in view of the fact that histories in clinical ecology characteristically consist of multiple complaints, histories are best recorded chronologically, discarding the "chief complaint." A format also discarding the "past history" and its anatomical subdivisions is also helpful in emphasizing the holistic features of clinical ecology and brings out the characteristic alternation between stimulatory and withdrawal manifestations affecting multiple organ systems.

Physical Examination

Although the physical examination is an integral part of all medical investigations, "examination of the environment" of a patient tends to be relatively more rewarding — once the decision has been made to investigate a patient ecologically. Details of the physical examination help in making this decision.

Specific Environmental Test Exposures

The basic diagnostic principle in clinical ecology is the demonstration of cause and effect relationships between a person and his or her intake capable of inducing and maintaining specific adaptation. The most commonly incriminated environmental exposures of advanced and complicated cases responding to diagnostic measures employed by clinical ecologists are: foods and environmental chemicals, including drugs (about equal frequency), tobacco smoke and particulate inhalants (pollens, molds, dusts, mites, and danders). These may either be avoided singly, in groups or multiply until the patient in question has recovered from the residual effects of the most recent specific exposure(s). After being first accentuated, withdrawal manifestations subside gradually as specifically adapted and maladapted responses become nonadapted. Re-exposures then induce acute test reactions.

Ecologically oriented medicine aims to establish a specific etiologic diagnosis by demonstrating a cause and effect relationship between nonpersonal exogenous agents and a given individual. Although such an etiology may be apparent in instances of nonadapted acute immediate reactions, the etiologic roles of common environmental substances are usually obscured in instances of adapted and maladapted chronic illnesses.

Under these circumstances, the most helpful and convincing diagnostic procedures are those which change chronic illnesses to acute illnesses as a result of avoidance followed by specific re-exposures. Although elimination diets and individual food ingestion tests accomplish this, the program of comprehensive environmental control under environmentally controlled hospital conditions is the program of choice, especially in advanced cases which tend to be characterized by polysymptomatic syndromes and a multiplicity of causative environmental exposures. Unfortunately, however, this program upon which these observations were largely based is no longer widely available because of economic reasons.

The greater the immediacy and advancement of such test responses, the greater the degree of individual susceptibility involved. Similar but generally less severe acute test responses also occur in provocative testing which are the diagnostico-therapeutic programs of choice to be employed on an outpatient basis.

Therapeutic Applications

The primary aim of treatment in clinical ecology is avoidance of incriminated environmental exposures and minimization of the tendency for individual susceptibility to develop to substituted materials. This program is most readily accomplished in food allergy by the use of the rotary diversified diet of test compatible foods. Treatment of demonstrated reactions to environmental materials by means of neutralization dosages is also widely used. These apparently maintain a specifically adapted relatively symptom-suppressed or symptom-reduced response. But in view of current differences of opinion as to what mechanisms may be involved in such injection or sublingual therapy, the modus operandi of this effect remains empirical.

Although avoidance of incriminated environmental exposures and/or specific therapy are the treatment programs of choice, acute reactions from accidental environmental exposures often require additional symptomatic therapy. Two of the most helpful measures, based on the application of the acid-anoxia-endocrine-enzymatic concept of allergic inflammation, consist of sodium and potassium bicarbonate (2/3 $NaHCO_3$ and 1/3 $KHCO_3$) orally or sodium bicarbonate intravenously when the oral route of administration is not available. Oxygen therapy is often helpful additionally.

Self-treatment, maintaining relatively constant specific exposures is easily the most frequently applied treatment of addictive responses to common foods, chemicals and other environmental exposures to which selected individuals are highly susceptible. This statement holds for both the adapted and maladapted phases, so exceedingly common is this physiologic mode of response, despite the fact that maladapted patients are often seeking medical management for their afflictions.

Although anthropocentric medicine with its emphasis on drug therapy may be moderately helpful in dealing with lesser reactions, it tends to be ineffective in more advanced and/or complicated cases. When dealing with progressive illness, many drugs either become ineffective with prolonged use or become illness-inducing or illness-perpetuating in their own right. The latter is a common eventuality in patients who are susceptible to various environmental chemicals because of the ease with which synthetically derived drugs also become incriminated. In short, traditional anthropocentric medical management not only tends to be ineffective when applied in ecologic disturbances, but often worsens them in the process. For these various reasons many chronically ill patients are abandoning the medical profession and attempting to seek help elsewhere.

In contrast to the relative gap or discontinuity between diagnostic and therapeutic measures both in medicine and often in conventional allergy, diagnosis and treatment in clinical ecology tends to be an ongoing continuum.

In view of these marked variations between different persons in respect to these personal and environmental variables and the dynamic nature of such interreactions, a relatively stereotyped statistically oriented approach from studying groups of patients tends to be misleading. Instead, the detailed case report, or a series of case reports, remains the format of choice in reporting these highly individualized responses to specific nonpersonal environmental exposures.

The overall results of ecologically oriented medical management are far superior, although the program of finding and avoiding specific environmental exposures may be disturbing initially. For instance, interference with patients' long ingrained eating habits and lifestyles which are precipitating their environmentally related illnesses tend to be disturbing and disruptive. Indeed, such accentuated withdrawals are difficult to endure when addicted persons are aware of how easily and simply they may be relieved by recourse to formerly habitual routines. Only after both stimulatory and withdrawal manifestations have cleared, as one reverts to the preadaptive (nonadapted) status, do such benefits become apparent.

Areas of medicine yielding to the application of this ecological approach include hyperactivity, obesity, alcoholism, drug addiction and epilepsy. This list also includes large areas of pediatrics, otolaryngology, ophthalmology, dermatology, psychiatry and psychosomatically interpreted illnesses as well as the subdivisions of internal medicine (allergy, rheumatology, cardiovascular disease, gastrointestinal and pulmonary medicine).

Prophylactic Applications

Rinkel's rotary diversified diet is the most important prophylactic measure to minimize both the original development of food allergy as well as to deter its spread to substituted foods once this problem has developed. This simply means that no primary food is to be eaten more often than once in four days or less frequently. Precluding eating a given food while the most recent intake of the same food is still in the gastrointestinal tract minimizes cumulative effects — the most important consideration leading to the development of specific individual susceptibility. In the presence of constipation, the interval between servings of a given food should be extended by a day or so. Unfortunately, this regimen is poorly appreciated. Although important enough to be taught in the primary grade, most college graduates are unfamiliar with it.

Avoidance of exposure to massive or sustained amounts of environmental chemicals is the most effective program to minimize the development of the chemical susceptibility problem. This means the absence of combustion of fossil fuels within living quarters, as well as eating food and drinking water known not to have been significantly contaminated by means of chemical additives and contaminants. Minimization of chemical air pollution is far more difficult. The best single measure is to choose living quarters on the windward side of metropolitan areas, removed as far as possible from the sources of industrial pollutants, as well as from automotive and railway traffic. For instance, one should not live on a major expressway, especially on a corner where traffic decelerates. Indeed, one may be chemically contaminated by traffic fumes if one can hear the sound of traffic.

Under no circumstances should the garage be interconnected with one's home — least of all located under the master bedroom. Installation of activated carbon filters aids in minimization of ambient levels of air pollution, whereas mechanical filters reduce the levels of pollens, dusts and other particulates. As a prophylactic measure, living quarters should never be shared with cats, dogs or other furry animals.

Discussion

Presentation of the historical background of clinical allergy in particular and medicine in general during the past century has aimed to acquaint the reader with what, retrospectively, appear to have been major medical errors of both commission and omission. These errors have not only led to an unfortunate polarization in the field of clinical allergy, but have also reflected in the generally poor ability of most physicians to handle many chronic illnesses.

The greatest error of commission by conventional allergists is the apriorism that allergy must be mediated by IgE antibodies or other immune mechanisms. An unfortunate corollary of this limitation is that if an illness cannot be diagnosed as IgE mediated or an allergen pinpointed by skin test or RAST, the illness simply does not exist as an allergy and the material in question is not an allergen. In contrast, the phenomena of clinical ecology tends to be mediated by both IgE and other antibodies or by additional mechanisms. Since such alternatives include the acid-anoxia-endocrine-enzymatic concept of inflammation, interpretations of which involve kinins and other mechanisms which are not skin sensitizing, skin tests of the wheal and flare type are of limited value in the diagnosis of specific environmental agents responsible for many ecologic illnesses.

Other errors of commission perpetuated by most physicians include the assumption that complex biological phenomena may be understood simply by an analysis of the factors entering into their production. Although such an excessively analytical and relatively stereotyped approach may be productive of new information and lends itself to computerization technology, it tends to obscure dynamic relationships between interacting biologic wholes. Symptomatic treatment instead of etiologic therapy tends to follow. When combined with anatomically demarcated specialties of medicine and anatomical diagnoses, these errors of commission may be referred to as the

ABCDs of MMM (Analytical, Bodily Centered and Drug related approaches of Modern Mass-applicable Medicine).

The greatest error of omission of conventional allergists as well as most other physicians is the relative neglect of many commonplace environmental excitants and perpetuants of chronic illness, the impact and the potency of which is magnified by individual susceptibility. Closely related is the tendency in modern medicine to interpret etiology in terms of bodily mechanisms rather than in respect to the selective impingement of environmental exposures.

An equally important sin of omission in modern medicine is failure to apply the basic concepts of adaptation to specific environmental exposures, to which individual susceptibility readily develops. This omission is reflected in the relatively poor performance of physicians in the management of many chronic illnesses and their poor understanding of the acute and chronic phases of the same illness in the same patients. These points will be summarized briefly.

In contrast to acute reactions resulting from the impingement of intermittent environmental exposures (in the presence of individual susceptibility) which are usually apparent, reactions from oft-repeated amounts of a substance tend to remain unrecognized. For instance, the repetitive intake of common foods, drugs, tobacco and environmental chemicals is rarely suspected because of their basically addictive features. But either the specifically adapted or maladapted phases of such chronic addictive behavior may be changed to nonadapted states as a result of temporary avoidance. Specific re-exposures then evoke acute immediate test reactions which demonstrate environmental etiology. The immediacy and severity of such experimentally induced acute responses also measures the degree of individual susceptibility which exists. Thus a working knowledge of specific adaptation, accentuated by individual susceptibility, is essential in the management of many chronic physical and mental illnesses.

Ecologic interpretations in medicine are concerned with multiple exogenous and endogenous materials capable of eliciting individual susceptibility and manifesting a wide range of interrelated physical and mental syndromes. Both naturally occurring biological materials and man-made or synthetically derived substances are involved in amounts generally regarded as safe (GRAS). However, such hydrocarbons are known to be toxic in greater concentrations, and in the presence of individual susceptibility, lesser doses may be responsible for inducing and maintaining advanced reactions. Thus there is no sharp distinction between clinical responses on the basis of susceptibility and toxicity in clinical ecology. Either, and more commonly both, manifest in polysymptomatic alternating responses consisting of stimulatory and withdrawal physical and mental syndromes — depending principally on factors of dosage and advancement of the process.

Resistance on the part of the majority of physicians to accept an environmentally focused etiologic interpretation of the production of symptoms is not only still present, but is not diminishing. If anything, this type of resistance within the medical profession seems to be increasing due to trends both within the medical profession, selected medical organizations and governmental agencies.

Sources of this resistance to the environmental etiology of illness center in academia, especially medical academia and the medical curriculum. And this is not improving. Indeed, there is even less emphasis on the demonstrable role of environmental factors in the etiology of clinical syndromes now than during recent decades due to an increasing focus on drug suppression of symptoms and minimal teaching of environmentally focused individualized medical care, individualized biological features of nutrition, clinical aspects of toxicology and adaptive interrelationships between individuals and their daily environmental exposures.

This unsatisfactory academic situation has been accentuated by the emphasis of traditional allergists responsible for most of the teaching in this field. Ecologically focused medical care is offered as a constructive alternative to the type of medical instruction currently emanating from medical academia.

Present governmental actions have also been disadvantageous for environmentally focused medicine. Not only has the human environment been allowed to deteriorate further, but Federal interference with the practice of medicine economically seems to be impairing the quality of medical care.

Although the application of ecologically focused medicine encounters compliance problems generally, intelligent and well read laymen are far more receptive than most physicians. The program that several of us initiated a decade ago of addressing medical problems to the general public has paid major dividends. I agree with Dr. Coca that this aspect of human health is far too important for major decisions to be left solely with the medical profession, especially in view of their present performance.

Positive public interest in the medical role played by the environment apparently stemmed from publication of ecologically focused popular books. These were initiated by Coca's semipopular books on allergy in the 1940s and his more popular presentations in the 1950s. Since this time many other popular presentations on ecologic aspects of allergy, including mine, have been published.

Presently, the decision to seek alternative medical care is most commonly initiated by unsatisfactorily treated chronically ill patients. Ecologically focused medicine is one of the more attractive alternatives because of long-standing environmental interests and concerns on the part of the public, and persistent relative neglect of individual-environmental interrelationships on the parts of medical academia and most practicing physicians.

Negative environmental interests of physicians apparently started with the beginnings of modern medicine. Medical academia's premature acceptance of the immunologic interpretation of allergy before the manifestations of clinical allergy had been described led to a series of disasters. The rigid defense of allergy in terms of immunologic mechanisms by traditional allergists precluded their appreciation of more advanced allergic manifestations, additional environmental excitants and/or alternative mechanisms.

It may be noted from this presentation that ecologically focused medical care not only developed outside of medical academia but has not been taught within the medical academic establishment. Although there has been a wide open opportunity for this to change, there is little evidence that this is occurring. Acceptance of these alternatives has virtually been precluded by full time academic teaching staffs of medical schools and placing medical residents in charge of teaching medical services seems to be precluding any likelihood of this happening within the foreseeable future.

When a practicing physician has a chronically ill patient in whom environmental exposures are suspected, the most effective way to demonstrate such relationships would be to refer this patient to a clinical ecologist. Should he wish to learn more about the alternative of ecologically focused medical care, he might attend basic instructional courses and annual meetings of the American Academy of Environmental Medicine — the Discipline of Clinical Ecology.

In view of the furor which clinical ecology has engendered amongst traditional allergists and the prospect that many other areas of medicine are becoming similarly involved, it is quite possible that this account of the development and present status of clinical ecology may well become the "most vigorously attacked" medical book of its time.

Many of the concepts and approaches which characterize present day medicine are anachronistic as far as the future of the health sciences in a repetitious polluted world is concerned. Fortunately, constructive alternatives embodied in the medical approach of clinical ecology exist. Application of this diagnostico-therapeutic program renders two useful functions: 1) An improved medical service for many illnesses, especially chronic physical and mental ills of presently obscure etiology. 2) In the long run, the greatest recipient of ecologically oriented medicine may be expected to be the human environment itself.

Appendix A

"Environmental Illness" by Steve McNamara

Environmental Illness

By Steve McNamara
(Pacific Sun, Week of August 5-11, 1983)
(P.O. Box 5553, Mill Valley, CA 94942)

On May 13, 1982, Vickie Moll was overcome by fumes from a cleaning fluid being used to remove stains from her living room carpet. These days she usually wears a respirator mask when she ventures outside her home and must avoid strong odors such as diesel exhaust, perfume and the soap aisle in supermarkets.

On March 12, 1981, Tom Ohlhausen became ill after three days of inhaling hot asphalt and diesel fumes while working for a Marin paving company. He is now confined to his Corte Madera home and says he must use oxygen from a tank in order to carry out any activity.

In January of 1978, Ruth Noe collapsed at her job as a kindergarten teacher at Short School in San Rafael. She says she was stricken by the cumulative effect of sprayed pesticides, a leaky gas heater and the chemicals used in an unventilated mimeograph room. Now a virtual prisoner in her Marinwood home, she is hoping that an expensive prospective hospital stay will reverse a rapid decline in her health.

These three people, and thousands like them throughout the world, believe that they suffer from environmental illness. They say it is an allergic reaction to foods and chemicals of the twentieth century, especially formaldehyde and phenols, which are harmless to most people. The three are under the care of three separate physicians who use a technique of treatment called clinical ecology. The clinical ecologists are less interested in hay fever and asthma than in illnesses with an environmental agent.

Conventional allergists say that all this is nonsense, that if people such as Moll, Ohlhausen and Noe are truly ill or merely crazy is a matter of substantial concern. If indeed they have been stricken by common chemicals in the environment, then a second question presents itself: are they a bizarre handful of unfortunate souls, or are they our canaries in the mine shaft?

Years ago miners would carry to work a canary in a cage. If the canary stopped singing, took sick or died, the miners knew to get out of the mine in a hurry — the bird's more sensitive system had detected lethal fumes before the miners had done so themselves. Does it follow that people who claim to have an environmental illness are sending a warning to the rest of us?

The mine shaft analogy merits special consideration because of the publication last month of the results of a study by Dr. Peter Budetti and Paul Newacheck at the University of California Medical School in San Francisco. They found that the percentage of youngsters to age 17 who suffer from chronic illnesses more than doubled from 1958 to 1981, from 1.7 percent to 3.8 percent. They have no ready explanation, but among the factors they will examine in their ongoing study are changes in the environment. Separately, Dr. Phyllis Saifer, a clinical ecologist, says that with more than 3000 new chemical compounds introduced each year, an increasing frequency of environmental ailments is inevitable.

So, is what we have here a significant but boring medical squabble which will be sorted out in the quiet corridors of science? Not exactly. The two sides freely charge each other with harboring charlatans who are ripping off patients for huge sums of money and providing inadequate care while selling their souls either to the drug and insurance industries or to quackery.

Says Dr. Joseph McGovern of Ross, a former hematologist on the faculty of the Harvard Medical School and now a clinical ecologist: "There are people in this fight who make Nixon look good."

Clinical ecologists say that much of the resistance by conventional allergists to their method of treatment stems from economic concerns — that their rivals fear losing business, especially the twice-a-week office visits for allergy shots which are the mainstay of conventional allergy practice. "Two days a week you give shots," says Dr. McGovern, himself a board certified allergist, "and the other three days you take deposits to the bank."

Conventional allergists counter in outraged tones that it is the clinical ecologists who are soaking their patients with expensive tests and treatment. Back comes Dr. McGovern: "There is a concerted effort on the part of conventional allergists to discredit clinical ecologists, to stop diagnostic testing, to prevent the use of sublingual testing. It is based on economic grounds only. We're starting a lawsuit with the Federal Trade Commission to enjoin them."

The total allergy field is huge. More than 20 million Americans are said to suffer from hay fever and asthma. There are only some 3000 boardcertified conventional allergists, but Hollister-Stier, the firm which provides anti-allergy substances to physicians, has some 25,000 professional customers, so a lot of allergy is being practiced with one technique or another.

It is interesting to note that this bitter struggle is not duplicated in Europe and Japan, where fee-for-service medicine is less common. In the United States the sums involved are enormous. There is money paid to doctors on either side for tests and treatment. There is money for drugs and for the oxygen which clinical ecology patients insist they need to function. There is

Appendix A — Environmental Illness

money on the line when an environmental illness victim sues for damages, as Vickie Moll is doing. And there is money paid in expert witnesses fees. Third-party insurers — private companies or government agencies such as Medicare — are anxious to deny liability for illnesses. It is even better if they can successfully deny that the illness exists. A run-of-the-mill expert witness can make $100,000 per year on cases such as these, while the superstars (most of them on the side of conventional allergy) can make more than $250,000. One Los Angeles area physician is said to charge $10 per minute for his services.

The case histories of Vickie Moll, Tom Ohlhausen and Ruth Noe cover most of the bases in this controversy:

Vickie Moll is married to a cabinetmaker and carpenter who works mainly in Marin. For several years Vickie had helped as a painter on her husband Larry's jobs. She says she had experienced no difficulties with paint fumes or other chemicals and she had not experienced mental problems.

In May of 1982 the Molls hired a firm to clean the carpets in their San Francisco home. The person who did the cleaning failed to raise up on blocks two chairs which had wooden cases. The result was stains where the carpet cleaning fluid acted on the wood. Larry complained to the cleaning firm, which on May 13 sent a young man to remove the stains. According to Larry, the stains refused to come up and the cleaning person, frustrated, dumped increasingly large quantities of cleaning fluid on the rug.

Vickie passed out. Her husband took her to the emergency room at Kaiser Hospital in South San Francisco, and also called the San Francisco Poison Control Center. Larry says the Poison Control Center seemed sufficiently concerned that he had his wife transferred to the Toxology Clinic at San Francisco General Hospital. There several blood tests were run (not the right test, say the Molls) and it was written in Vickie Moll's medical record that her problem was chiefly mental.

"I was devastated," says Moll. "I thought maybe I really was crazy, but then I knew that it wasn't possible." Her case was transferred back to Kaiser in South San Francisco where eventually she came under the care of Dr. Robert Sinaiko, a clinical ecologist. Dr. Sinaiko diagnosed her ailment as environmental illness triggered by cleaning fluid fumes.

Moll's system had been thrown into shock, she and Dr. Sinaiko say, and she could no longer tolerate soaps, perfumes, ink, corn, yeast, milk, sugar products and more than 40 other common substances. She says that when exposed to these things her face becomes numb, she gets sick to her stomach, her abdomen swells, she becomes dizzy and suffers from arthritic pains, and that once her right leg turned bright red. She says her discomfort is matched by her anger at those who claim that her trouble is all in her head: "I'd like for one of them to be around when I have a reaction, to see what I physically go through. And then I'd like them to tell me that it isn't happening to me. I dare any of them to say it's in my head. How dare they say that this isn't happening!

"If there is anything mental it is the fear of never knowing when you are going to be bombarded with petrochemicals. It has taken its toll on me and my family. It is really difficult to deal with. I cry a lot. I'm stubborn about it; I refuse to be cooped up in my home. But I'm forced to use my oxygen unit in order to be halfway healthy and to not suffer from tremendous arthritic pains and mental confusion."

Moll's case illustrates two of the main dividing points between conventional allergists and clinical ecologists: Conventional allergists believe that if Moll's blood does not show an elevated level of Immunoglobulin-E, then she does not have an IgE mediated-allergy and is not sick. Clinical ecologists observe Moll's symptoms and note that she is plainly sick. They say the fact that she does not have an IgE-mediated classic allergy is irrelevant.

Conventional allergists say that the fact that she was around paint fumes for years with no ill effect shows that she is not susceptible to petrochemicals. Clinical ecologists reverse that position: They say that Moll's long-term exposure to petrochemical fumes laid the groundwork for the reaction to the overdose of cleaning fluid.

Dr. Sinaiko believes that his views are reasonable: "I'm a very conventional allergist for the most part. I don't do sublingual testing or food allergy testing. But I also know that chemical exposures are really important in causing people a lot of serious illnesses. Also, I see food allergy as a major cause of a wide variety of symptoms, including headaches, muscle aches, fatigue-ability, loss of the enjoyment of life. I've seen people get better when you get them on the right diet.

"Because the conventional allergists are complacently ignorant, I can't throw my lot in with them at all. The people that I consider interesting in my field are the clinical ecologists."

Tom Ohlhausen grew up in west Texas, in the Buffalo Gap area south of Abilene. He says he has spent much of his life exposed to hydrocarbon gases, heavy metals, formaldehyde, asbestos, asphalt, diesel fumes and other threatening substances. In 1979 he moved to Marin and went to work for a paving company. He says that after several bouts with acute systemic poisoning, he finally succumbed in March of 1981.

The doctors he first encountered didn't share Ohlhausen's view of the problem and, he says, he let them know about it: "I said, 'Look, doctor, I'm going fast and you people are swarming all over me. I can't stand all this exposure. I'd be better off out on the grass in front of your place. I know I sound drunk, I know I sound like a dope addict, I know I sound like everything else; but I'm not. I'm a poison victim and I want to be treated as such.' Then they went and wrote it up like it was an emotional reaction; the typical psychiatric discredit."

Ohlhausen, 47, now stays shut up in his home, attended by a housekeeper. He is feisty, to put it mildly, and for two years he has fought running battles with various federal and state agencies for benefits, which he appears to have won, and for payments for oxygen, which seem endangered. Ohlhausen's physician now is Dr. Jeffrey Anderson of Mill Valley, who for years had a family practice with an emphasis on nutritional medicine, then switched to clinical ecology: "I developed more and more of an interest in immunology," says Anderson. "Clinical ecology seemed to be where most of the interesting new things were being done. Is Tom making all this up? No, he's got too much physical data to support the fact that he suffers from hypersensitivity. But I won't deny that he's very aggressive and opinionated."

Ruth Noe grew up in Arizona, where she says she led an active life with no history of mental or physical illness. She says her mother lives in Mill Valley and still drives a car at age 87. In 1962 Noe moved West and in 1965 began teaching in the San Rafael city schools. In 1974 she became the kindergarten teacher at Short School near Albert Park.

It was not, she says, a healthy environment: "It was a modern 'pod' school with windows that didn't open; the air-conditioning system was often in need of repair, the gas heater in my pod leaked fumes and the mimeograph equipment was kept in the teachers' lounge, which was not ventilated. But the worst part was the pesticide sprayings; once a month a service came and sprayed for rodents, inside and out. The residue collected on the rugs and when you teach kindergarten you are down on the rug a lot."

Noe, who is being treated by Dr. McGovern, says that of the seven staff members at the school during this period, five have become ill. (One of them is also seeing Dr. McGovern.) Her collapse, she says, came in January of 1978. Conventional allergists said that her illness was mental and so she obligingly entered therapy. Much later her psychiatrist, Dr. Seymour Boorstein of Kentfield, suggested that she go to Oakland to see Dr. McGovern, who in 1975 had bought one of the larger allergy practices in the West.

Says Dr. Boorstein, "Ruth has emotional problems but her primary disease is of a chemical sensitivity nature. She used to be a very fine school teacher; very highly thought of. We see more and more of this sort of thing. For example, many of the airtight new buildings are causing people to have headaches. I got interested in this because for 18 years I was a consultant at the Sunny Hills home for disturbed youngsters and also for the state Department of Corrections at Vacaville. By looking at the data I became convinced that a lot of the learning disabilities — dyslexia, hyperactivity, minimal brain dysfunction — are sensitivities to various substances. The ramifications are staggering."

After becoming a patient of Dr. McGovern, Noe journeyed to the Environmental Care Unit of Carrollton Hospital in Dallas, which is run by Dr. William Rea. After a period of "de-toxification," her condition improved. Then, some months ago, it began to deteriorate. She now has isolated herself from all outside contact, drinks special water and says she sleeps on a cot in her bathroom because the porcelain walls exude fewer chemicals than the walls in the rest of the house.

How does she feel when conventional allergists insist that her problems are mental? "I feel saddened. I feel that physicians who are healers haven't the privilege to be narrowminded. There should be a respect for the fact that no physician can have all the knowledge, and new knowledge is coming to the fore all the time. People's lives and well-being are at stake. It is very important that the medical community be open-minded."

Is she crazy?

Neither Dr. Boorstein nor Dr. McGovern thinks so. Nor do Robert Vandre and the Mormon Church. Noe is a member of the Church of Jesus Christ of Latter Day Saints (LDS). She appealed to it for help in financing a second trip to Dr. Rea's Environmental Control Unit in Texas. Vandre, a dentist at Letterman Hospital in San Francisco, is the bishop of Noe's LDS "stake" (jurisdiction) in San Rafael. A stay at Dr. Rea's hospital costs $3000 per week and Noe wants to stay more than three weeks.

"That's an awful lot of money," says Vandre. "I had to refer the question to Salt Lake City. We wanted to make sure that Ruth's problems weren't psychosomatic. The treatment might cost the same but it would be a quite different treatment. The case was handled by the brother who deals with catastrophic illnesses. He took all the documents in her case — it was a stack about four inches thick — and had them reviewed by some old-school physicians. They said she was being taken for a ride by a bunch of charlatans. Then we looked into it further and realized this wasn't the case at all. It was a fight between the old-school allergists and some new ones who haven't published in Annals of Allergy. Now we're behind Ruth. She's a very sharp woman. It's just a sad thing to be held a prisoner in your own home, especially during these beautiful days we've been having."

The practice of allergy is said by most physicians to have begun one day in 1911 when an English physician and hay fever sufferer, Dr. L. Noon, had an inspired hunch. He took a pinch of what he suspected was the offending grass pollen, diluted it greatly and injected it into himself. His hay fever symptoms cleared up.

It is an irony of the struggle between rival allergists over scientific credentials that in the 72 years since Dr. Noon's shot, little has been learned about why it worked then and works now. The basis of conventional allergy — anti-allergy shots consisting of an increasingly strong solution of the offending substance — is both quite effective and quite unscientific. It has the disquieting echo — disquieting to the allergists — of homeopathy.

However — and this is an enormously important "however" — there has been a parallel development in allergy which is scientific. In the 1920s it was determined that there was a substance in the blood of allergy victims which signaled their condition. The field was redefined in terms of an immunological mechanism. The mysterious substance was called Reagin. Its presence could be confirmed by transferring blood from an allergy victim to a nonvictim, then getting a reaction from the newly created "victim."

As the years ticked by, allergy and allergists suffered from a low ranking in the medical pecking order. It was a field in search of a science. Brain surgeons might send their sneezing wives to an allergist, but the field was thought to be decidedly unscientific and rather flaky. Then, in 1967, the sun burst through the clouds. Teruko and Kimishige Isizaka, a husband and wife team at Yale Medical School with M.D. and Ph.D. degrees, identified the chemical composition of Reagin and named it Immunoglobulin-E (IgE). Suddenly allergy was a medical science. Its practitioners could hold up their heads at medical society meetings and country club dances. In following years more Ig factors were added: A, D, G and N.

There remained, however, an embarrassing fact: anti-allergy shots continued to work and IgE experiments continued to flourish but the two refused to be tied closely together. For example, shots would deal with an allergy victim's symptoms rather quickly, but the IgE level would stay elevated for a year or more. Some sufferers from asthma — one of the worst allergies — had mysteriously normal IgE levels. It was observed that allergy had embraced a beautiful piece of science which was not entirely relevant to the prescribed treatment.

Dr. Sinaiko says, "They've chosen this one very easy thing to do and they do it day in and day out. But they don't confront the hard questions. If the patient doesn't satisfy their definition of allergy, then they throw the patient out of the office and say they don't have an allergy: 'It must be in your head because I can't explain it with my little IgE mechanism.'"

Thirty years ago, into this field delicately balanced between science and guesswork, walked a physician named Theron Randolph. He had been trained at Massachusetts General Hospital, the Marine Corps of training hospitals, and was a respected, board-certified allergist on the faculty of Northwestern University Medical School in Chicago. Dr. Randolph kept careful patient records. He noticed that one of his allergy patients, who was also a schizophrenic, seemed to get worse after eating sugar. He had her sugar supply analyzed and determined that it was beet sugar. Then he took a daring step, one from which the field of allergy has never recovered. He confined her to bed in the hospital, set up a movie camera, turned it on and fed her through a tube a solution of beet sugar.

She went crazy and stayed crazy for four days. Then, abruptly, she snapped out of it.

"It was the first motion picture tying a reaction of the brain to foods," Dr. Randolph recalls from his Chicago office. "I showed it to the medical school staff. It didn't fit into their way of viewing things. I was charged more or less officially with having faked the picture. I was 'excommunicated' from the medical school staff in 1950 and denied privileges at Wesley Memorial Hospital the following year.

"The allergists are stuck in a trap of their own making. There are numerous mechanisms in allergy. It's ridiculous to limit the concept of hypersensitivity to the one mechanism of IgE. They're trying to make it an exclusive practice. They won't give up. Why? Because they're blockheads."

Thus, with Dr. Theron Randolph and his beet sugar movie, began the field of clinical ecology. Today it and conventional allergy are divided in every way imaginable. Here are four ways:

Methods

When you visit a conventional allergist you will be tested by having perhaps 100 drops of different substances in solution put on your skin. The drops will be scratched into the skin and an identification made of the bumps, or wheals, which appear. For treatment you will return to the allergist's office twice a week for shots of increasingly strong dilutions of the offending substances. In three to four years some 75 percent of your fellow patients will be substantially better, many of them symptom free, by a mechanism nobody clearly understands.

If you visit a clinical ecologist you will be tested not with a scratch test of 100 drops, but with either an intradermal (into the skin) or sublingual (under the tongue) test of fewer than ten substances, one at a time. It might be found that a No. 6 dilution of acacia tree pollen provokes the symptoms of asthma. Then stronger or weaker doses are given until the symptoms disappear. Nobody knows why this works, any more than they know why the shots of the conventional allergist work. The clinical ecologist will then send the patient home with a vial of the treatment dose of tree pollen solution and advise that drops be placed under the tongue twice a week. Clinical ecologists call this technique "provocative (first shot) and neutralization (second shot) therapy."

Science

In the area of published double-blind scientific studies, conventional allergists are miles ahead of the clinical ecologists, and they make much of it. Marin allergist Dr. William Sawyer says flatly, "There is no true scientific research showing the validity of what most clinical ecologists are talking about. There are people in clinical ecology whom I respect as persons for their sincerity, but they find results that aren't there. It's almost like a religion to these people."

Clinical ecologists offer several explanations for the scarcity of double-blind tests supporting their technique. First, they say, it is a young field and they have been too busy making people feel better to do much scientific research. Second, clinical ecologists say that because they believe in the cumulative effect of many environmental factors, their studies are more difficult to design than those of the single-cause, single-symptom conventional allergists. Third, they say that when clinical ecology studies are done they are invariably turned down for publication by the two major allergy journals controlled by conventional allergists: Annals of Allergy and the Journal of Allergy and Clinical Immunology.

Dr. Sawyer, a conventional allergist, was asked about a double-blind study on sublingual testing done by David King of Langley Porter Psychiatric Institute. Dr. Sawyer dismissed it, unseen, as probably flawed.

Dr. Randolph, who has published some 300 papers, says he has now given up on conventional allergy publications: "I decided that it was impossible to change academia and so I decided to go public. I write for the public and virtually all the writings for the public are from our group. The subject is going around the traditional allergists and they are too stuffy to do anything about it."

This may look like a harmless rock-throwing session, but for somebody caught in the middle it can be painful. Tom Ohlhausen received payments for oxygen from Medi-Cal because of what is now said to be a mistake. In order to determine finally whether Ohlhausen is entitled to oxygen, Medi-Cal wants him examined by an expert of their choice: Dr. Sawyer. Dr. Sawyer says he is a special examiner for Workers' Compensation cases and does a good deal of expert opinion work. He says he doesn't prejudge a case but merely looks at the facts and makes an independent judgment.

Doesn't Dr. Sawyer generally find that the position of clinical ecologists is in error? "Yes, I generally do," he says. Has he ever given an opinion that favored the position of clinical ecology? "I find that question offensive," replies Dr. Sawyer. "I'm not going to answer it."

Dr. Anderson, Ohlhausen's clinical ecologist, says that asking a conventional allergist for a second opinion on environmental illness is "like asking a proctologist to do brain surgery. They don't belong there. They're not qualified to comment on it. All they're qualified to do is put on a skin test and say, 'Yes, his skin is allergic to mold' or 'No, it isn't.' "

Motivation

Conventional allergists say their bitter resistance to the inroads of clinical ecology stems from concern for the health of patients and devotion to scientific truth. Dr. Abba Terr, with his Stanford position, flourishing practice at 450 Sutter and heavy schedule as an expert witness, is acknowledged by all to be an exceptionally shrewd medical politician and forceful spokesperson for conventional allergy. He has said:

"I am confident that examination of patients diagnosed as having 'cerebral allergy' etcetera by a competent allergist, internist, psychiatrist or other appropriate medical specialist would reveal that the procedures, diagnoses and treatments recommended by the clinical ecologist have no standing whatsoever in modern-day medicine."

Clinical ecologists say that they are motivated by developments in an exciting and important new field of medicine. They charge that their opponents are motivated by a desire to protect their lucrative practices and expert witness fees. Dr. Alan Levin was a nationally noted immuno-pathologist in the cancer field before switching to clinical ecology.

"I got into clinical ecology because it made scientific sense to me based on my knowledge of immunology," says Levin. "Then I found out there was a controversy. That was shocking. So I wondered, 'What's going on here?' I looked at the conventional allergist's science and it was bad science — a seven-year-old abstract, a 12-year-old letter to the editor, a study for which the data was so manipulated that the statistician quit. So you wonder why these so-called scientists are basing such weighty hypotheses on such poor science? What's the advantage of knocking out these poor people in clinical ecology who say that food can cause problems and that sometimes drugs can cause problems?

"You begin to see that it is the drug industry that has a vested interest. The American Academy of Allergy [AAA] is working for the drug industry to wipe out anybody who recognizes that food additives are a problem, anybody who does not use chemotherapy. The AAA's life breath depends on funding from the drug industry. So they phony up all these studies, and believe me they are phony. This is my profession and I am insulted."

Numbers

Dr. Abba Terr heads an advisory panel of the California Medical Association in the field of allergy. The panel consists only of conventional allergists. Dr. Sawyer was asked if the panel might profit by including a variety of views, inasmuch as the field is so divided. "It isn't all that divided," he replied. "The number of people on this clinical ecology is very, very small. Not even 10 percent of the total; maybe 2 or 3 percent. They are a small but very vocal group; they are prone to preach their sermons."

Dr. Randolph has a different count, and so, apparently, does Dr. Abba Terr, who is concerned that "their impact seems to be increasing." Says Dr. Randolph, "There are 500 dues-paying members of the Society for Clinical Ecology; there are 900 ear, nose and throat physicians who use the techniques of clinical ecology; there are 300 members of the Pan-American Allergy Society plus another 200 physicians handling problems in this field. I'd say there are nearly 2000 of us as opposed to 3000 to 3500 conventional allergists. When there were only a few of us we were treated as gadflys. Now that we are 40 percent of the total we are perceived as a real threat and dealt with accordingly."

The Society for Clinical Ecology and the Pan-American Allergy Society have retained San Francisco attorney Richard Spohn, director of the California Department of Consumer Affairs under Governor Jerry Brown. Members of the two groups point to a 1982 U.S. Supreme Court decision concerning a struggle between psychiatrists and psychologists over qualifications. The court ruled that if agreement between competing medical practitioners were to be required, then nothing would ever get done.

Spohn is preparing for a lawsuit and is also assembling a massive presentation of the case for clinical ecology. It will have three parts: the legal case done by Spohn, the scientific case done by Dr. Alan Levin and Dr. Iris Bell, and the case for cost effectiveness done by a medical economist. "We will have a document," says Spohn, "that will say, 'Here's what we do, here's why it works and here's why it saves you guys money and, by the way, here are some legal problems you are going to have if you persist in ignoring this data and acting in collusion with our competitors.'

Appendix A — Environmental Illness

"Why is it that the CMA has refused to sit down and dialogue with the clinical ecologists?" asks Spohn. "It seems like a hell of a way to conduct an honest, open, intellectual inquiry. It's the antithesis of science. One of the things that really upsets me is the lack of interest by the antagonists of clinical ecology in really having a look at the data. They just say, 'Well, it was published in journals that we don't think are the best.' Just a lot of back-of-the-hand stuff. If they are really interested in healing people rather than in protecting their own hegemony, then why are they unwilling to do that?"

In the end we come to the same two questions with which we began:

1. Are people who claim to suffer from environmental illness really ill, or are they crazy?

2. Is the problem restricted to a few unfortunates, or does it have ominous overtones for many of us?

Steve Lerner is research director of Commonweal in Bolinas, a highly regarded research and treatment facility dealing with the effects of food and chemicals on behavior. Says Lerner, "Are they sick or crazy? The answer is both; some are sick and some are crazy. Those who are crazy may have been that way before and they may have been made worse by their illness. A lot of people who recover will talk about the earlier time when they were very befuddled."

Dr. Anderson says, "It's common for environmental illness patients to be in an uncomfortable psychological set. They're disoriented, spacy, isolated, frustrated and angry. The stress increases immune disregulation. It's a snowball effect, a vicious circle. But the fact remains that most of the people I see are not hypochondriacs and not malingerers."

Says another close observer: "I've been heavily involved with clinical ecology research and I really do believe that these people are on to something important, although some of their claims are excessive. It is also an important fact that clinical ecology patients are in many instances really very interested in their own ailments. They will often talk about them for long periods of time. For many of them it's an important part of their identity. They have found a label which allows them to have a fairly positive self-image. Also, they are very intensive users of health care, much more so than people who often have more severe illnesses, such as lupus, heart disease or cancer. There are clinical ecology patients who are intelligent, reasonable people who happen to have a sensitivity, and there are others who tend to talk a great deal about their illnesses and who can be extraordinarily difficult patients."

And the larger question: the effect of a changing environment on all of us? Says Michael Lerner, president of Commonweal:

"That is an issue that has absolutely fascinated me for ten years. About five years ago I looked at the issue and thought that America was experiencing what I called bio-social decline. There was an increase in the total stress load on the population, things that were never measured together, such as the increase in chronic illness, the increase in suicides among young adults, the increase in birth defects and handicaps among children. It appeared that a form of cultural disintegration was taking place. It has biological as well as social roots. It wasn't just permissive education or violence on television that was causing it. You needed to look at the total environmental stress load: the effect of cigarette smoking, of dioxin, of a high-fat diet and so forth.

"But I've changed my mind. I realized that you have to take into consideration the fact that in some ways the environment is getting healthier. Good food, exercise, avoiding stress, people rediscovering true spiritual values —all these are very positive reponses. So what I talk about now is bio-social transformation.

"That's where the clinical ecology patients are telling us something. They are stressed out, unhappy, sick individuals. They are responding to the increase in key environmental stresses — to the nuclear revolution, the petrochemical revolution, the introduction of television into every home, changes in the American diet, enormous crowding, depressed times economically.

"When all these interact, these are the people in the population who are going to be in trouble first. Next, all of us could begin to function at a lower level than we have every right to expect. But there are also these positive responses at work. There are a lot of very healthy people around; a lot of very healthy family units; a lot of people who say quietly, 'We're going to do it, here.'

"I like to quote Yogi Berra; 'It's not over till it's over.' We have a good shot at pulling it off in a positive way."

Appendix B

Training Fellowships in Clinical Ecology for Physicians
Training Fellowships in Clinical Ecology for Nurses

RECIPIENTS OF TRAINING FELLOWSHIPS IN CLINICAL ECOLOGY FOR PHYSICIANS

John P. Rollins, M.D.
Clyde K. Walter, M.D.
Ryuzo Tomidokoro, M.D.
Kendall A. Gerdes, M.D.
Stephen N. Wilson, M.D.
L. Neil Enkema, M.D.
Roger L. Fife, M.D.
David S. Buscher, M.D.
William M. Orr, M.D.
Bhaskar D. Power, M.D.
Karl E. Humiston, M.D.
Robert T. Marshall, M.D.
George F. Kroker, M.D.

Gary Campbell, D.O.
Robert C. Filice, M.D.
Richard S. Wilkinson, M.D.
Richard B. Lyman, M.D.
Edna B. Pretila, M.D.
Laurine P. Ludwig, M.D.
George Juetersonke, D.O.
Jeanette S. Schoonmaker, M.D.
Donald J. Boon, M.D.
Paul E. Dart, M.D.
John W. Crayton, M.D.
Myrna T. Maniulit, M.D.
John L. Bowman, M.D.

RECIPIENTS OF TRAINING FELLOWSHIPS IN CLINICAL ECOLOGY FOR NURSES

Marilyn Kruse, R.N.
Irmgard Dix, R.N.
Ilene K. Buchholz, R.N.
Roberto Guillet, R.N.
Renee Sannes Puza, R.N.
Helen Lewis, R.N.

Appendix C

*Comments on Position Paper on Clinical Ecology
by the American Adademy of Allergy and Immunology
by Doris J. Rapp, M.D.*

Comments on the Position Paper on Clinical Ecology by the American Academy of Allergy and Immunology

By Doris J. Rapp, M.D.
September 15, 1985

Leonard Bernstein, M.D.
Chairman, Practice Standards
Committee AAAI
611 Wells Street
Milwaukee, WI 53202

Dear Leonard:

You asked for comments on your Position Paper on Clinical Ecology. As a board-certified specialist in allergy who practiced traditional allergy for 18 years, and clinical ecology for the past 10 years, I feel impelled to respond to your request.

Your *background* review is excellent and of the caliber I would expect from what I recall about you. My only objection is that you tend to refer to the "tip of the iceberg"-type person, and not the type who are seen routinely in my and most other offices where this type of medicine is practiced. The changes in the home environment that are now recommended are a bit different from what I routinely suggested during the first 18 years. We now simply are aware of more home factors that can cause illness. The diets are somewhat restrictive, but in many ways they are more varied and sometimes nutritionally better.

The *comments* in your paper, however, fall far short of what I would have expected from the aspect of clarity, fairness and fact. The newer applications and variations of the basic tenets of allergy require more time to confirm, but much evidence of efficacy is already in print. More is presently awaiting publication.

The published positive data of the ecologists is strongly suggestive that these newer approaches are most helpful.[1-13] The published negative review articles about our studies, unfortunately, consistently fail to review *most* of the positive articles which were in print at the time of the review publications.[14-17] When one looks critically at the so-called negative articles which disprove ecologic methods, any academic scientist would have to admit that the latter articles leave a lot to be desired, from many aspects, research methodology not withstanding.[18-26]

Let me give you only one example, but be assured, many more are available. Let us compare the study of Dr. David King which is rarely quoted, versus the articles published by Food Allergy Committee which are so widely quoted.[19,20]

King's (1981) article[1] is double-blind, is internally and externally controlled, and shows the efficacy of sublingual testing with stock allergy extracts in 30 patients. The study is valid, reliable, and includes repetitive pre and post base rate trials, pre and post-screening trials with placebos, multiple randomized challenges with 18 placebos and antigens, and agreement checks for reliability of inter-rater interpretations. Sound statistical analysis provided a basis for the conclusions.

After studying it, please give me the name of any article, by any traditional allergist which compares to it in the quality of the research.

Let us now discuss a widely quoted paper: The final 1973 Food Allergy Committee Report of the American College of Allergists, included two *experienced and seven "purposely" inexperienced* investigators.[19] The patients had suspected but *unproven allergies* at the time of the study. Fifty percent of the data from the 1973 study was discarded "because the design of the study was faulty and gave an inordinate number of positive (i.e., favorable) responses."[27] The statistician for the study referred to the "faulty protocol" and stated that "because of the design of the study, one must be careful about drawing conclusions."[28] In spite of this admonition, sublingual testing was labelled as neither reliable nor sensitive in determining food allergy in this article.

The 1974 Food Allergy Committee Report failed to correct several of the admitted limitations in the 1973 study.[20] Part of the conclusions of the 1974 study, however, were based on the admittedly flawed 1973 report.[19] In both studies there were flagrant inconsistencies in data found in the tables versus that in the text of the article, confounding variable, and obvious errors in design and execution.

So far there are 4 single-blind[29-32] and 8 double-blind diet studies.[33-40] There are 9 double-blind studies related to provocation-neutralization.[1-9] There are double-blinded studies which indicate that some patients with colitis,[41] hyperactivity,[2-7,9,33-38,40,42,43,45,51] asthma,[11,44] migraine,[48,49] epilepsy,[46] muscle ache,[2,31,47] cardiac problems,[13] arthralgia,[7] depression,[1,31,47] and conjunctivitis[4] can have food-related problems.

While the ecologists' studies have some weaknesses, such as some being only single well-conducted case reports,[3,4,6,32] the allergists often quote abstracts[24] or committee reports, never fully in print,[18,20,25] or letters to the editor.[18] Articles by Rippere[54] and Weiss[55] are rarely quoted. Is it because they negate the conclusions of key accepted articles that hyperactivity is unrelated to food coloring?

New, but as yet unpublished, studies by Boris, Crayton, Eggers and Brostoff are completed and clearly indicate that scientific studies do document the efficacy of these newer approaches, not only for foods, but for inhalants. Why did you accept the initial article by Boris,[11] and 18 months later reject it? What did you have to gain? What did the patients have to lose? His research has been replicated and expanded since his initial investigation. Is there nothing that a traditional allergist can learn from such quality research?

All of this evidence cannot be called anecdotal.

In addition, I personally have innumerable edited and unedited movies and video tapes, blinded, in infants and toddlers, clearly documenting what appears to be "cause and effect relationships." We have, also, demonstrated EEG changes, and changes in immunological parameters, during testing. At the very least, data such as this raises further research questions.

Your statements in relation to life style/restrictions are totally unfair and inaccurate. You are *not* describing the "average" patient that is seen in our offices. You are describing relatively a few patients who are severely incapacitated by illness which is totally, or in part, ecologically-related. The 12,000 letters which were sent to the FDA should be required reading by the disbelieving allergists. Perhaps they would begin to realize that what we are doing is helping a desperate group of patients who can be helped.

At this point in time, based on the existent research, it is ludicrous that you state that what ecologists do is not allergy. I am doing the same things, for example, that I did for the first 18 years, but much better. I use the same extracts to test and treat. What I do, however, requires much more time, and the overhead is discouragingly increased. But, the rewards are that patients, not helped by others, or previously not helped by myself, often get well quickly.

At this point it is almost ridiculous that traditional allergists continue to say that our work is unproven. It is controversial only because the decision-making leaders in allergy appear much more interested in drug therapy, drug sponsorship, and their old ways rather than in patients. Unfortunately, I have too many files of patients seen by well-trained, often prominent allergists, who did what I did for the first 18 years and were unable to determine why the patients had the typical and the atypical allergy symptoms, and how to help them. We must learn from each other before the field of allergy is absorbed by the other specialists who are recognizing the validity of these newer methods.

The statement is repeatedly made that what we are doing is not allergy.
- These patients have positive personal histories of classical allergy.
- These patients have positive family histories of allergy.
- These patients have symptoms from eating certain foods, or after specific exposures, which are relieved by avoidance.

- These patients often can have their specific symptoms precipitated by single-blind testing, with single antigens in 1:5 concentrations. (Have you personally tried it? If not, ask yourself seriously during a quiet reflective time, why not?)
- These patients' symptoms subside when they receive the correct dilution of specific antigens, initially, and on a long-term basis.
- Eventually many of these patients can stop their individualized allergy extract made from standard stock supply house antigens, and they can ingest the previously problematic foods, and be exposed to inhalants, pollens, molds, or epithelials which previously caused symptoms.
- They often can stop using regular allergy medications and merely use subcutaneous or sublingual therapy until that time. Of course, every patient is not equally responsive. However, if it did not help, the patients would simply not buy more extract or return to us when symptoms recur.

Think about the above and consider if allergy is not at least part of the problem, and what research can be designed to further prove and refine these methods that have been shown to help so many, so quickly.

Think about the credibility of allergists who have adamantly stated we are wrong. How does a mother feel after she *sees* the symptoms her child has complained about for years, blindly reproduced, and then eliminated in an ecologist's office because we attempted to confirm or negate her anecdotal observations. This happens on a daily basis in *many* of our offices. For your patients' sakes, spend a week with one of us. Find out what really goes on. There is a whole new world of help which you are denying your patients on the basis of hearsay.

Yes, we do need more studies to help explain some of the modalities we use for therapy, but this is true with any advancement. That takes time. More immunologic studies are needed, but the large number of studies ALREADY IN PRINT does more than suggest these methods help.[53,56-69]

Try not to wedge the academy in a corner with statements that will haunt allergy in the years to come. If the thinking, leading allergists do not listen, and soon, the whole specialty of allergy will be lost. Use your mighty caches of money, and brains to help elucidate our impressive observations and successes. Help us refine what we are doing. Not only will you gain, but the patients, who should be the bottom line of whatever we discuss, will be helped.

I know you are a caring concerned physician. Please think about what I have written. Call if you have questions. 716-875-5578

Sincerely,
Doris J. Rapp, M.D., FAAA, FAAP

1. King DS. Can allergic exposure provoke psychological symptoms? A double-blind test. Biological Psychiatry 16:3, 1981.
2. Miller JB. A double-blind study of food extract injection therapy: A preliminary report. Annals of Allergy 38:185, 1977.
3. Rapp DJ. Double-blind confirmation and treatment of milk sensitivity. Medical Journal of Australia 1:571, 1978.
4. Rapp DJ. Weeping eyes in wheat allergy. Transactions of the American Society of Ophthalmology and Otolaryngology 18:149, 1978.
5. Rapp DJ. Food allergy treatment for hyperkinesis. Journal of Learning Disabilities 12:42, 1979.
6. Rapp DJ. Possible new way to treat herpes progenitalis. NYS Journal of Medicine 78:693, 1978. (Letter to the Editor)
7. Mandell M, Conte A. The role of allergy in arthritis, rheumatism, and polysymptomatic cerebral, visceral, and somatic disorders: A double-blind study. Journal of the International Academy of Preventive Medicine, July 5-6, 1982.
8. Monro J. Food Allergy in migraine. Proc Nutr Society 42:241, 1983.
9. O'Shea JA, Porter SF. Double-blind study of children with hyperkinetic syndrome treated with multi-allergen extract sublingually. Journal of Learning Disabilities 14:189, 1981.
10. Green M. Sublingual provocation testing for foods and F, D, & C dyes. Annals of Allergy 33:274, 1974.
11. Boris M, et al. Bronchoprovocation blocked by neutralization therapy. Journal of Allergy and Immunology 71:92, 1983. (Abstract)
12. Boris M. Other studies completed but not yet in print.
13. Rea WJ, Podell RN, Williams M, Fenyves I, Sprague DE, Johnson AR. Elimination of oral food challenge reaction by injection of food extracts. Archives of Otolaryngology 110:248, 1984.
14. Golbert TM. A review of controversial diagnostic and therapeutic techniques employed in allergy. Journal of Allergy and Clinical Immunology 56:170, 1975.
15. Reisman R. American Academy of Allergy: Position Statements — controversial techniques. Journal of Allergy and Clinical Immunology 67:333, 1981.
16. Grieco MH. Controversial practices in allergy. Journal of the American Medical Association 247:3106, 1982.
17. Van Metre TE. Critique of controversial and unproven procedures for diagnosis and therapy of allergic disorders. Pediatric Clinics of North America 30:807, 1983.
18. Kailin E, Collier R. "Relieving" therapy for antigen exposure. Journal of the American Medical Association 217:78, 1971. (Letter to the Editor)
19. Breneman JC, Crook WC, Deamer W, Exline L, Gerrard JW, Heiner D, Hurst A, Leney FL. Report of the Food Allergy Committee on the sublingual method of provocative testing for food allergy. Annals of Allergy 31:382, 1973.
20. Breneman JC, et al. Final report of the Food Allergy Committee of ACA on clinical evaluation of sublingual provocative testing method for diagnosis of food allergy. Annals of Allergy 33:164, 1974.
21. Lehman CW. A double-blind study of sublingual provocative food testing: A study of its efficacy. Annals of Allergy 45:144, 1980.
22. Health Care Technology Report: Summary of Assessments. Journal of the American Medical Association 246(14):1499, 1981.
23. Hirsch SR, et al. Rinkel injection therapy: A multicenter controlled study. Journal of Allergy and Clinical Immunology 68:133, 1981.
24. Bronsky EA, Burkley DP, Ellis EF. Evaluation of the provocative food skin test technique. (Abstract) Journal of Allergy 47:104, 1971.

25. Crawford LV, et al. A double-blind study of subcutaneous food testing sponsored by the Food Committee of the Amerian Academy of Allergy. Journal of Allergy and Clinical Immunology 57:236, 1976.
26. Draper WL. Food testing in allergy: Intradermal provocation versus deliberate feeding. Archives of Otolaryngology 95:169, 1972.
27. Breneman, p. 383, 1973. Reference #19.
28. Breneman, p. 385, 1973. Reference #19.
29. Crook WC, Harrison WE, Crawford SE, Emerson BS. Systemic manifestations due to allergy. Pediatrics 27:790, 1961.
30. Green, M. Sublingual provocation testing of foods and F, D, and C dyes. Annals of Allergy 33:274, 1974.
31. Finn R, Cohen HN. Food allergy: fact or fiction? Lancet 1:426, 1978.
32. O'Banion D, Armstrong B, Cummings RH. Disruptive behavior: a dietary approach. Journal of Autism and Childhood Schizophrenia 8:325, 1978.
33. Atherton DJ, Soothill JF, Sewell W, Wells RS. A double-blinded controlled crossover trial of an antigen avoidance diet in atopic eczema. Lancet 1:401, 1978.
34. Levy F, et al. Hyperkinesis and diet: A double-blind crossover trial with a tartrazine challenge. Medical Journal of Australia 1:61, 1978.
35. William IJ, Cram DM, Douglas M, Tausig FT, Webster E. Relative effects of drugs and diet on hyperactive behaviors: an experimental study. Pediatrics 61:811, 1978.
36. Trites RW, Tryphonas H, Ferguson B. Treatment of hyperactivity in a child with allergies to food. In Case Studies in Hyperactivity, ed. Marvin J. Fine. Springfield, IL. Charles C. Thomas, 1978. (In Press)
37. Rapp DJ. Does diet affect hyperactivity? Journal of Learning Disability 11:56, 1978b.
38. Rapp DJ. Food allergy treatment for hyperkinesis. Journal of Learning Disability 11:79, 1978c.
39. Rapp DJ. Weeping eyes in wheat allergy. Transactions of American Society of Ophthalmology and Otolaryngology 18:149, 1978d.
40. Shaywitz BA, Goldenring JR, Wool RS. Effects of chronic administration of food colorings on activity levels and cognitive performance in normal and hyperactive developing rat pups. Annals of Neurology 4:196, 1978.
41. Soothill JF, Jenkins HR, Pincott JR, Milla PG, Harris JT. Food allergy: The major cause of infantile colitis. Archives of Disease in Childhood 59:326, 1983.
42. Weiss B. Behavioral responses to food coloring. Science 207:1487, 1980.
43. Swanson JM. Food dyes in pair performance of hyperactive children on a laboratory learning test. Science 207:1485, 1980.
44. Selner JC, Gerdes KA. Bronchospasm following IV dextrose. Annals of Allergy 44:48, 1980.
45. Prinz RJ, et al. Dietary correlates of hyperactive behavior in children. Journal of Consult Clinical Psychology 48:760, 1980.
46. Crayton JW, et al. Epilepsy precipitated by food sensitivity: Report of a case with double-blind placebo-controlled assessment. Clinical Electroencephalography 12:192, 1981.
47. Radcliffe MJ, et al. Food allergy in polysymptomatic patients. The Practitioner 225:1651, 1981.
48. Monro J. Food allergy in migraine. Proc Nutr Soc 42:241, 1983.
49. Egger J, Carter CM, Wilson J, Turner MW, Soothill JF. Is migraine food allergy? A double-blind controlled trial of oligoantigenic diet treatment. Lancet 865, 1983.
50. Rea WJ, Podell RN, Williams M, Fenyves I, Sprague DE, Johnson AR. Elimination of oral food challenge reaction by injection of food extracts. Archives of Otolaryngology 110:248, 1984.

51. Egger J, Carter CM, Graham PJ, Gumley D, Soothill JF. Controlled trial of oligoantigenic treatment in the hyperkinetic syndrome. Lancet 1:540, 1985.
52. Gerrard JR. Just food intolerance. Lancet 2:413, 1984.
53. Sandberg DH, Bernstein CW, McIntosh RM, Carr R. Severe steroid-responsive nephrosis associated with hypersensitivity. The Lancet, February 19, pp. 388-391, 1977.
54. Rippere V. Food additives and hyperactive children: A critique of Conners. British Journal of Clinical Psychology 22:19, 1983.
55. Weiss B. Color me hyperactive. American Health, May/June pp. 68-73, 1982.
56. McGovern JJ, Lazaroni JA, Hicks MF, Adler JC, Cleary P. Food and chemical sensitivity: Clinical and immunologic correlates. Archives of Otolaryngology 109:292, 1983.
57. Brostoff J, Carini C, Wraith DG, Jones P. Production of IgE complexes by allergen challenge in atopic patients and the effect of sodium cromoglycate. Lancet, June 16, pp. 1268-1270, 1979.
58. Little CH, Stewart AG, Fennessy MR. Platelet serotonin release in rheumatoid arthritis: a study in food-intolerant patients. Lancet, August 6, p. 297, 1983.
59. McGovern JJ. Clinical evaluation of the major plasma and cellular measures of immunity. The Journal of Orthomolecular Psychiatry 12(1):60, 1983.
60. McGovern JJ, Gardner RW, Painter K, Rapp DJ. Natural foodborne aromatics induce behavioral disturbances in children with hyperkinesis. The International Journal for Biosocial Research 4(1):40, 1983.

Appendix D

Bibliographies of Clinical Ecologists

Arthur F. Coca, M.D.	*1875-1959*
Hal M. Davison, M.D.	*1891-1958*
William W. Duke, M.D.	*1883-1946*
French K. Hansel, M.D.	*1893-1981*
Francis W. E. Hare, M.D.	*1858-1928*
Herbert J. Rinkel, M.D.	*1896-1963*
Albert H. Rowe, M.D.	*1889-1970*
Warren T. Vaughan, M.D.	*1893-1944*
Michael Zeller, M.D.	*1900-1977*
Theron G. Randolph, M.D.	*1906-*

BIBLIOGRAPHY

of

ARTHUR FERNANDEZ COCA (1875-1959)

1. Coca AF. The significance of the "fibroglia" fibrils: an embryological study. U Penn Med Bull 19:90, 1906.
2. Coca AF. Die Bedeutung der Fibroglia-Fibrillen, eine embryologische Studie. Virchow's Arch. für pathol. Anat 186:297, 1906.
3. von Dungern E, Coca AF. Ueber Hämolyse durch Schlangengift. Münch med Wochenschrift 54:2317, 1907.
4. von Dungern E, Coca AF. Spezifische Hämolyse der durch Osmium fixierten Blutkörperchen. Berlin klin Wochenschrift 44:1471, 1907.
5. Coca AF. An improved method of staining fibroglia, myoglia, myofibrillae of striped muscle, etc. U Penn Med Bull 20:60, 1908.
6. von Dungern E, Coca AF. Ueber Hämolyse durch Schlangengift. Biochem Zeitschrift 12:417, 1908.
7. Coca AF. Beitrag zur Antikörperstehung. Biochem Zeitschrift 14:125, 1908.
8. von Dungern E, Coca AF. Ueber spezifische Hämolyse durch isotonische Salzlösungen. Münch med Wochenschr 55:14, 1908.
9. von Dungern E, Coca AF. Ueber Hämolyse durch Kombination von Oelsaure oder öelsaurem Natrium und Kobragift. Münch med Wochenschr 55:105, 1908.
10. Coca A.F. The cause of sudden death following the intravenous injection of the blood corpuscles of foreign species. U Penn Med Bull 243, 1909.
11. Coca AF. The bacterial substances in fibrin. Philippine J Science B 4:171, 1909.
12. Coca AF, Gilman PK. The specific treatment of carcinoma. Philippine J Science B 4:391, 1909.
13. Coca AF. Die Ursache des plötzlichen Todes bei intravenöser Injektion artfremder Blutkörperchen. Virchow's Arch Pathol Anat 196:92, 1909.

14. Coca AF. Ueber die von Beitzke und Neuberg angenommene synthetische Wirkung der Antifermente. Zeitschrift Immunitätsforsch, Jena (Orig) 2:1, 1909.
15. von Dungern E, Coca AF. Ueber Hasensarcoma die in Kaninchen wachsen und über des Wesen der Geschwulstimmunität. Zeitschrift Immunitätsforsch exp Therap, Jena 2:391, 1909.
16. Weil R, Coca AF. An experimental study of anti-anaphylaxis. Proc Soc Exper Biol & Med 9:114, 1911.
17. Coca AF. "Vaccination" in cancer. I. Vaccination in human cancer in the light of the experimental data upon normal tissue and tumor immunity. Zeitschrift Immunitätsforsch exp Therap, Jena (Orig) 13:324, 1911.
18. Coca AF, Dorrance GM, Labredo MG. "Vaccination" in cancer. II. A report of the results of the vaccination therapy as applied in 79 cases of human cancer. Zeitschrift Immunitätsforsch exp Therap 13:543, 1911.
19. Coca AF, Dorrance GM, Labredo MG. "Vaccination" in cancer: a report of the results of the vaccination therapy as applied in 79 cases of human cancer. Mulford Digest Phila 1:96, 1912.
20. Coca AF. The separation of protozoan species by means of immunity reactions. Zeitschrift Immunitätsforsch exp Therap 12:127, 1912.
21. Coca AF, L'Esperance ES. A modification of the technic of the Wassermann reaction. Zeitschrift Immunitätsforsch exp Therap 14:139, 1912.
22. Coca AF. The plurality of the toxic substances of snake venoms. Zeitschrift Immunitätsforsch exp Therap 12:134, 1912.
23. von Dungern E, Coca AF. Some remarks upon the publication of Preston Kyes entitled "Venom Hemolysis." J Infect Dis 10:57, 1912.
24. Weil R, Coca AF. The nature of anti-anaphylaxis. Zeitschrift Immunitätsforsch exp Therap (Orig) 17:141, 1913.
25. Coca AF, L'Esperance ES. A Modification of the technic of the Wassermann reaction. Arch Int Med 11:84, 1913.
26. Coca AF. The site of reaction in anaphylactic shock. Zeitschrift Immunitätsforsch exp Therap 20:123, 1914.
27. Coca AF, Cooke RA. Studies on the blood of two cases of paroxysmal hemoglobinuria, II. The multiplicity of complement. Zeitschrift Immunitätsforsch exp Therap 21:632, 1914.
28. Coca AF. Toxins and the side-chain theory. J Infect Dis 17:351, 1915.
29. Coca AF. A rapid and efficient method of producing hemolytic amboceptor against sheep corpuscles. J Infect Dis 17:361, 1915.
30. L'Esperance ES, Coca AF. Further experiences with the isolated organ lipoids as "antigen" in the Wassermann test. J Immunol 1:129, 1916.
31. Coca AF. A study of some diagnostic reactions for malignant tumors. J Cancer Res Balto 2:61, 1917.
32. Coca AF. Sero-diagnostic methods in cancer; reactions of Freund and of von Dungern. Proc Pan American Scient Congress Washington 10:520, 1917.
33. Cooke RA, Flood EP, Coca AF. Hay fever: the nature of the process and of the mechanism of the alleviating effect of specific treatment. J Immunol 2:217, 1917.
34. Coca AF. The examination of the blood preliminary to the operation of blood transfusion. J Immunol 3:93, 1918.
35. Coca AF. The perfusion experiment in the study of anaphylaxis. J Immunol 4:209, 1919.
36. Coca AF. The mechanism of the anaphylaxis reaction in the rabbit. J Immunol 4:219, 1919.
37. Coca AF. Hypersensitiveness, in *Tice's Practice of Medicine* 1:109, 1920.
38. Coca AF, Kosakai M. Studies in anaphylaxis. I. On the quantitative reaction of partially neutralized precipitin in vitro and in vivo. J Immunol 5:297, 1920.

39. Coca AF, Kosakai M. Studies in anaphylaxis. II. On antisensitization. J Immunol 5:310, 1920.
40. Coca AF, Kosakai M. Studies in anaphylaxis. III. Experiments with specific precipitates. J Immunol 5:316, 1920.
41. Coca AF, Kelly MF VI. A serological study of the bacillus of Pfeiffer. J Immunol 6:87, 1921.
42. Coca AF, Russell EF, Bauchman WH. The reaction of the rat to diphtheria toxin, with observations on the technic of the Roemer method of testing diphtheria toxin and antitoxin. J Immunol Baltimore 6:387, 1921.
43. Coca AF. Studies in specific hypersensitiveness. V. The preparation of fluid extracts and solutions for use in the diagnosis and treatment of the allergies, with notes on the collection of pollens. J Immunol Baltimore 7:163, 1922.
44. Coca AF. Studies on specific hypersensitiveness. VII. The age incidence of serum disease and of dermatitis venenata as compared with that of the natural allergies. J Immunol 7:193, 1922.
45. Coca AF, Diebert O, Menger EF. Studies in specific hypersensitiveness. VIII. On the relative susceptibility of the American Indian race and the white race to the allergies and to serum disease. J Immunol 7:201, 1922.
46. Coca AF, Cooke RA. On the clasification of the phenomena of hypersensitiveness. J Immunol 8:163, 1922.
47. Coca AF, Klein H. Hitherto unidentified pair of isoagglutination elements in human beings. J Immunol 8:477, 1923.
48. Coca AF, Diebert O. A study of the occurrence of the blood groups among the American Indians. J Immunol 8:487, 1923.
49. Coca AF, Grove EF. Studies in hypersensitiveness. XIII. A study of the atopic reagins. J Immunol 10:445, 1925.
50. Grove EF, Coca AF. Studies in specific hypersensitiveness. XV. On the nature of the atopens of pollens, house dust, horse dander, and green peas. J Immunol 10:471, 1925.
51. Coca AF, Milford EL. Studies in specific hypersensitiveness. XVII. The preparation of fluid extracts and solutions for use in the diagnosis and treatment of atopic conditions. J Immunol 10:55, 1925.
52. Coca AF. Anaphylactic and atopic oversensitiveness. Med Klinik Berlin 21:57, 1925.
53. Coca AF. Relation of atopic hypersensitiveness (hay fever, asthma) to anaphylaxis. A review of recent literature. Arch Pathol Lab Med 1:96, 1926.
54. Levine P, Coca AF. Studies in hypersensitiveness. XX. A quantitative study of the interaction of atopic reagin and atopen. J Immunol 11:411, 1926.
55. Levine P, Coca AF. Studies in hypersensitiveness. XXI. A quantitative study of the atopic reagin in hay fever. The relation of skin sensitivity to reagin content of serum. J Immunol 11:435, 1926.
56. Levine P, Coca AF. Studies in hypersensitiveness. XXII. On the nature of the alleviating effect of the specific treatment of atopic conditions. J Immunol 11:449, 1926.
57. Coca AF. Food poisoning, with special reference to food idiosyncrasy (hypersensitiveness). Dietary Admin & Therapy 5:35, 1927.
58. Coca AF. Studies in hypersensitiveness. XXIX. On the influence of heredity in atopy. J Lab Clin Med 12:1135, 1927.
59. Coca AF. Ueber die Beziehungen der Atopie (Asthma und Heufiebergruppe) zur Anaphylaxie. Centralblatt Bakt Parasitenk Infektionskrankh Erste Abteilung: Referate 88:515, 1928.
60. Coca AF, Milford EL. Studies in specific hypersensitiveness. XXX. Additional notes on the preparation of the fluid extracts and solutions for use in the diagnosis and treatment of atopic conditions. J Immunol 15:1, 1928.
61. Clarke JA, Donnally HH, Coca AF. Studies in specific hypersensitiveness. XXXII. On the influence of heredity in atopy. J Immunol 15:9, 1928.

62. Coca AF. The skin as a shock tissue. Bull NY Acad Med 2nd ser 5:223, 1929.
63. Coca AF. Grounds for etiologic classification of phenomena of specific sensitiveness. J Allergy 1:74, 1929.
64. Coca AF. Atopy in The Newer Knowledge of Bacteriology and Immunology, 1004-1005, Jordan EO, Falk IS, eds. U Chicago Press, 1929.
65. Coca AF. Specific sensitiveness as a cause of symptoms in disease. Essential hematuria and localized retinal edema as possibly allergic symptoms. Bull NY Acad Med 6:593, 1930.
66. Coca AF. On dialyzability of proteins. J Immunol 19:405, 1930.
67. Coca AF, Walzer M, Thommen AA. *Asthma and Hay Fever in Theory and Practice.* CC Thomas, Springfield, 1931.
68. Coca AF. Slide method of titrating blood grouping sera. J Lab Clin Med 16:405, 1931.
69. Zinsser H, Coca AF. Remarks concerning Landsteiner's discovery of isoagglutination and blood groups, with special reference to a paper by JA Kennedy. J Immunol 20:259, 1931.
70. Brown A, Milford EL, Coca AF. Studies in contact dermatitis. I. The nature and etiology of pollen dermatitis; preliminary report. J Immunol 20:393, 1931.
71. Brown A, Milford EL, Coca AF. Studies in contact dermatitis; nature and etiology of pollen dermatitis. J Allergy 2:301, 1931.
72. Coca AF. Principles of diagnosis and treatment of allergic diseases. J Amer Med Assn 97:1201, 1931.
73. Coca AF. Immunity in human beings to pneumococcus. Proc Soc Exper Biol & Med 29:1042, 1932.
74. Coca AF. Classification of allergic diseases, with reference to diagnosis and treatment. J Lab Clin Med 18:219, 1932.
75. Coca AF. On the plan of standardization of pollen extracts proposed by Cooke and Stull. J Allergy 4:354, 1933.
76. Coca AF. Critical review of investigations of allergic diseases. Ergeb Hyg Bakt Immunitätsforsch exp Therap 14:538, 1933.
77. Coca AF. New definition of Noon pollen unit. J Allergy 5:345, 1934.
78. Coca AF. Specific diagnosis and treatment of allergic diseases of the skin; present status. J Amer Med Assn 103:1275, 1934.
79. Coca AF. Study of pneumococcus toxins. J Immunol 30:1, 1936.
80. Coca AF. Milford EL. Chemical standardization of pollen extracts. J Allergy 7:337, 1936.
81. Townsend IM, Coca AF. On some practical consequences of influence of temperature upon isoagglutination. J Lab Clin Med 21:729, 1936.
82. Blank JM, Coca AF. Study of prophylactic action of extract of poison ivy in control of rhus dermatitis. J Allergy 7:552, 1936.
83. Coca AF, Little PA, Lyon BM, Voigt EF. Study of pneumococcic toxins and antitoxins in animals. J Immunol 32:271, 1937.
84. Strauss HW, Coca AF. Studies in experimental hypersensitiveness in the rhesus monkey. III. On the manner of development of the hypersensitiveness in contact dermatitis. J Immunol 33:215, 1937.
85. Coca AF. Selection of donors for blood transfusion, with special reference to preliminary blood tests and use of universal donor. Amer J Med Technol 4:28, 1938.
86. Coca AF. Tratamienta de algunas enfermedada alergicas. Día Méd 12:820, 1940.
87. Coca AF. Familial nonreaginic food allergy as predisposing cause of common cold. J Lab Clin Med 26:757, 1941.
88. Coca AF. Familial nonreaginic food allergy: its specific diagnosis and treatment. J Lab Clin Med 26:1878, 1941.

89. Coca AF. *Familial Nonreaginic Food Allergy.* CC Thomas, Springfield, IL, 1942 [1st edition]
90. Coca AF. Brief critical review of fundamental knowledge concerning allergic diseases. Ann Allergy 1:120, 1943.
91. Coca AF. Normal adult human pulse rate. Ann Allergy 1:212, 1943.
92. Coca AF. Art of interpreting pulse-diet record in familial nonreaginic food allergy. Ann Allergy 2:1, 1944.
93. Coca AF. Sympathectomy as aid in relief of familial nonreaginic food allergy. Ann Allergy 2:213, 1944.
94. Coca AF. Adjuvant effect of aerosol upon germicidal action of cadmium chloride. J Lab Clin Med 29:689, 1944.
95. Coca AF. Dermatologic manifestations of familial nonreaginic food allergy. Ann Allergy 3:101, 1945.
96. Coca AF. Successful therapy of dermatologic syndrome with L. casei factor (folic acid); preliminary report. Ann Allergy 3:443, 1945.
97. Wodehouse RP, Coca AF. Progress in allergy; pollen antigens; critical review of literature. Ann Allergy 4:58, 1946.
98. Coca AF. Sympathectomy as antiallergic measure. Ann Allergy 5:95, 1947.
99. Coca AF. Influence of idioblaptic cigarette sensitivity. Ann Allergy 5:458, 1947.
100. Coca AF. New method for detection of food allergy treatment. Polski tygodnik tek 2:593, 1947.
101. Coca AF. Hereditary nature of food-allergy: notes for study in human genetics. J Heredity 39:157, 1948.
102. Coca AF. Environmental excitants of idioblaptic allergy (inhalants). Ann Allergy 6:501, 1948.
103. Coca AF. Dust-Seal: its use in avoidance of "house dust" by dust sensitive persons. Ann Allergy 6:506, 1948.
104. Coca AF. Various kinds of eczematoid eruptions attributable to exposure to foods. J Invest Dermatol 13:17, 1949.
105. Coca AF. Idioblaptic diseases of alimentary tract. Rev Gastroenterol 16:786, 1949.
106. Coca AF. Antiallergic action of sympathectomy. Ann NY Acad Sciences 50:807, 1949.
107. Locke AP Jr, Coca AF. Idioblaptic allergy as implementing background factor in anterior poliomyelitis; exploratory study. Ann Allergy 8:26, 1950.
108. Coca AF. *Familial Nonreaginic Food Allergy.* CC Thomas, Springfield, IL. 1953. [3rd ed.]
109. Meyer MG, Johnston A, Coca AF. Is multiple sclerosis manifestation of idioblaptic allergy? Psychiat Quart 28:57, 1954.
110. Coca AF. Overweight and underweight as manifestations of idioblaptic allergy. Arch Immun ter dośw 3:91, 1955 (Warszawa)
111. Coca AF. *The Pulse Test.* New York: Arc Book, Inc., 1956.
112. Arthur Fernandez Coca, 1875-1959. J Immunol 85:330, 1960.

BIBLIOGRAPHY

of

HAL McCLUNY DAVISON (1891-1958)

1. Davison HM. A case of asthma caused by sensitiveness to dog hair. J Med A Georgia 11:68, 1922.
2. Davison HM. Typhus in Siberia; observations upon myself and others. J Med A Georgia 11:258, 1922.
3. Davison HM. The diagnosis and treatment of non-seasonal hay fever. J Med A Georgia 14:21, 1925.
4. Davison TC, Davison HM. Basal metabolic rate in toxic goitre. J Med A Georgia 15:424, 1926.
5. Davison HM, Wood RH, Paullin JE. Incidence of syphilitic infection among negroes in South, its influence in causation of disability, and methods which are being used to combat this infection. Boston M & S J 197:345, 1927.
6. Davison HM, Thoroughman JC. Study of heart disease in negro race. South M J 21:464, 1928.
7. Davison HM. Asthma in children; diagnosis and treatment; report of 72 cases. J M A Georgia 18:122, 1929.
8. Davison HM, Lowance MI. Allergy; its relation to general medicine. J South Carolina M A 26:88, 1930.
9. Davison HM, Lowance MI. Neuroses and their relation to internal medicine. South Med & Surg 92:715, 1930.
10. Davison HM, Lowance MI, Durhan OC. Principal hay fever plants and pollens of the South. South M J 27:529, 1934.
11. Davison HM, Lowance MI, Barnett CF. Hyperpyrexia; evaluation of its use in office practice. M Rec 143:253, 1936.

12. Davison HM, Lowance MI, Barnett CF. Chronic arthritis and fibrositis. J M A Georgia 25:427, 1936.
13. Davison HM, Lowance MI, Barnett CF. Allergy in general medicine. Journal-Lancet 57:102, 1937.
14. Davison HM, Lowance MI, Barnett CF. Neuroses; discussion from standpoint of internist. Tr Am Therapy Soc (1935) 35:98, 1936.
15. Davison HM, Lowance MI, Crowe WR. Allergy: its relation to surgical specialities. South Surgeon 7:233, 1938.
16. Davison HM, Lowance MI, Barnett, CF. Urticaria in female; results of endocrine therapy. Tr Am Therapy Soc (1937) 37:49, 1938.
17. Davison HM, Lowance MI, Crowe WR. Arthritis: treatment by hyperpyrexia. J M A Georgia 25:245, 1939.
18. Davison JM, Lowance MI, Crowe WR. Evaluation of hyperpyrexia in general medicine. Arch Phys Therapy 21:728, 1940.
19. Davison HM, Thoroughman JC, Peschau JB. Intestinal infestations. J South Carolina M A 37:294, 1941.
20. Davison HM, Thoroughman JC, Peschau JB, Jr. Urology and internal medicine. Uron & Cutan Rev 46:79, 1942.
21. Davison HM, Thoroughman JC, Peschau JB, Jr. Management of asthmatic patient. M Times 70:335, 1942.
22. Davison HM, Thoroughman JC, Bowcock H. Cardiovascular allergy. South MK J 36:560, 1943.
23. Davison HM, Thoroughman JC, Peschau JB, Jr. Management of asthmatic patient. J Nat M A 36:45, 1944.
24. Davison HM, Bowcock H, Vogt E. Hypertension; examination of patients. J M A Georgia 33:201, 1944.
25. Davison HM. Present status of allergy. J Nat M A 37:16, 1945.
26. Davison HM. Cerebral allergy. South M J 42:712, 1949.
27. Arp CR, Davison HM, Atwater JS. Cartoid sinus syndrome. J M A Georgia 39:196, 1950.
28. Davison HM. Allergy of nervous system. Quart Rev Allergy 6:157, 1952.
29. Peacock LB, Davison HM. Observations on iodide sensitivity. Ann Allergy 15(2):158, 1957.
30. Davison HM. Clinics owned by doctors. J M Georgia 46(4):183, 1957.
31. Davison HM. Perspective. J M Georgia 45(6):239, 1957.
32. Davison HM. Medical ethics. J M Georgia 47(9):423, 1958.
33. Thomas JW. Hal McCluny Davison, 1891-1958. Annals of Allergy 16:467, 1958.

BIBLIOGRAPHY

of

WILLIAM WADDELL DUKE (1883-1946)

1. Howell WH, Duke WW. Experiments on the isolated mammalian heart to show the relation of the inorganic salts to the action of the accelerator and inhibitory nerves. J Physiol, London 35:131, 1907.
2. Howell WH, Duke WW. The effect of vagus inhibition on the output of potassium from the heart. Am J Physiol, Boston 21:51, 1908.
3. Howell WH, Duke WW. Note upon the effect of stimulation of the accelerator nerve upon calcium, potassium, and nitrogen metabolism of the isolated heart. Am J Physiol, Boston 23:174, 1909.
4. Duke WW. The relation of blood platelets to hemorrhagic disease; description of a method for determining the bleeding time and coagulation time and report of three cases of hemorrhagic disease relieved by transfusion. JAMA 55:1185, 1910.
5. Duke WW. The rate of regeneration of blood platelets. J Exper M, Lancaster, PA 14:265, 1911.
6. Duke WW. A simple instrument for determining the coagulation time of the blood. Arch Int Med 9:258, 1912.
7. Duke WW. The behavior of the blood platelets in toxemias and hemorrhagic disease: a preliminary report. Johns Hopkins Hosp Bull, Baltimore 23:144, 1912.
8. Duke WW. Tuberculin treatment. J Missouri M Assn, St. Louis 10:229, 1914.
9. Duke WW. Status lymphaticus and status hypoplasticus and their relationship to the glands of internal secretion. J Missouri M Assn, St. Louis 11:207, 1915.
10. Duke WW. A note on the differential diagnosis of tympanites and pneumoperitonitis. J Missouri M Assn 12:70, 1915.
11. Duke WW. Variation in the platelet count, its causes and clinical significance. JAMA 65:1600, 1915.

12. Duke WW. Pyorrhea alveolaris in relation to systemic disease. Internat J Orthodontia 4:189, 1918.
13. Duke WW. *Oral Sepsis in its Relation to Systemic Disease.* St. Louis: Mosby, 1918.
14. Duke WW. Multiple infections: a study of the relation of one infection to another. JAMA 71:1703, 1918.
15. Duke WW. Multiple infections: a study of the relation of one infection to another. N Jersey Dent J 7:379, 1918.
16. Duke WW, Diveley RL. Multiple infections; a study of relationship between infections. Dental Cosmos 61:1059, 1919.
17. Duke WW, Diveley RL. Multiple infection; a study of the relationship between infections. J Missouri M Assn 17:4, 1920.
18. Diveley RL, Duke WW. Oral infection in its relation to systemic diseases. J Radiol, Iowa City 1:326, 1920
19. Duke WW. Transfusion in the treatment of pernicious anemia. Southwest J M & S, El Reno 28:177, 1920.
20. Duke WW. Ice water bath in complement fixation for the Wassermann reaction; a shortened technic. J Lab & Clin M, St. Louis 6:392, 1921.
21. Duke WW. Ice water bath in complement fixation for the Wassermann reaction; a shortened technic. Am J Syph, St. Louis 5:312, 1921.
22. Duke WW. Food Allergy as a cause of abdominal pain. Arch Int Med, Chicago 28:151, 1921.
23. Duke WW, Stofer DD. A comparison of capillary and venous blood in pernicious anemia. Arch Int Med, Chicago 30:94, 1922.
24. Duke WW. Food Allergy as a cause of abdominal pain. South M J Birmingham 15:599, 1922.
25. Duke WW. A practical trocar and cannula for removing ascitic fluid. JAMA 79:134, 1922.
26. Duke WW. Food Allergy as a cause of bladder pain. Ann Clin Med 1:117, 1922.
27. Duke WW. Specific tests in the diagnosis of allergy. Arch Int Med 32:298, 1923.
28. Duke WW. Food allergy as a cause of illness. JAMA 81:886, 1923.
29. Duke WW. Food allergy as a cause of irritable bladder. J Urol, Baltimore 10:173, 1923.
30. Duke WW. Specific hypersensitiveness as a common cause of illness. Med Herald 42:181, 1923.
31. Duke WW. The quantitative Wasserman in relation to the diagnosis and treatment of syphilis. Ann Clin Med 2:137, 1923.
32. Duke WW. Meniere's syndrome caused by allergy. JAMA 81:2179, 1923.
33. Duke WW. Food allergy as a cause of irritable bladder. South M J 16:776, 1923.
34. Duke WW. Details in the treatment of hay fever, asthma, and other manifestations of allergy. Am J M Sc 166:645, 1923.
35. Duke WW. Specific hypersensitiveness or allergy as a common cause of illness. Nebraska MJ 8:241-279, 1923.
36. Duke WW, Stofer DD. Allergic shock as a result of blood transfusion. Med Clin N Am 7:1255, 1924.
37. Duke WW, Durham OC. A botanic survey of Kansas City, Missouri and neighboring rural districts; with reference to the flora responsible for hay fever, asthma, and dermatoses. JAMA 82:939, 1924.
38. Duke WW. The treatment of leukemia by irradiation of the chest. Radiology 1:98, 1923.
39. Duke WW. Medical aspect of oral diagnosis. Internat J Orthodontia 10:494, 1924.
40. Duke WW. Urticaria caused specifically by the action of physical agents (light, cold, heat, burns, mechanical irritation, and physical and mental exertion). JAMA 83:3, 1924.

41. Duke WW. Epinephrin in the treatment of the pain of herpes zoster. JAMA 83:1919, 1924.
42. Duke WW. Physical Allergy; preliminary report. JAMA 84:736, 1925.
43. Duke WW. Light sensitiveness. Radiology 4:279, 1925.
44. Duke WW. *Allergy, Asthma, Hay Fever, Urticaria and Allied Manifestations of Reaction.* St. Louis: Mosby, 1925.
45. Duke WW. The present status of our knowledge of allergy in its relationship to otolaryngology. Arch Otolaryng 2:460, 1925.
46. Duke WW. *Allergy, Asthma, Hay Fever, Urticaria and Allied Manifestations of Reaction.* St. Louis: Mosby, 1926. [2nd edition]
47. Duke WW. Physical allergy as a cause of dermatoses. Arch Dermat & Syph 13:176, 1926.
48. Duke WW. Transfusion in the treatment of anemia. Kansas City Clin Soc Month Bull 2(8):29, 1926.
49. Duke WW. Advancement in our knowledge of allergy as related to otolaryngology during the past two years. Arch Otolaryng 4:430, 1926.
50. Duke WW. Estimation of circulating hemoglobin volume; simple practical method for clinical purposes. Tr Sect Path & Physiol, AMA 35-49, 1926.
51. Duke WW. Allergy as related to otolaryngology. Ann Otol Rhin & Laryng 36:820, 1927.
52. Duke WW. Mental and neurologic reactions of asthma patient. J Lab & Clin Med 13:20, 1927.
53. Duke WW. Advancement in knowledge of allergy as related to otolaryngology. Arch Otolaryng 6:450, 1927.
54. Duke WW. Physical therapy in treatment of illnesses caused by light, heat, cold, mechanical irritation and by mental or physical exertion. Arch Physical Therapy 9:193, 1928.
55. Duke WW, Durham OC. Pollen count of air; relationship to symptoms and treatment of hay fever, asthma and eczema. JAMA 90:1529, 1928.
56. Duke WW. Subnormal temperature in perennial asthmatic patient. J Lab & Clin Med 13:1010, 1928.
57. Duke WW. Chronic use of adrenalin in treatment of asthma. J Lab & Clin Med 13:1012, 1928.
58. Duke WW. Aplastic anemia. JAMA 91:720, 1928.
59. Duke WW. Palm color test; simple, practical clinical method for diagnosis of anemia and plethora. Arch Int Med 42:533, 1928.
60. Duke WW. Diagnosis and treatment of anemias. Ann Int Med 2:463, 1928.
61. Duke WW. Advances in field of allergy as related to otolaryngology during years 1927 and 1928. Arch Otolaryng 8:573, 1928.
62. Duke WW. Pollen content of still air relationship to symptoms and treatment of hay fever and asthma. JAMA 91:1709, 1928.
63. Duke WW. Allergy as encountered by general practitioner. Bull NY Acad Med 5:939, 1929.
64. Duke WW. Advances in field of allergy as related to otolaryngology, during years 1928 and 1929. Arch Otolaryng 10:521, 1929.
65. Duke WW. Our idiosyncrasies to food. Hygeia 7:19, 1929.
66. Duke WW. Heat and effort sensitiveness; cold sensitiveness; relationship to heat prostration, effort syndrome, asthma, urticaria, dermatoses, noninfectious coryza and infections. Arch Int Med 45:206, 1930.
67. Duke WW. New method of administering pollen extract for purpose of preventing reactions; subcuticular method. JAMA 94:767, 1930
68. Duke WW. Allergy as recounted by general practitioner. J Iowa M Soc 20:240, 1930.
69. Duke WW. Deformity of face caused by nasal allergy in childhood. Arch Otolaryng 12:493, 1930.

70. Duke WW. Advances in field of allergy as related to otolaryngology during years 1929 and 1930; summary of bibliographic material available in field of otolaryngology. Arch Otolaryng 12:651, 1930.
71. Duke WW. Allergy as cause of gastrointestinal disorder. South M J 24:363, 1931; also Am J Surg 12:249, 1931.
72. Duke WW. Allergy as cause of gastrointestinal disorder. Tr Am Gastro-Enterol A (1930) 33:107, 1931.
73. Duke WW. Dropping out of pollen seasons; relation of pollen content of air to hay fever, asthma, and dermatitis of exposed parts. J Allergy 2:471, 1931.
74. Duke WW. Advances in field of allergy as related to otolaryngology during years 1930 and 1931; summary of bibliographic material available in field of otolaryngology. Arch Otolaryng 14:623, 1931.
75. Duke WW. Clinical manifestations of heat and effort sensitiveness, and cold sensitiveness; relationship to heat prostration, effort syndrome, asthma, urticaria, dermatoses, noninfectious coryza, and infections. J Allergy 3:257, 1932.
76. Duke WW. Specific hypersensitiveness to own breast milk; interesting therapeutic result in treatment of galactorrhea. JAMA 98:1445, 1932.
77. Duke WW. Treatment of physical allergy; treatment of heat and effort sensitiveness and cold sensitiveness and treatment of contact urticaria caused by light, cold and scratches. J Allergy 3:408, 1932.
78. Duke WW. Type of drug allergy caused by contact sensitiveness to ether and to chlorine. J Allergy 3:495, 1932.
79. Duke WW. Sensitiveness to corn shuck. JAMA 99:468, 1932.
80. Duke WW. Advances during years 1931 and 1932 in field of allergy as related to otolaryngology; summaries of bibliographic material available in field of otolaryngology. Arch Otolaryng 16:721, 1932.
81. Duke WW. Rapid and more accurate method of determining pollen content of air. JAMA 99:1686, 1932.
82. Duke WW. Relationship of heat and effort sensitiveness and cold sensitiveness to functional cardiac disorders including angina pectoris, tachycardia, and ventricular extrasystoles. J Allergy 4:38, 1932.
83. Duke WW. Medical clinic. Journal-Lancet 53:33, 1933.
84. Duke WW. Aspirin allergy; method of testing sensitiveness and method of avoiding aspirin catastrophes. J Allergy 4:426, 1933.
85. Duke WW. Dawn of speciality in medicine; allergy and physical allergy. South Med & Surg 95:128, 1933; also Journal-Lancet 53:237, 1933.
86. Duke WW. Soybean as possible important source of allergy. J Allergy 5:300, 1934.
87. Duke WW. Dawn of speciality in medicine; allergy and physical allergy. Wisconsin M J 33:265, 1934.
88. Duke WW. Allergy as related to otolaryngology; summary of bibliographic material available in the field of otolaryngology. Arch Otolaryng 20:712, 1934.
89. Duke WW. Wheat hairs and dust as common cause of asthma among workers in wheat flour mills. JAMA 105:957, 1935.
90. Duke WW. Wheat millers' asthma. J Allergy 6:568, 1935.
91. Duke WW. Allergy as related to otolaryngology; summaries of bibliographical material available in field of otolaryngology. Arch Otolaryng 22:638, 1935.
92. Duke WW. Advances in field of allergy as related to otolaryngology during years 1936 and 1937; summaries of bibliographic material available in field of otolaryngology. Arch Otolaryng 26:739, 1937.

93. Duke WW, Kohn CM. Advances in field of allergy as related to otolaryngology during year 1938; summaries of bibliographic material available in field of otolaryngology. Arch Otolaryng 28:1003, 1938.
94. Duke WW, Kohn CM. Advances in field of allergy as related to otolaryngology during year 1939; summaries of bibliographic material available in field of otolaryngology. Arch Otolaryng 31:687, 1940.
95. Duke WW, MacQuiddy EL. Review of allergy for 1940; summaries of bibliographic material available in field of otolaryngology. Arch Otolaryng 34:1178, 1941.
96. Duke WW. R.C. Lowdermilk, 1872-1945. Ann Allergy 3:395, 1945.
97. William Waddell Duke, 1883-1946. JAMA 130:1185, 1946.

BIBLIOGRAPHY

of

FRENCH KELLER HANSEL (1893-1981)

1. Hansel FK. Vasomotor rhinitis. JAMA 82:15-17, 1924.
2. Hansel FK. Mastoiditis without apparent otitis media. Arch Otolaryng 3:433-437, 1926.
3. Hansel FK. Otalgia from abscess of tongue controlled by cocainization of nasal ganglion. Arch Otolaryng 7:165-166, 1928.
4. Hansel FK. Trigeminal disturbances of otitic origin. Ann Otol Rhin & Laryng 38:335-350, 1929.
5. Hansel FK. Clinical and histopathologic studies of nose and sinuses in allergy. J Allergy 1:43-70, 1929.
6. Hansel FK. Allergy as etiologic factor in paranasal sinus disease. J Missouri M A 27:275-277, 1930.
7. Hansel FK. Allergy and its relation to inflammatory diseases of paranasal sinuses. Ann Otol Rhin & Laryng 39:510-526, 1930.
8. Hansel FK. Diagnosis and treatment of allergic disease of nose and paranasal sinuses. Tr Am Acad Ophth 35:343-356, 1930.
9. Hansel FK. Allergy and its relation to inflammatory diseases of paranasal sinuses. Tr Am Laryng Rhin & Otol Soc 36:403-408, 1930.
10. Hansel FK. Histopathologic changes in nose and paranasal sinuses in allergy. Tr Am Laryng Rhin & Otol Soc 37:180-186, 1931.
11. Hansel FK. Treatment of allergic disease of nose and paranasal sinuses. Tr Am Laryng Rhin & Otol Soc 37:492-502, 1931.
12. Hansel FK. Allergy and its relation to acute and chronic inflammatory diseases of nose and paranasal sinuses in children. Tr Am Laryng Rhin & Otol Soc 38:511-518, 1932.
13. Hansel FK. Clinical analysis of 125 cases of nasal allergy. Tr Am Laryng Rhin & Otol Soc 39:193-207, 1933.

14. Hansel FK. Allergy as related to otolaryngology and pediatrics. J Pediat 3:516-522, 1933.
15. Hansel FK. Observations on cytology of secretions in allergy of nose and paranasal sinuses. J Allergy 5:357-366, 1934.
16. Hansel FK. Allergy as related to otolaryngology and ophthalmology; literature for 1935. J Allergy 7:164-179, 1936.
17. Hansel FK. Panel discussion of allergic rhinitis; surgical treatment of nose in allergy. Tr Am Laryng Rhin & Otol Soc 42:409-419, 1936.
18. Hansel FK. *Allergy of the Nose and Paranasal Sinuses.* St. Louis: Mosby, 1936.
19. Hansel FK. Allergy as related to otolaryngology and ophthalmology; literature for 1936. J Allergy 8:196-210, 1937.
20. Hansel FK. Respiratory allergy; incidence of other manifestations. Journal-Lancet 57:83-87, 1937.
21. Hansel FK. Allergy as related to otolaryngology and ophthalmology; literature for 1937. J Allergy 9:189-198, 1938.
22. Hansel FK. Status of ionization in nasal allergy. Arch Phys Therapy 19:489-498, 1938.
23. Hansel FK. Diagnosis of nasal allergy and its relation to other manifestations. South M J 31:1003-1010, 1938.
24. Hansel FK. Allergy as related to otolaryngology and ophthalmology; literature for 1938. J Allergy 10:187-198, 1939.
25. Hansel FK. Allergy in otolaryngology and its relation to other manifestations; general considerations. Ann Otol Rhin & Laryng 48:54-72, 1939.
26. Hansel FK. Allergy in otorhinolaryngology and ophthalmology; review of recent current literature. Laryngoscope 49:323-373, 1939.
27. Hansel FK. Allergy in otolaryngology and its relation to other manifestations; diagnosis and treatment with case reports. Ann Otol Rhin & Laryng 48:359-374, 1939.
28. Hansel FK. Allergy as related to rhinology and ophthalmology; literature for 1939. J Allergy 11:178-194, 1940.
29. Hansel FK. Allergy of nose and paranasal sinuses; principles of diagnosis and treatment. Nebraska M J 25:41-46, 1940.
30. Hansel, FK. Allergy in otorhinolaryngology and ophthalmology. Review of recent current literature. Laryngoscope 50:201-221, 1940.
31. Hansel FK. Hay Fever; value of daily atmospheric counts of pollen grains and mold spores in diagnosis and treatment. J Missouri M A 37:241-246, 1940.
32. Hansel FK. Cytologic observations on secretions of nose and paranasal sinuses in allergy. Tr Am Laryng A 62:293-339, 1940.
33. Hansel FK. Allergy of upper and lower respiratory tracts in children. Ann Otol Rhin & Laryng 49:579-627, 1940.
34. Hansel FK. Allergy in otorhinolaryngology and ophthalmology. Review of recent current literature. Laryngoscope 51:221-240, 1941.
35. Hansel FK. Treatment of allergy of nose and paranasal sinuses by hyposensitization with dust extracts. Tr Am Laryng Rhin & Otol Soc 46:156-179, 1940.
36. Hansel FK. Coseasonal intracutaneous treatment of hay fever. J Allergy 12:457-469, 1941.
37. Hansel FK. Diagnosis of allergy of nose and paranasal sinuses, with particular attention to atypical types. Arch Otolaryng 34:1152-1162, 1941.
38. Hansel FK. Diagnosis of allergy of nose and paranasal sinuses, with particular attention to atypical types. Tr Sect Laryng Otol & Rhin, AMA, pp. 202-218, 1941.
39. Hansel FK. Allergy in otolaryngology and ophthalmology. Review of recent current literature. Laryngoscope 52:242-252, 1942.

40. Hansel FK. Recent advances in otolaryngologic allergy. Ann Otol Rhin & Laryng 51:1025-1049, 1942.
41. Hansel FK. Allergy in otolaryngology and ophthalmology. Review of recent current literature. Laryngoscope 53:210-220, 1943.
42. Hansel FK. Principles of Diagnosis and Treatment of allergy as related to otolaryngology. Laryngoscope 53:260-275, 1943; also J Iowa M Soc 33:211-217, 1943.
43. Hansel FK. Nethamine hydrochloride and theophylline isobutanolamine in treatment of nasal allergy and asthma. Ann Allergy 1:199-207, 1943.
44. Hansel FK. Allergy in otolaryngology and opthalmology. Review of recent current literature. Ann Allergy 2:165-172, 1944.
45. Hansel FK. Recent advances in otolaryngologic allergy. Tr Am Laryng Rhin & Otol Soc 48:260-283, 1942.
46. Hansel FK. Allergy in relation to otolaryngology and ophthalmology. Review of recent current literature. Laryngoscope 54:238-252, 1944.
47. Hansel FK. Some experience with small dosage dust and pollen therapy. South M J 38:608-613, 1945.
48. Hansel FK. Allergy in otorhinolaryngology and ophthalmology. Review of recent current literature. Laryngoscope 56:121-134, 1946.
49. Hansel FK. Treatment of certain specific types of headache with histamine. Ann Otol Rhin & Laryng 56:152-160, 1947.
50. Hansel FK. Nethaphyl in treatment of nasal allergy and bronchial asthma. Ann Allergy 5:397-401, 1947.
51. Hansel FK. Allergy in relation to ophthalmology and otolaryngology. Tr Indiana Acad Ophth & Otolaryng 31:64-87, 1948.
52. Hansel FK. Diagnosis and treatment of hay fever with particular reference to ragweed type. Laryngoscope 58:380-395, 1948.
53. Hansel FK. Allergy in otolaryngology. Laryngoscope 58:652-672, 1948.
54. Hansel FK. Treatment of headache, with particular reference to use of cafergone (ergotamine tartrate and caffeine) for relief of attacks. Ann Allergy 7:155-161, 1949.
55. Hansel FK. Penicillin sensitivity. Ann Allergy 7:619-624, 1949.
56. Hansel FK. Symposium: allergy; allergy in otolaryngology. Tr Am Acad Ophth 54:287-298, 1950.
57. Hansel FK. Nethaprin in treatment of respiratory allergy. Ann Allergy 8:745-776, 1950.
58. Hansel FK. Allergy in otolaryngology: historical review. Tr Am Acad Ophth (Supp) pp. 3-8, 1951.
59. Hansel FK. Methods of immunization employed in treatment of sinusitis. Ann Allergy 10:131-162, 1952.
60. Hansel FK. Allergy of external ear. Tr Am Acad Ophth 56:197-200, 1952.
61. Hansel FK. Use of staphylococcus toxoid and extracts of pathogenic molds in otolaryngology and ophthalmology. Tr Am Acad Ophth 56:267-271, 1952.
62. Hansel FK. Vascular headaches and related phenomena. Tr Am Acad Ophth 57:447-464, 1953.
63. Hansel FK. Pharyngeal and laryngeal phenomena of vascular origin. Ann Otol Rhin & Laryng 62:431-438, 1953.
64. Hansel FK. *Clinical Allergy*. St. Louis: Mosby, 1953.
65. Hansel FK. Allergic and other untoward reactions to antibiotics and drugs. Tr Am Acad Ophth 58:73-88, 1954.
66. Hansel FK. Use of staphylococcus toxoid in otolaryngology. Ann Otol Rhin & Laryng 63:324-345, 1954.

67. Hansel FK. Effects of tobacco smoking upon respiratory tract. South M J 47:745-749, 1954.
68. Barrett BM, Hansel FK. Reserpine new adjunct in management of resistant headache patterns: preliminary report. Ann NY Acad Soc 61:250-266, 1955.
69. Hansel FK, Youngerman WM, McGannon FL, and Hampsey JW. Panel discussions: otolaryngologic allergy; treatment of allergy. Tr Am Acad Ophth 59:659-662, 1955.
70. Hansel FK, and others. Panel discussions: otolaryngologic allergy; respiratory allergy. Tr AM Acad Ophth 59:650-658, 1955.
71. Hansel FK, Derlacki EL, Davidson RW, Maloney WH, Goldman JL. Respiratory allergy. Trans Am Acad Ophthalmo 59:650, 1955.
72. Hansel FK. Vascular headaches and related phenomena. Angiology 7:457-465, 1956.
73. Hansel FK. Optimal dosage therapy in allergy and immunity. Ann Otol Rhino 66:729-742, 1957.
74. Hansel FK. Management of vascular headaches. Amer Practit 11:215-218, 1960.
75. Hansel FK. Cytologic diagnosis in respiratory allergy and infection. Ann Allergy 24:564-569, 1966.
76. Prince HE. French Keller Hansel, 1893-1981. Annals of Allergy 48:54, 1982.
77. Anderson JR. French Keller Hansel, 1893-1981. Journal of Allergy and Clinical Immunology 69(1):77, 1982.
78. Shambaugh GE, Jr. French Keller Hansel and the History of Allergy. Arch Otolaryng 109:126, 1983.

BIBLIOGRAPHY

of

FRANCIS WASHINGTON EVERARD HARE (1858-1928)

1. Hare FWE. Typhoid fever. Australas M Gaz, Sydney, 6:161-164, 1887.
2. Hare FWE. A case of typhoid treated by cold baths. Australas M Gaz, Sydney, 7:60, 1887.
3. Hare FWE. Tracheotomy in diptheria; nine cases; five recoveries. Australas M Gaz, Syndey, 7:211, 1888.
4. Hare FWE. Five consecutive cases of tracheotomy for diptheria. Brit M J, London, 2:76, 1888.
5. Hare FWE. The cold bath treatment of typhoid fever. Intercolon. M Cong Tr, Melbourne, 2:179-185, 1889.
6. Hare FWE. The treatment of perforation in typhoid. Australas M Gaz, Sydney, 8:132, 1889.
7. Hare FWE. The influence of the cold bath treatment on the hospital mortality of typhoid. Australas M Gaz, Sydney, 7:265-268, 1889.
8. Hare FWE. The true value of quinine in continued fevers. Australas M Gaz, Sydney, 9:157-160, 1890.
9. Hare FWE. The cold bath treatment in typhoid fever. Practitioner, London, 44:161-184, 1891.
10. Hare FWE. Quinine as a cardiac stimulant. Lancet, London, 1:930, 1891.
11. Hare FWE. The flap-splitting operation for rupture of the perineum into the bowel. Australas M Gaz, Sydney, 12:249-251, 1893.
12. Hare FWE. Some remarks on the operatative treatment of perforation in typhoid. Intercolon. Q J M & S, Melbourne, I:323-332, 1895.
13. Hare FWE. The cold bath treatment of typhoid; a reply. Australas M Gaz, Sydney, 15:223-226, 1896.

14. Hare FWE. Some speculations as to the pathology of relapsing typhlitis. Australas M Gaz, Sydney, 15:290, 1896.
15. Hare FWE. Peroxide of hydrogen in acute septic arthritis. Intercolon. M J, Australas, Melbourne, 1:477-480, 1896.
16. Hare FWE. The cold bath treatment of typhoid. Australas M Gaz, Sydney, 15:325-327, 1896.
17. Hare FWE. Ten years experience of the cold bath treatment of typhoid at the Brisbane Hospital. Intercolon. M J Australas, Melbourne, 2:130-135, 1897; also Med Rec NY, 51:656, 1897.
18. Hare FWE. Ten years experience of the cold bath treatment of typhoid fever at the Brisbane Hospital. Practitioner, London, 59:254-259, 1897.
19. Hare FWE. The 1897 epidemic of dengue in North Queensland. Australas M Gaz, Sydney, 17:98-107, 1898.
20. Hare FWE. The cold bath treatment of typhoid fever, the experience of a consecutive series of nineteen-hundred and two cases treated at the Brisbane Hospital. London, 1898, Macmillan & Co. p. 207.
21. Hare FWE. The cold bath treatment of typhoid. Brit M J, London, 2:1810, 1898.
22. Hare FWE. Mechanism of the paroxysmal neuroses. Australas M Gaz, Sydney, 22:283-394, 1903.
23. Hare FWE. The treatment of haemoptysis. Australas M Gaz, Sydney, 23:65, 1904.
24. Hare FWE. Paroxysmal neuroses. Med Times & Hosp Gaz. London, 32:82-242, 1904.
25. Hare FWE. Cauterisation of the nasal mucosa and the paroxysmal neuroses. Australas M Gaz, Sydney, 23:209-212, 1904.
26. Hare FWE. The medical treatment of deep-seated haemorrhage. Lancet, London, 2:522-942, 1904.
27. Hare FWE. Amylnitrite v. adrenalin in haemoptysis. Lancet, London, 2:1446, 1904.
28. Hare FWE. The treatment of typhoid fever. Lancet, London, 1:253, 1905.
29. Hare FWE. The treatment of haemoptysis. Clin J, London, 25:366, 1904; also (abstract): Am Med, Philadelphia, 9:528, 1905.
30. Hare FWE. The hepatic factor in biliousness. Brit M J, London, 1:817-819, 1905.
31. Hare FWE. The mechanism of asthma. Med Press & Circ, London, n.s., 79:395-398, 1905.
32. Hare FWE. The carbon factor in gout; hyperpyremia. Med Rec NY, 67:921-925, 1905.
33. Hare FWE. The mechanism of the pain in migraine. Med Press & Circ, London, n.s., 79:583-586, 1905.
34. Hare FWE. The mechanism of angina pectoris. Med Press & Circ, London, n.s. 80:23l-234, 1905.
35. Hare FWE. The food factor in uricemia. Med Rec, NY, 68:366-369, 1905.
36. Hare FWE. Pathological variations of physiological vasomotor action, with special reference to the malarial paroxysmal neuroses. Practitioner, London, 75:145-155, 1905.
37. Hare FWE. The food factor in asthma: hyperpyraemia. NY M J (etc.), 82: 573-647, 1905.
38. Hare FWE. Angina pectoris and allied conditions. Lancet, London, 2:991, 1905.
39. Hare FWE. The food factor in migraine: hyperpyraemia. Clin J, London, 26:412-416, 1905.
40. Hare FWE. *The Food Factor in Disease*, being an investigation into the humoral causation, meaning, mechanism, and rational treatment, preventive and curative, of the paroxysmal neuroses (migraine, asthma, angina pectoris, epilepsy, etc.), bilious attacks, gout, catarrhal and other affections, high blood-pressure, circulatory, renal and other degenerations. 2.v. London: Longmans, Green & Co., 1905. p. 1054.
41. Hare FWE. The vasomotor factor in the pain of migraine. Clin J, London, 27:237-240, 1906.

42. Hare FWE. The food factor in the paroxysmal neuroses. Practitioner, London, 76:179-191, 1906.
43. Hare FWE. The medical treatment of inebriety. Brit J Inebr, London, 3:196-201, 1906.
44. Hare FWE. The vasomotor factor in asthma. NY M J (etc.), 82:701-704, 1906.
45. Hare FWE. Angina pectoris: its mechanism and treatment. Med Rec, NY, 70:601-605, 1906.
46. Hare FWE. Amylnitrite in haemoptysis and in other haemorrhages: recent results. Lancet, London, 2:1435, 1906.
47. Hare FWE. Amylnitrite in haemoptysis. Lancet, London, 1:189, 1907.
48. Hare FWE. Treatment of haemoptysis by nitrite of amyl. Brit J Tuberculosis, London, 1:55-59, 1907.
49. Hare FWE. The mechanism of the asthmatic dyspnoea. Clin J, London, 30:171-176, 1907.
50. Hare FWE. The dietetic treatment of certain diseases in the well-nourished and corpulent. Med Mag, London, 16:722-731, 1907.
51. Hare FWE. The mechanism of the asthmatic dyspnoea. Brit M J, London, 1:307, 1909.
52. Hare FWE. The vasomotor theory of epilepsy. Lancet, London, 1:1275, 1909.
53. Hare FWE. Some points in the management of the chronic alcoholist. Brit J Inebr, London, 7:21-23, 1909.
54. Hare FWE. The withdrawal of narcotics from habitues. Brit J Inebr, London, 8:86-90, 1911.
55. Hare FWE. Menorrhagia in virgins; a medicinal treatment. Brit M J, London, 2:110, 1911.
56. Hare FWE. Alcohol epilepsy. Brit J Inebr, London, 9:70-76, 1912.
57. Hare FWE. *On Alcoholism, Its Clinical Aspects and Treatment.* London: J & A Churchill, 1912.
58. Hare FWE. The management of the male inebreiate. Brit J Inebr, London, 11:192-197, 1914.
59. Hare FWE. Alcohol and delirium tremens. Brit M J, London, 1:446, 1915.
60. Mercier CA, Hare FWE, et al. The rum ration. Brit M J, London, 1:489-491, 1915.
61. Hare FWE. The meaning and mechanism of menstruation. Clin J, London, 45:105-146, 1916.
62. Hare FWE. The causation and treatment of delirium tremens. Clin J, London, 45:293-301, 1916.
63. Hare FWE. Zamia paralysis. North Am Vet, Chicago, 2:478-480, 1921.
64. Hare FWE. Treatment of alcoholism. Practitioner, London, 113:295-316, 1924.
65. Francis Washington Everard Hare, 1858-1928. Lancet 2:1319, 1928; also M J Australia 1:365-366, 1929.

BIBLIOGRAPHY

of

HERBERT JOHN RINKEL (1896-1963)

1. Balyeat RM, Rinkel HJ. Symptomatology of asthma in children. J Oklahoma M A 24:145, 1931.
2. Rinkel HJ, Balyeat RM. Secondary factors in seasonal hay fever. J Oklahoma M A 24:240, 1931.
3. Balyeat RM, Rinkel HJ. Allergic migraine in children. Am J Dis Child 42:1126, 1931.
4. Balyeat RM, Rinkel HJ. Distribution and importance of paper mulberry (Papyrius papyrifera Kuntze) as cause of hay fever and asthma in United States. J Allergy 3:7, 1931.
5. Balyeat RM, Rinkel HJ. Further studies in allergic migraine: based on series of 202 consecutive cases. Ann Int Med 5:713, 1931.
6. Rinkel HJ. Contact dermatitis and eczema with report of 9 cases. South M J 25:621, 1932.
7. Rinkel HJ, Balyeat RM. Occupational dermatitis due to lettuce. JAMA 98:137, 1932.
8. Balyeat RM, Rinkel HJ. Headache due to specific hypersensitiveness; history taking and etiology. Southwestern Med 16:5, 1932.
9. Balyeat RM, Rinkel HJ. Urinary retention due to use of ephedrine. JAMA 98:1545, 1932.
10. Balyeat RM, Rinkel HJ. Episcleritis due to allergy. JAMA 98:2054, 1932.
11. Rinkel HJ, Balyeat RM. Pathology and symptomatology of headaches due to specific sensitization. JAMA 99:806, 1932.
12. Rinkel HJ, Balyeat RM. Localization and specificity of cellular sensitization. J Allergy 3:567, 1932.
13. Balyeat RM, Rinkel HJ, Stemen TR. Contact dermatitis (venenata); distribution and importance of heleniums as cause of contact dermatitis in United States. Am J M Soc 184:547, 1932

14. Rinkel HJ. Gastrointestinal allergy, consideration of general and clinical features. J Kansas M Soc 34:53, 1933.
15. Rinkel, HJ. Migraine: some considerations of allergy as factor in familial recurrent headache. J Allergy 4:303, 1933.
16. Rinkel HJ. Clinical allergy — some factors governing diagnostic and therapeutic procedures. J Oklahoma M A 27:49, 1934.
17. Rinkel HJ. Gastrointestinal allergy, II. Concerning mimicry of peptic ulcer syndrome by symptoms of food allergy. South M J 27:630, 1934.
18. Rinkel HJ. Skin testing in allergy. J Missouri M A 31:382, 1934.
19. Rinkel HJ. Coseasonal Hay Fever Therapy. With special reference to low pollen doses. Clin Med & Surg 42(9):426, 1935.
20. Rinkel HJ. Leukopenic index in allergic diseases. J Allergy 7:356, 1936.
21. Rinkel HJ. Treatment of seasonal hay fever, II. The value of a prescribed diet as an adjunct to specific pollen therapy. Clin Med & Surg 433:211, 1936.
22. Rinkel HJ. Food allergy. J Kansas M Soc 37:177, 1936.
23. Rinkel HJ. Leukopenic index, II. Concerning nature of food sensitization in intractable allergic diseases. J Lab & Clin Med 21:814, 1936.
24. Rinkel HJ. Diagnostic measures in atopic infantile eczema. South M J 29:507, 1936.
25. Rinkel HJ, Gay LP. Leukopenic index; technic and interpretation. J Missouri M A 33:182, 1936.
26. Rinkel HJ. Infantile eczema; relation of specific ingesta to puritus and eczema. Arch Pediat 53:559, 1936.
27. Rinkel HJ. Perennial vasomotor rhinitis. J Iowa M Soc 27:93, 1937.
28. Rinkel HJ. Allergy in children. Arch Pediat 54:349, 1937.
29. Rinkel HJ. Food allergy; concerning diagnostic problems and procedures. J Kansas M Soc 38:374, 1937.
30. Rinkel HJ. Respiratory allergy; diagnosis and treatment. J Missouri M A 34:327, 1937.
31. Rinkel HJ. Food allergy; definition, diagnosis and clinical importance. Tri-State M J 11:2272, 1939.
32. Rinkel HJ. Food allergy. J Missouri M A 37:428, 1940.
33. Rinkel HJ. Food allergy, I. Role of food allergy in internal medicine. Ann Allergy 2:115, 1944.
34. Rinkel HJ. Food allergy, II. Technique and clinical application of individual food tests. Ann Allergy 2:504, 1944.
35. Rinkel HJ. Food allergy, III. A brief outline of general facts of food sensitization. Clin Med 54:147, 1947.
36. Rinkel HJ. Food allergy, IV. Function and clinical application of rotary diversified diet. J Pediat 32:226, 1948.
37. Rinkel HJ. Allergy problems. J Missouri M A 46:91, 1949.
38. Rinkel HJ. Inhalant allergy, I. Whealing response of skin to serial dilution testing. Ann Allergy 7:625-650, 1949.
39. Rinkel HJ. Inhalant allergy, II. Factors modifying whealing response of skin. Ann Allergy 7:631, 1949.
40. Rinkel HJ. Inhalant allergy, III. Coseasonal application of serial dilution testing (titration). Ann Allergy 7:639, 1949.
41. Rinkel HJ. Management of food allergy. Trans Am Acad Ophth and Otolaryng (Supp):18, 1951.
42. Rinkel HJ. Thermal allergy; clinical evaluation and management. South M J 44:1067, 1951.

43. Rinkel HJ, Randolph TG, Zeller M. *Food Allergy.* Springfield, IL: Thomas, 1951.
44. Rinkel HJ. Food allergy; clinical features of food sensitization. Trans Am Acad Ophth and Otolaryng 60:475, 1956.
45. Rinkel HJ. Symposium: allergy of the ear; the diagnosis of allergy. Tr Am Acad Ophth Otolar 61(1):82, 1957.
46. Rinkel HJ. The management of clinical allergy, I. General considerations. Arch Otolaryng (Chicago) 76:491, 1962.
47. Rinkel HJ. The management of clinical allergy, II. Etiologic factors and skin titration. Arch Otolaryng (Chicago) 77:42, 1963.
48. Rinkel HJ. The management of clinical allergy, III. Inhalant allergy therapy. Arch Otolaryng (Chicago) 77:205, 1963.
49. Rinkel HJ. The management of clinical allergy, IV. Food and mold allergy. Arch Otolaryng (Chicago) 77:302, 1963.
50. Rinkel HJ, Lee CH, Brown DW, Jr, Willoughby JW, Williams JM. The diagnosis of food allergy. Arch Otolaryng (Chicago) 79:71, 1964.
51. Herbert J. Rinkel, 1896-1963. J Allergy 34:556, 1963.
52. Williams Rl. Herbert J. Rinkel, M.D. Arch Otolaryng (Chicago) 79:1, 1964.

BIBLIOGRAPHY

of

ALBERT HOLMES ROWE (1889-1970)

1. Rowe AH. On the creatin-splitting enzyme of the parathyroids and the adrenals. Am J Physiol, Boston, 31:169-173, 1912.
2. Tranter CL, Rowe AH. The refractometric determination of albumin, globulin, and nonprotein in normal human blood serum. JAMA Chicago, 65:1433, 1915.
3. Falconer EH, Rowe AH. A case of sprue with necropsy. Am J Trop Dis (etc.), New Orleans, 3:400-405, 1916.
4. Rowe AH. The effect of venous statsis on the proteins of human blood serum. J Lab & Clin M, St. Louis, 1:485-489 1916.
5. Rowe AH. An automatic pipette. J Lab & Clin M, St. Louis, 1:439-441, 1916.
6. Rowe AH. The albumin and globulin content of human blood serum in health, syphilis, pneumonia, and certain other infections, with the bearing of globulin on the Wassermann reaction. Arch Int Med, Chicago, 18:455-473, 1916.
7. Rowe AH. Refractometric studies of serum proteins in nephritis, cardiac decompensation, diabetes, anemia, and other chronic diseases. Arch Int Med, Chicago, 19:354-366, 1917.
8. Rowe AH. The effect of muscular work, diet and hemolysis on the serum proteins together with comment on the technic and clinical usefulness of Robertson's micro-refractometric method. Arch Int Med, Chicago, 19:499-506, 1917.
9. Rowe AH. Diagnosis and treatment of acidosis, especially in diabetes. Calif State J M, San Francisco, 15:451-456, 1917.
10. Rowe AH. Important phases of the Allen treatment for diabetes. Northwest Med, Seattle, 17:85-88, 1918.
11. Rowe AH. Clinical aspects of the fasting treatment of diabetes. Calif State J M, San Francisco, 16:433-438, 1918.

12. Rowe AH. A review of the pneumonia of last year. Calif State J M, San Francisco, 17:244-247, 1919.
13. Rowe AH. Basal metabolism in thyroid disease, as an aid to diagnosis and treatment, with notes on the utility of the modified Tissot apparatus. Calif State J M, San Francisco, 18:332-336, 1920.
14. Rowe AH, Eakin Margaret. Monthly fluctuations in the normal metabolic rates of men and women. Calif State J M, San Francisco, 19:320-325, 1921.
15. Rowe AH. Recent advances in the diagnosis and treatment of hay fever and asthma. Calif State J M, San Francisco, 20:94-98, 1922.
16. Rowe AH. Focal infection from internist's point of view. Northwest Med, Seattle, 22:51-56, 1923.
17. Rowe AH. Insulin treatment of diabetes mellitus. Calif State J M, San Francisco, 21:204-208, 1923.
18. Rowe AH. The diagnosis and treatment of thyroid disease as controlled by the metabolic rate. Endocrinology, Los Angeles, 7:256-272, 1923.
19. Rowe AH. The insulin control of diabetes mellitus and its complications. Endocrinology, Los Angeles, 7:670-688, 1923.
20. Rowe AH. The treatment of bronchial asthma. J Am M Assn, Chicago, 84:1902-1906, 1925.
21. Rowe AH, Rogers H. Allergic dermatitis. Calif & West Med, San Francisco, 23:1589, 1925.
22. Rowe AH. Bronchial asthma in children and in young adults. Am J Dis Child, Chicago, 31:51-57, 1926.
23. Rowe AH. Hypothyroidism as a complication of diabetes mellitus. Endocrinology, Los Angeles, 10:499, 1926.
24. Rowe AH. Food allergy, a common cause of abdominal symptoms and headache. Food Facts 3:9, 1927.
25. Rowe AH. Housedust in etiology of bronchial asthma and of hay fever. Arch Int Med 39:498-507, 1927.
26. Rowe AH, Rogers H. Value of carbohydrate tolerance tests in diagnosis of diabetes mellitus. California & West Med 26:64-70, 1927.
27. Rowe AH, Rogers H. Carbohydrate tolerance in normal persons and in nondiabetic patients. Arch Int Med 39:330-342, 1927.
28. Rowe AH. House dust in etiology of bronchial asthma and of hay fever. Arch Int Med 40:396, 1927.
29. Rowe AH. Allergy in etiology of disease. J Lab & Clin Med 13:31-40, 1927.
30. Rowe AH. Food allergy: Its manifestations, diagnosis and treatment. JAMA 91:1623, 1928.
31. Rowe AH. Study of atmospheric pollen and botanic flora of east shore of San Francisco Bay. J Lab & Clin Med 13:416-439, 1928.
32. Rowe AH. Botanical survey of San Francisco. California & West Med 30:173-175, 1929.
33. Rowe AH. Food allergy and pollen hay fever. J Allergy 1:531-535, 1930.
34. Rowe AH. Allergic toxemia and migraine due to food allergy; report of cases. California & West Med 33:785-793, 1930.
35. Rowe AH, Richet C, Jr. Manifestations nerveuses chroniques de l'anaphylaxie alimentaire. J méd franc 19:170-177, 1930. (Chronic nervous manifestations of alimentary anaphylaxis.)
36. Rowe AH. *Food Allergy: Its Manifestations, Diagnosis and Treatment.* Philadelphia: Lea & Febiger, 1931.
37. Rowe AH. Elimination diets for diagnosis and treatment of food allergy. J Allergy 2:92-105, 1931.
38. Rowe AH. Bronchial asthma — its etiology and control. J Lab & Clin Med 16:1047-1055, 1931.

39. Rowe AH. Desensitization to foods. J Allergy 3:69-75, 1931.
40. Rowe AH. Gastrointestinal allergy. JAMA 97:1440-1445, 1931.
41. Rowe AH. Food allergy in differential diagnosis of abdominal symptoms. Am J M Soc 183:529-537, 1932.
42. Rowe AH. Botanical survey of San Joaquin County in central California. J Allergy 3:375-388, 1932.
43. Rowe AH. Allergic migraine. JAMA 99:912-915, 1932.
44. Rowe AH. Uterine Allergy. Am J Obst & Gynec 24:333-338, 1932.
45. Rowe AH, Garrison OH. Lipodystrophy; atrophy and tumefaction of subcutaneous tissue due to insulin injections. JAMA 99:16-18, 1932.
46. Rowe AH. Present status of food allergy. Northwest Med 32:217-224, 1933.
47. Rowe AH. Roentgen studies of patients with gastrointestinal food allergy. JAMA 100:394-400, 1933.
48. Rowe AH. Evaluation of skin reactions in food sensitive patients. J Allergy 5:135-147, 1934.
49. Rowe AH. Challenge of allergy in medical practice. California & West Med 40:352-358, 1934.
50. Rowe AH. Camomile (Anthemis cotula) as skin irritant. J Allergy 5:383-388, 1934.
51. Rowe AH. Revised "elimination diets" for diagnosis and treatment of food allergy. Am J Digest Dis & Nutrition 1:387-392, 1934.
52. Rowe AH. Food allergy; common problem in practice. South M J 28:261-267, 1935.
53. Rowe AH. Nasal and bronchial allergy in childhood. Arch Otolaryng 21:653-662, 1935.
54. Rowe AH. Protection of nutrition during use of "elimination diets." Am J Digest Dis & Nutrition 2:306-307, 1935.
55. Rowe AH. Allergy; its recognized causes. Northwest Med 34:371-434, 1935.
56. Graeser JB, Rowe AH. Inhalation of epinephrine for relief of asthmatic symptoms. J Allergy 6:415-420, 1935.
57. Rowe AH, Howe JW. Botanical survey of northwestern California. J Allergy 6:494-503, 1935.
58. Rowe AH. Gastrointestinal allergy. Journal-Lancet 56:120-126, 1936.
59. Rowe AH, Graeser JB. Treatment of pollen and other inhalant allergy. Southwestern Med 20:297-302, 1936.
60. Graeser JB, Rowe AH. Inhalation of epinephrine hypochloride for relief of asthma in children. Am J Dis Child 52:92-99, 1936.
61. Graeser JB, Rowe AH. Administration of epinephrine by inhalation. J Lab & Clin Med 21:1134-1136, 1936.
62. Rowe AH. *Clinical Allergy*. Philadelphia: Lea & Febiger, 1937.
63. Rowe AH. Dietary problem of food sensitive patient. Am J Digest Dis & Nutrition 4:787-789, 1938.
64. Rowe AH. Nasal allergy. Arch Otolaryng 28:98-105, 1938.
65. Rowe AH. Bronchial asthma; its diagnosis and treatment. JAMA 111:1827-1834, 1938.
66. Rowe AH. Bronchial Asthma. Thomas Nelson & Son. Loose Leaf Medicine, Chapter 6, 1938.
67. Rowe AH. Contact allergy to cocklebur (Xanthium spinosum); preliminary report. Arch Dermat & Syph 39:149, 1939.
68. Rowe AH. Pine pollen allergy. J Allergy 10:377-378, 1939.
69. Rowe AH, Fong J. Specificity of graminae pollens as evidenced by precipitin reactions. Proc Soc Exper Biol & Med 40:570-572, 1939.

70. Rowe AH. Elimination diets in diagnosis and treatment of food allergy. J Am Dietet A 16:193-198, 1940.
71. Rowe AH. *Elimination Diets and the Patient's Allergies.* Philadelphia: Lea & Febiger, 1941.
72. Rowe AH, Mauser CL. Cereal-free elimination diets and soybean emulsion for study and control of infantile eczema. J Allergy 13:166-169, 1942.
73. Rowe AH. Chronic ulcerative colitis-allergy in its etiology. Ann Int Med 17:83-100, 1942.
74. Rowe AH. Elimination diets for study and treatment of food allergy. Journal-Lancet 62:307-311, 1942.
75. Mauser CL, Rowe AH, Michael PPE. Intercapillary glomerulosclerosis. Ann Int Med 17:101-105, 1942.
76. Rowe AH. Clinical allergy in nervous system. J Nerv & Ment Dis 99:834-841, 1944.
77. Rowe AH. *Elimination Diets and the Patient's Allergies.* Philadelphia: Lea & Febiger, 1944. [2nd edition]
78. Rowe AH. Dermatitis of hands due to atopic allergy to pollen. Arch Dermat & Syph 53:437-453, 1946.
79. Rowe AH. Delayed healing of abdominal wound due to food allergy. West J Surg 54:313-316, 1946.
80. Rowe AH. Atopic dermatitis of hands due to food allergy. Arch Dermat & Syph 54:683-703, 1946.
81. Rowe AH, Rowe A, Jr. Bronchial asthma in patients over age of 55 years; diagnosis and treatment. Ann Allergy 5:509-518, 1947.
82. Rowe A, Jr, Rowe AH. Local cutaneous allergy (Arthus phenomenon) from epinephrine. J Allergy 19:62-67, 1948.
83. Rowe AH. Fever due to food allergy. Ann Allergy 6:252-259, 1948.
84. Rowe AH, Rowe A, Jr. Bronchial asthma in infants and children; its diagnosis and treatment. California Med 69:261-268, 1948.
85. Rowe AH. Chronic ulcerative colitis — an allergic disease. Ann Allergy 7:727, 1949.
86. Rowe AH, Rowe A, Jr. Prognosis of Allergic Conditions. J Ins Med IV(2), 1949.
87. Rowe AH. Pruritus ani et vulvae due to allergy. JAMA 140:644-645, 1949. (Comment on Hailey's article.)
88. Rowe AH. Allergic Toxemia and Fatigue. Ann Allergy 8:72-79, 1950.
89. Rowe AH. Management of food allergy. Postgrad Med 8:52-55, 1950.
90. Rowe AH. Elimination diets (Rowe) for study and control of food allergy. Quart Rev Allergy 4:227-237, 1950.
91. Rowe AH, Rowe A, Jr. Bronchial Asthma in Adult Patients — 15 to 55 years — Its Causes and Treatment. California Med 5:72, 1950.
92. Rowe A, Jr, Rowe AH. Atopic dermatitis in infants. Pediat. 39:80-86, 1951.
93. Rowe AH. Botanical survey of northern California. Ann Allergy 10:605-614, 1952.
94. Rowe A, Jr, Rowe AH. Cortisone and corticotropin in allergic disease. California Med 77:387-390, 1952.
95. Rowe AH. Bronchial Asthma Due to Food Allergy. Progress in Allergy. (Fortschritte Der Allergielehre) S. Karger, Basel, Schweiz, 3:222-275, 1952.
96. Rowe AH, Rowe A, Jr. Allergy and infections. JAMA 1151:846-847, 1953. (Comment on Chobot's article.)
97. Rowe AH, Rowe A, Jr, Uyeyama K. Regional enteritis — its allergic aspects. Gastroenterology 23:554-571, 1953.

98. Rowe A, Jr, Tufft RW, Mechanick PG, Rowe AH. Phenylbutazone (butazolidin) in musculo-skeletal diseases, bronchial asthma and other allergic diseases. Am Pract & Digest Treat 4:390-394, 1953.
99. Rowe AH, Rowe A, Jr. Chronic ulcerative colitis and regional enteritis — their allergic aspects. Ann Allergy 12:387-402, 1954.
100. Rowe AH, Rowe A, Jr, Uyeyama K. Allergic epigastric syndrome. J Allergy 25:464-471, 1954.
101. Rowe AH. Food allergy; reasons for delayed recognition and control by physicians. Quart Rev Allergy 8:391-403, 1954.
102. Rowe A, Jr, Rowe AH. Strained meat formulas in allergic diseases of infants and children. California Med 81:279-280, 1954.
103. Rowe AH, Rowe A, Jr, Uyeyama K. Chronic ulcerative colitis due to pollen allergy with 6 case reports. Acta Med Scandinav 152:139-151, 1955.
104. Rowe H, Rowe A, Jr, Uyeyama K, Young EJ. Diarrhea caused by food allergy. J Allergy 27:424-436, 1956.
105. Rowe AH. La alergia alimenticia y sus manifestaciones clinicas. (Food allergy and its clinical manifestations). Rev Clin Españ 62(6):366-373, 1956.
106. Rowe AH. Atopic Dermatitis of the Hands Due to Food Allergy. Arch and Syphilology 54:683-703, 1956.
107. Rowe AH. La alergia alimenticia y sus manifestaciones clinicas. (Nutritional allergies and their clinical manifestations). Día méd, Buenos Aires 29(41):1339-1342, 1957.
108. Rowe AH, Rowe A, Jr. Seasonal and geographic influences on food allergy. Internat Arch Allergy, Basel 13(3-4):233-244, 1958.
109. Rowe AH. Allergic fatigue and toxemia. Ann Allergy 17(1):9-18, 1959.
110. Rowe AH, Rowe A, Jr, Young EJ. Bronchial asthma due to food allergy alone in ninety-five patients. J Am M Assn 169(11):1158-1162, 1959.
111. Rowe AH, Rowe A, Jr. Chronic Ulcerative Colitis and Regional Enteritis Responding to Anti-Allergic Therapy. Gastoenterologia, Basel 91:(6), 1959.
112. Rowe AH, Rowe A, Jr. Bronchial asthma — its treatment and control JAMA 172:1734-1743, 1960.
113. Rowe AH. Chronic ulcerative colitis — atopic allergy in its etiology. Amer J Gastroent 34:49-66, 1960.
114. Rowe A, Jr, Rowe AH. Unusual extra-respiratory manifestations of pollen allergy. Ann Allergy 19:1004-1009, 1961.
115. Rowe AH. The Manifestations and Control of Food Allergy. Read by invitation at the First Congress on Food Allergy, Vichy, France, July 2, 1963.
116. Rowe AH, Rowe A, Jr. *Bronchial Asthma. Its Diagnosis and Treatment.* Springfield, IL: Thomas, 1963.
117. Rowe AH, Young EJ, Uyeyama K. Atopic Allergy in Chronic Ulcerative Colitis. (To the editor) JAMA 184:429, 1963.
118. Rowe AH, Rowe A, Jr. Eczema of the hands due to food and pollen allergy. Ann Allergy 23:385-388, 1965.
119. Rowe AH, Rowe A, Jr. Perennial nasal allergy due to food sensitization. J Asthma Res 3:141-154, 1965.
120. Rowe AH, Rowe A, Jr. Food allergy — its role in emphysema and chronic bronchitis. Dis Chest 48:609-612, 1965.
121. Rowe AH, Rowe A, Jr, Sinclair C. Immunology in ulcerative colitis. JAMA 200:1133, 1967.
122. Rowe AH, Rowe A, Jr, Sinclair C. Bronchial asthma in infants and children due to food and inhalant allergies J Asthma Res 4:189-195, 1967.

123. Rowe AH, Rowe A, Jr, Sinclair C. Food allergy — its role in the symptoms of obstructive emphysema and chronic bronchitis. J Asthma Res 5:11-20, 1967.

124. Rowe AH, Rowe A, Jr. Bronchial asthma: food, inhalant, drug and chemical allergies in its etiology. Minn Med 50:1321-1325, 1967.

125. Lietze A, Rowe AH, Rowe A, Jr. Hypoallergenic Breads. California Med 107:500-503, 1967.

126. Rowe AH, Rowe A, Jr, Sinclair C. Asma bronquial debida a alergia a alimentos e inhalatoria con una discusión de alergia atópica en enfisema obstructivo y bronquitis crónica. Prense Med Argent 53:949-953, 1966. (Spanish)

127. Lietze A, Rowe AH, Rowe A, Jr, et al. An empirical test for corn allergy. Ann Allergy 26:587-590, 1968.

128. Lietze A, Rowe AH, Rowe A, Jr. Die Bedeutung unlöslicher Nahrungspartikel für die Allergie. 1. Stürke-Antikörper beim Menchen Allerg Asthma (Leipzig) 15:11-17, 1969. (German)

129. Lietze A, Rowe AH, Rowe A, Jr. Die Bedeutung unlöslicher Nahrungspartikel für die Allergie. II. Antikörper gegen lösliche und unlöslicher Weizenantigene in Seren von Nahrungsmittelallergikern. Allerg Asthmna (Leipzig) 15:17-22, 1969. (German)

130. Lietze A, Sinclair C, Rowe AH. An intrinsic inaccuracy of radial immunodiffusion measurements of incomplete antigens. Clin Biochem 3:335-338, 1970.

131. Young EJ. Albert Holms Rowe, 1889-1970. Ann Allergy 29:167, 1971.

132. Rowe AH, Rowe A, Jr. *Food Allergy (Its Manifestations and Control and the Elimination Diets) — A Compendium.* Springfield, IL: Thomas, 1972.

133. Albert H. Rowe, Sr. 1889-1970, Ann Allergy 29:167-168, 1971.

134. Albert Holmes Rowe (1889-1970), J Allergy & Clin Immunol 48:183-184, 1971.

BIBLIOGRAPHY

of

WARREN TAYLOR VAUGHAN (1893-1944)

1. Vaughan WT. Influenza, in past and in future. Virginia M Monthly 47:361, 1920.
2. Vaughan WT. Influenza. An Epidemiologic Study. The Journal of Hygiene Monographic Series 1:260.
3. Vaughan WT. Philosophy of medical diagnosis. Virginia M Monthly 48:94, 1921.
4. Vaughan WT. Hepatic function. Virginia M Monthly 48:625, 1922.
5. Vaughan WT. Duodenal ulcer and cholecystitis. Virginia M Monthly 49:34, 1922.
6. Vaughan WT. Diseases associated with protein sensitization. Virginia M Monthly 49:316, 1922.
7. Vaughan WT. Lead poisoning from drinking "moonshine" whisky. JAMA 79:966, 1922.
8. Vaughan WT. Discussion of medical features of postoperative treatment of duodenal ulcer. S Clinics N America 2:1229, 1922.
9. Vaughan WT. Cooperation between internist and dentist. Virginia M Monthly 49:465, 1922.
10. Vaughan WT, Horsley JS. Surgical treatment of gastric and duodenal ulcers. JAMA 78:1371, 1922.
11. Vaughan WT, Van Dyke NH. Postoperative dietotherapy. Am J M Soc 163:272, 1922.
12. Vaughan WT, Dodson AI, Horsley JS. Direct transfusion of blood. Arch Surg 5:301, 1922.
13. Vaughan WT. Specific treatment of hay fever during attack. JAMA 80:245, 1923.
14. Vaughan WT. Lumbar puncture in routine treatment of syphilis. Virginia M Monthly 50:75, 1923.
15. Vaughan WT. Recent advances in diagnosis and treatment of syphilis. Virginia M Monthly 50:372, 1923.
16. Vaughan WT. Group medicine; critical survey. Southern M J 16:724, 1923.
17. Vaughan WT. Simplified apparatus for alveolar carbon dioxide determination. JAMA 81:830, 1923.

18. Vaughan WT. System for use of insulin with diabetic diet in general practice. Virginia M Monthly 50:683, 1924.
19. Vaughan WT. Eczema as an allergic phenomenon. Southern M J 17:749, 1924.
20. Vaughan WT. Interaction of specific and nonspecific factors in allergic disease. Virginia M Monthly 51:472, 1924.
21. Vaughan WT. Unilateral hyperhidrosis and erosion of teeth following parotid abscess. JAMA 84:583, 1925.
22. Vaughan WT. Rational treatment of pneumonia. Virginia M Monthly 52:88, 1925.
23. Vaughan WT, Graham WR. Functional capacity of kidneys; significance of slight deviation from normal. Virginia M Monthly 52:239, 1925.
24. Vaughan WT. Pollinosis; constitutional and local factors. Arch Int Med 40:386, 1927.
25. Vaughan WT. Reaction of omentum to germ substance. Warthin Ann Vol, p. 503, 1927.
26. Vaughan WT. Allergic Migraine. JAMA 88:1383, 1927.
27. Vaughan WT. Clinical study of hypotension. Virginia M Monthly 54:757, 1928.
28. Vaughan WT. Role of specific and nonspecific factors in allergy and allergic equilibrium. J Lab & Clin Med 13:633, 1928.
29. Vaughan WT. Some causes for failure in specific treatment of allergy. J Lab & Clin Med 13:955, 1928.
30. Vaughan WT. Allergic factor in mucous colitis. South M J 21:894, 1928.
31. Vaughan WT. Interpretation of borderline allergic reactions. J Lab & Clin Med 14:433, 1929.
32. Vaughan WT. Food allergy in specialities and in general medicine. Virginia M Monthly 56:725, 1930.
33. Vaughan WT. Effect of allergic reactions on course of nonallergic disease. J Lab & Clin Med 15:726, 1930.
34. Vaughan WT. Food allergens: genetic classification, with results of group testing. J Allergy 1:385, 1930.
35. Vaughan WT, Graham WR. Hypotension in South. South M J 23:1140, 1930.
36. Vaughan WT. Food allergens; trial diets in elimination of allergenic foods. J Immunol 20:313, 1931.
37. Vaughan WT, Hawke EK. Angioneurotic edema with some unusual manifestations. J Allergy 2:125, 1931.
38. Vaughan WT, Beck R, Shelton TS. Primary Bacillus pyocyaneus meningitis; report of case with recovery. Arch Int Med 47:155, 1931.
39. Vaughan WT. *Allergy and Applied Immunology: A Handbook for Physician and Patient on Asthma, Hay Fever, Urticaria, Eczema, Migraine and Kindred Manifestations of Allergy.* St. Louis: Mosby, Kimpton, 1931.
40. Vaughan WT. Perennial pollen desensitization. JAMA 97:90, 1931.
41. Vaughan WT. Diagnostic program in food allergy. Am J M Sc 182:459, 1931.
42. Vaughan WT. Bacterial allergy and chronic arthritis. Virginia M Monthly 59:7, 1932.
43. Vaughan WT. Specificity of bacterial allergy. Am J Clin Path 2:179, 1932.
44. Vaughan WT. Control of perennial allergic patient. South Med & Surg 94:350, 1932.
45. Vaughan WT. Improved coseasonal therapy. J Allergy 3:542, 1932.
46. Vaughan WT. Improved marking for precision syringes. J Lab & Clin Med 18:79, 1932.
47. Vaughan WT. Control of pollen allergy. J Lab & Clin Med 18:240, 1932.
48. Vaughan WT, Crockett RW. Assay of goldenrod as cause of hay fever. Ann Int Med 6:789, 1932.
49. Vaughan WT. Some rhinologic aspects of allergy. J Allergy 4:127, 1933.

50. Vaughan WT. Atypical and borderline allergic manifestations as important factor in general medicine. South Med & Surg 95:15, 1933.
51. Vaughan WT. Allergic migraine: analysis of followup after 5 years. Am J M Sc 185:821, 1933.
52. Vaughan WT. So-called urinary proteose in individuals allergic to ragweed. J Allergy 4:385, 1933.
53. Vaughan WT. Food allergy as common problem. J Lab & Clin Med 19:53, 1933.
54. Vaughan WT, Cooley LE. Air-conditioning as means of removing pollen and other particulate matter and of relieving pollinosis. J Allergy 5:37, 1933.
55. Vaughan WT, Graham WR, Crockett RW. Hay fever pollen prevalences in Virginia: review of 6-year survey. Virginia M Monthly 60:158, 1933.
56. Vaughan WT. Minor allergy: its distribution, clinical aspects and significance. J Allergy 5:184, 1934.
57. Vaughan WT. Nasal pathology as nonspecific factor in treatment of inhalant allergy. Virginia M Monthly 60:598, 1934.
58. Vaughan WT. *Allergy and Applied Immunology: A Handbook for Physician and Patient on Asthma, Hay Fever, Urticaria, Eczema, Migraine and Kindred Manifestations of Allergy*. St. Louis: Mosby, 1934. [2nd edition]
59. Vaughan WT. Some observations on food allergy. Am J Digest Dis & Nutrition 1:384, 1934.
60. Vaughan WT. Food allergens: leukopenic index, preliminary report. J Allergy 5:601, 1934.
61. Vaughan WT. Further studies on leukopenic index in food allergy. J Allergy 6:78, 1934.
62. Sulzberger MB, Vaughan WT. Experiments in silk hyposensitivity and inhalation of allergen in atopic dermatitis (neurodermatitis disseminatus). J Allergy 5:554, 1934.
63. Vaughan WT. Analysis of allergic factor in recurrent paroxysmal headaches. Tr A Am Physicians 49:348, 1934; also J Allergy 6:365, 1935.
64. Vaughan WT. Clinical photography with leica camera. J Lab & Clin Med 20:550, 1935.
65. Vaughan WT. Allergic headache. South M J 28:267, 1935.
66. Vaughan WT. Food idiosyncrasy as factor in digestive leukocyte response. J Allergy 6:421, 1935.
67. Vaughan WT, Fowlkes RW. Allergic reactions associated with cohabitation. JAMA 105:955, 1935.
68. Vaughan WT. Theory concerning mechanism and significance of allergic response. J Lab & Clin Med 21:629, 1936; also M Papers, Christian Birthday Vol, p. 711, 1936.
69. Vaughan WT. Modern methods in study of migraine. Journal-Lancet 56:127, 1936.
70. Vaughan WT. Leukopenic index as diagnostic method in study of food allergy with discussion of its reliability. J Lab & Clin Med 21:1278, 1936.
71. Vaughan WT. Allergy in rhinologist's waiting room. Tr Am Acad Ophth 40:255, 1935.
72. Vaughan WT, Pipes DM. On probable frequency of allergic shock. Am J Digest Dis & Nutrition 3:578, 1936.
73. Vaughan WT, Pipes DM. Is there correlation between food dislikes and food allergy? J Allergy 8:257, 1937.
74. Vaughan WT, Sullivan CJ. On possibility of allergic factor in essential hypertension. J Allergy 8:573, 1937.
75. Sullivan CJ, Vaughan WT. Blood surface tension, sedimentation rate and hypertensive blood pressure responses following ingestion of allergenic foods. J Allergy 9:48, 1937.
76. Vaughan WT. Food idiosyncrasy as factor of importance in gastroenterology and in allergy. Rev Gastroenterol 5:1, 1938.
77. Grubb GD, Vaughan WT. Evidence of group specific and species specific sensitization to pollen. J Allergy 9:211, 1938.

78. Vaughan WT. *Practice of Allergy.* St. Louis: Mosby, 1939.
79. Sullivan CJ, Vaughan WT. Highly concentrated pollen extracts and their deterioration in various media. J Allergy 10:551, 1939.
80. Vaughan WT. Why we eat what we eat. Scient Monthly 50:148, 1940.
81. Vaughan WT, Pipes DM. Subcutaneous tissue pressure studies in urticaria and angioneurotic edema. J Allergy 11:349, 1940.
82. Vaughan WT. Future of allergy; presidential address. J Allergy 11:584, 1940.
83. Vaughan WT. *Strange Malady: The Story of Allergy.* Doubleday, Doran, 1941.
84. Vaughan WT. Treatment of bronchial asthma. Proc Interst Postgrad M A North America (1940) 129, 1941.
85. Vaughan WT. Problems of allergy in war time. Mil Surgeon 89:737, 1941.
86. Vaughan WT, Derbes, VJ. Comparative study of incidence of acute infectious diseases in allergic and nonallergic persons. J Allergy 12:477, 1941.
87. Vaughan WT. Social revolution and the physician. J Lab & Clin Med 27:279, 1941; also Univ Hosp Bull, Ann Arbor 8:37, 1942.
88. Vaughan WT, Derbes VJ. Further modifications of nasal contact test for allergy. Ann Otol Rhin & Laryng 50:1141, 1941.
89. Vaughan WT, Graham WR. Death from asthma; warning. JAMA 119:556, 1942.
90. Vaughan WT. *Primer of Allergy: A Guidebook for Those Who Must Find Their Way Through the Mazes of This Strange and Tantalizing State.* St. Louis: Mosby, 1943. [2nd edition]
91. Vaughan WT. What price glory? Science 97:183, 1943.
92. Vaughan WT. Palindromic rheumatism among allergic persons. J Allergy 14:256, 1943.
93. Vaughan WT. On lacquer dermatitis. Virginia M Monthly 70:193, 1943.
94. Vaughan WT. On contact dermatitis. South M J 36:380, 1943.
95. Vaughan WT, Perkins RM, Derbes VJ. Epinephrine and ephedrine analogues and their clinical assay. J Lab & Clin Med 28:255, 1942.
96. Vaughan WT. Alergia en el nuevo mundo. Día méd 16:54, 1944.
97. Peterson WF, Vaughan WT. Weather and death in asthma. J Allergy 15:97, 1944.
98. Thomas JW. Warren Taylor Vaughan, 1893-1944. Annals of Allergy 2:184, 1944.

BIBLIOGRAPHY

of

MICHAEL ZELLER (1900-1977)

1. Zeller M, Murphy LJ. Resume of study of serology and blood chemistry of 100 consecutive cases. Illinois M J 61:42, 1932.
2. Zeller M. Hypersensitivity to ephedrine and ephetonie. JAMA 101:1725, 1933.
3. Zeller M. Allergic management of vasomotor rhinitis. Illinois M J 66:255, 1934.
4. Zeller M. Leukopenic index in intractable asthma. Illinois M J 69:54, 1936.
5. Zeller M. Leukopenic index in vasomotor rhinitis. J Lab & Clin Med 21:1274, 1936.
6. Zeller M. Leukopenic index with reference to normal white blood cell variations. Am J M Sc 193:652, 1937.
7. Zeller M. Dermatitis due to permafix: case report. J Am Dent A 25:1719, 1938.
8. Zeller M. Oral ragweed pollen therapy. J Allergy 10:579, 1939.
9. Zeller M, Feinberg SM, et al. Oral pollen therapy in ragweed pollinosis; cooperative study. JAMA 115:23, 1940.
10. Zeller M, Edlin JV. Allergy in insane. J Allergy 14:564, 1943.
11. Zeller M. Influence of hypnosis on passive transfer and skin tests. Ann Allergy 2:515, 1944.
12. Zeller M. Penicillin urticaria. Ann Allergy 3:360, 1945.
13. Zeller M. Unusual effect of aminophylline on intestinal tract; case report. Ann Allergy 3:369, 1945.
14. Zeller M. Temporal arteritis; case report. Ann Allergy 6:148, 1948.
15. Zeller M. Nasal pyribenzamine for relief of hay fever. Ann Allergy 7:103, 1949.
16. Zeller M. Rheumatoid arthritis — food allergy as factor. Ann Allergy 7:200, 1949.
17. Zeller M. Adrenocorticotropic hormone (ACTH) effect on ragweed hay fever. Ann Allergy 9:603, 1951.

18. Rinkel HJ, Randolph TG, Zeller M. *Food Allergy.* Springfield, IL.: Thomas, 1951.
19. Davis TA, Zeller M. Multiple peptic ulcers with massive hemorrhage during oral cortisone therapy; report of case. JAMA 150:31, 1952.
20. Grupper C, Zeller M. Poikilodermie cervico-faciale: maladie de Thomson ou atrophodermie? Cas pour diagnostic. Bull Soc fr derm syph 65(1):60, 1958.
21. Zeller M, Hetz HH. Rupture of a pancreatic cyst into the portal vein. Report of a case of subcutaneous nodular and generalized fat necrosis. JAMA 195:869, 1966.
22. Zeller M, Thier L. Gasbrand bei einem Bullen (Clostr perfringens-Infektion). Berlin Munchen Tieraerztl Wschr 78:243, 1965. (German)
23. Zeller M. Bruno Klopfer. J Anal Psychol 19(1):97, 1974.
24. Zeller M. Die Simulation des instationären thermischen Verhaltens Klimatisierter Räume mit einem elektrischen Analogie modell nach Beuken. Gesund Ing 96(9):218, 1975. Gesundheits-Ingenieur (Munchen).
25. Zeller M. Haltungsschaden-eine Zivilisationskrankheit Zentralbl Arbeitsmed Arbeitsschutz Prophyl Ergonomie 32(9):324, 1982.
26. Michael Zeller, 1900-1977. Chicago Tribune, Chicago, IL: Section 3, page 11, November 8, 1977.

BIBLIOGRAPHY

of

THERON GRANT RANDOLPH

My bibliography differs somewhat from previous ones — the formats of which have been listed in the Quarterly Cumulative Index Medicus. The numbering represents the order in which publications occurred. The popular presentations have been excluded.

Several of my references include preliminary published abstracts and program presentations that aid in establishing priority in contents and techniques. Those individuals interested may obtain a complete personal updated bibliography upon request.

1. Sheldon JM, Randolph TG. Allergy in Migraine-like Headaches. Amer J Med Sci 190:232, 1935.
2. Rackemann FM, Randolph TG, Guba EF. Specificity of Fungous Allergy. J Allergy 9:447-453, 1938.
3. Randolph TG, Rackemann FM. Blood Histamine Level in Asthma and in Eosinophilia. J Allergy 12:450-456, 1941.
4. Randolph TG. The Squibb Diabetic Diet Calculator. New York: ER Squibb & Sons, 1941.
5. Randolph TG, Squier TL. The Seasonal Incidence of Atmospheric Mold Spores in Relation to Inhalant Allergy. Proc Cen Soc Clin Research 14:61, 1941. (Abstract)
6. Randolph TG. Diet Calculator for Simplifying Diet Prescription in Diabetes Mellitus. Amer J Med Sci 204:111-117, 1942.
7. Randolph TG. Chart for Converting Carbohydrate Percentage of Fruits and Vegetables. J Amer Dietet Assn 18:523, 1942.
8. Randolph TG, Squier TL. Incidence of Atmospheric Mold Spores in Relation to Climatic Conditions in Milwaukee, 1935-1941. Wis Med J 41:987-991, 1942.
9. Randolph TG. Enumeration and Differentiation of Leukocytes in Counting Chamber with Propylene Glycol-Aqueous Stains. Proc Soc Experimental Biol & Med 52:20-22, 1943.
10. Randolph TG, Gibson EB. The Presence in Allergic Disease of Atypical Lymphocytes and Symptoms Suggesting the Recovery Phase of Infectious Mononucleosis. Proc Cen Soc Clin Research 16:41, 1943. (Abstract)
11. Randolph TG, Mikell RF. Carbol Fuchsin in Propylene Glycol for Rapid Staining of Tubercle Bacillus, Preliminary Report. Amer Rev Tuberculosis 49:109, 1944.
14. Randolph TG. Blood Studies in Allergy: I. Direct Counting Chamber Determination of Eosinophils by Propylene Glycol-Aqueous Stains. J Allergy 15:89-96, 1944.
15. Randolph TG, Gibson EB. Blood Studies in Allergy: II. Presence in Allergic Disease of Atypical Lymphocytes and Symptoms Suggesting Recovery Phase of Infectious Mononucleosis. Amer J Med Sci 207:638-643, 1944.
16. Randolph TG. Differentiation of Leukocytes in Counting Chamber by Propylene Glycol-Aqueous Stains; Screen for Detection of Major Blood Abnormality. Amer J Clin Path, Tech Sect 8:48-53, 1944.
17. Randolph TG, Rawling FFA. Bronchial Asthma as Manifestation of Sulfonamide Sensitivity. JAMA 126:166-167, 1944.

18. Randolph TG. Allergic Headache; Unusual Case of Milk Sensitivity. JAMA 126:430-432, 1944.
19. Randolph TG, Stanton CL. Comparison of Counting Chamber and Stained Film Differential Counts; Further Use of Propylene Glycol-Aqueous Stain. Proc Cen Soc Clin Research 17:44, 1944. (Abstract)
20. Randolph TG, Hettig RA. The Coincidence of Allergic Disease, Unexplained Fatigue and Lymphadenopathy; Possible Diagnositc Confusion with Infectious Mononucleosis. Proc Cen Soc Clin Research 17:44-45, 1944. (Abstract)
21. Randolph TG, Rawling FFA. Blood Studies in Allergy: III. Cellular Reactions in Sulfonamide Sensitivity. J Allergy 16:17-29, 1945.
22. Randolph TG. The Direct Enumeration of Eosinophils in the Counting Chamber by Means of Propylene Glycol-Aqueous Stains. 7th Ann Forum on Allergy, Pittsburgh, PA: 1945. (Exhibit).
23. Randolph TG, Hettig RA. Coincidence of Allergic Disease, Unexplained Fatigue and Lymphadenopathy; Possible Diagnostic Confusion with Infectious Mononucleosis. Amer J Med Sci 209:306-314, 1945.
24. Randolph TG, Stanton CL. Comparison of Differential Counts from Stained Film and Counting Chamber Using Propylene Glycol-Aqueous Stain. Amer J Clin Path, Tech Sect 9:17-22, 1945.
25. Randolph TG. Fatigue and Weakness of Allergic Origin (Allergic Toxemia) To Be Differentiated from "Nervous Fatigue" or Neurasthenia. Ann Allergy 3:418-430, 1945.
28. Randolph TG. Clinical Allergy. 105th Ann Meeting, IL State Med Soc, Chicago: 1946. (Exhibit)
29. Randolph TG, Rawling FFA. Blood Studies in Allergy: V. Variations of Total Leukocytes Following Test Feeding of Foods; An Appraisal of Individual Food Test. Ann Allergy 4:163-178, 1946.
30. Randolph TG. Fatigue and Weakness of Allergic Origin to be Differentiated from "Nervous Fatigue" or Neurasthenia. Proc Institute Med of Chicago 16:145-146, 1946. (Abstract)
31. Randolph TG. Blood Studies in Allergy: IV. Variations in Eosinophils Following Test Feeding of Foods. J Allergy 18:199-211, 1947.
32. Randolph TG. Fatigue and Weakness. Clin Med 54:223-224, 1947.
33. Randolph TG. Clinical Allergy. 12th Ann Miss Valley Med Soc Meeting, Burlington, IA: 1947. (Exhibit)
34. Randolph TG. Allergy as Causative Factor of Fatigue, Irritability and Behavior Problems of Children. J Pediat 31:560-572, 1947.
35. Randolph TG. Masked Food Allergy as Factor in Development and Persistence of Obesity. J Lab & Clin Med 32:1547, 1947. (Abstract)
36. Randolph TG, Yeager LB. Incidence of Allergy to Major Foods. J Lab & Clin Med 32:1547-1548, 1947. (Abstract)
37. Randolph TG. Gelatin as an Allergen. J Lab & Clin Med 32:1548-1549, 1947. (Abstract)
38. Randolph TG. Food Allergy. Med Clin N Amer 32:245-263, 1948.
39. Randolph TG. Fatigue Syndrome of Allergic Origin. Miss Valley Med J 70:105-108, 1948.
40. Randolph TG. Cornstarch as Allergen, Sources of Contact in Food Containers. J Amer Dietet Assn 24:841-846, 1948.
41. Randolph TG. Acute Torticollis Due to Food Allergy. J Lab & Clin Med 33:1614, 1948. (Abstract)
42. Randolph TG. Muscular Symptoms of Allergic Origin. Proc Amer Federation Clin Research 4:19, 1948. (Abstract)

43. Randolph TG. Ingredients of Bread. Testimony in the Matter of a Definition and Standard of Identity for Bread and Related Products, Docket No FDC-31(B), Before the Administrator, Federal Security Agency, 1949, pp. 14593-14628.
44. Randolph TG. Allergy; Whence, Whether and Whither. Quart Bull Northwestern U Med School 23:93-99, 1949.
45. Randolph TG, Yeager LB. Corn Sugar as an Allergen. Ann Allergy 7:651-661, 1949.
46. Randolph TG. How Do You Know Your Patient is Food Sensitive? J Mich Med Soc 48:1253-1276, 1949.
47. Randolph TG, Markson DE, Rollins JP. The Eosinophil Response in Adrenocorticotrophic Hormone (ACTH) Therapy. J Lab & Clin Med 34:1740-1741, 1949. (Abstract)
48. Randolph TG. Allergy as a Cause of Nuchal Myalgia and Associated Headache. Arch Otolaryng 50:745-758, 1949.
49. Randolph TG. Differentiation and Enumeration of Eosinophils in the Counting Chamber with Glycol Stain; Valuable Technique in Appraising ACTH Dosage. J Lab & Clin Med 34:1696-1701, 1949.
50. Randolph TG, Rollins JP, Walter CK. Allergic Reactions Following the Intravenous Injection of Corn Sugar (Dextrose or Glucose). J Lab & Clin Med 34:1741, 1949. (Abstract)
51. Randolph TG, Rollins JP. Eosinophil Observations in Adrenocorticotrophic Hormone (ACTH) Therapy. Proc 1st Clin ACTH Conference (Mote JR, Ed.), Philadelphia: Blakiston, 1950, pp. 1-13.
53. Randolph TG, Rollins JP. Relief of Allergic Diseases by ACTH Therapy. Proc 1st Clin ACTH Conference (Mote JR, Ed.), Philadelphia: Blakiston, 1950, pp. 479-490.
54. Randolph TG, Rollins JP, Zeller M. Adrenocorticotropic Hormone (ACTH), Concentrated Adrenal Cortex Extract (ACE) and Cortisone in Allergy. 6th Ann Meeting Amer College of Allergists, St. Louis, MO: 1950. (Exhibit)
55. Randolph TG, Sisk WN. Cottonseed Protein vs Cottonseed Oil Sensitivity; Case of Cottonseed Oil Sensitivity. Ann Allergy 8:5-10, 1950.
56. Randolph TG, Rollins JP. Adrenocorticotropic Hormone (ACTH); Its Effect in Bronchial Asthma and Ragweed Hay Fever. Ann Allergy 8:149-162, 1950.
57. Zeller M, Randolph TG, Rollins JP. Adrenocorticotropic Hormone (ACTH): Gross and Histologic Effects on Skin Tests and Passive Transfer. Ann Allergy 8:163-168, 1950.
58. Randolph TG, Rollins JP. Concentrated Adrenal Cortex Extract (ACE): Its Effect in Bronchial Asthma and Gastrointestinal Allergy. Ann Allergy 8:169-187, 1950.
60. Randolph TG, Rollins JP. Effect of Cortisone on Bronchial Asthma. J Allergy 21:288-295, 1950.
61. Randolph TG. House Dust Allergy. Arch Otolaryng 52:40-57, 1950.
62. Randolph TG. Allergy to So-called "Inert Ingredients" (Excipients) of Pharmaceutical Preparations. Ann Allergy 8:519-529, 1950.
63. Randolph TG, Rollins JP. Allergic Reactions from Ingestion or Intravenous Injection of Cane Sugar (Sucrose). J Lab & Clin Med 36:242-248, 1950.
64. Randolph TG, Rollins JP. Beet Sensitivity; Allergic Reactions from Ingestion of Beet Sugar (Sucrose) and Monosodium Glutamate of Beet Origin. J Lab & Clin Med 36:407-415, 1950.
65. Randolph TG, Rollins JP, Walter CK. Allergic Reactions Following Intravenous Injection of Corn Sugar (Dextrose). Arch Surg 61:554-564, 1950.
67. Randolph TG. Allergy as a Cause of Acute Torticollis. Amer Practitioner & Digest Treat 1:1062-1067, 1950.
68. Clark HG, Randolph TG. The Acid-Anoxia-Endocrine Theory of Allergy. J Lab & Clin Med 36:811-812, 1950. (Exhibit)

69. Clark HG, Randolph TG. The Clinical Application of the Acid-Anoxia-Endocrine Theory of Allergy. J Lab & Clin Med 36:811, 1950. (Abstract)
70. Randolph TG. Symposium on Food Allergy; Concepts of Food Allergy Important in Specific Diagnosis. J Allergy 21:471-477, 1950.
72. Randolph TG. Specific Food Allergens in Alcoholic Beverages. J Lab & Clin Med 36:976-977, 1950. (Abstract)
73. Randolph TG. The Mechanism of Chronic Alcoholism. J Lab & Clin Med 36:978, 1950. (Abstract)
74. Randolph TG. Allergic Factors in the Etiology of Certain Mental Symptoms. J Lab & Clin Med 36:977, 1950. (Abstract)
75. Rinkel HJ, Randolph TG, Zeller M. *Food Allergy.* Springfield, IL: Thomas, 1951.
76. Clark HG, Randolph TG. Mode of Action of ACTH. Proc 2nd Clin ACTH Conference (Mote JR, Ed.), 1(Research):525-527, Philadelphia: Blakiston, 1951. (Discussion)
77. Randolph TG, Rollins JP. Adrenocorticotropic Hormone (ACTH): Its Effect in Atopic Dermatitis. Ann Allergy 9:1-18, 1951.
78. Randolph TG. The Diagnosis of Food Allergy. Trans Amer Acad Ophthalmol & Otolaryng, (Supp), pp.13-17, 1951.
79. Randolph TG. An Experimentally Induced Acute Psychotic Episode Following the Intubation of an Allergenic Food. 7th Ann Congress Amer College of Allergists, Chicago: 1951. (A Motion Picture)
81. Randolph TG, Interlandi J. Allergy and Alcoholism. 7th Ann Congress Amer College of Allergists, Chicago: 1951. (Exhibit)
82. Randolph TG. Unlabeled Allergenic Constituents of Commercial Foods and Drugs; Critique of Food, Drug and Cosmetic Act. Ann Allergy 9:151-165, 1951.
83. Randolph TG. Allergic Myalgia. J Mich Med Soc 50:487-494, 1951.
84. Randolph TG, Interlandi J. Allergy and Alcoholism. 110th Ann Meeting, IL State Med Soc, Chicago: 1951 (Exhibit) — Awarded Bronze Medal for Originality.
85. Randolph TG. Allergic Ills Limiting Student Performance. Proc 29th Ann Meeting Amer College Health Assn 31:46-48, 1951.
86. Clark HG, Randolph TG. The Acid Anoxia Endocrine Theory of Allergy. Abstracts of Lectures and Communications, 1st Internat Congress of Allergology 1:61, Zurich, Switzerland: 1951. (Abstract)
87. Randolph TG. Food Addiction (Advanced Chronic Food Allergy) in the Etiology of Obesity and Chronic Alcoholism. Abstracts of Lectures and Communications. 1st Internat Congress of Allergology 1:95-96, Zurich, Switzerland: 1951.
88. Randolph TG. Sensitivity to Petroleum; Including its Derivatives and Antecedents. J Lab & Clin Med 40:931-932, 1952. (Abstract)
89. Randolph TG. The Alternation of the Symptoms of Allergy and Those of Alcoholism and Certain Mental Disturbances. J Lab & Clin Med 40:932, 1952. (Abstract)
90. Randolph TG. Food Allergy and Food Addiction. 9th Ann Congress Amer College of Allergists, Chicago, 1953. (Exhibit)
91. Randolph TG. Food Allergy and Food Addiction. 112th Ann Meeting, IL State Med Soc, Chicago, 1953. (Exhibit)
92. Randolph TG. Food Allergy and Food Addiction. Ann Meeting, Indiana State Med Soc, French Lick, IN: 1953. (Exhibit)
93. Randolph TG. Experimentally Induced Cerebral Syndrome Following the Test Feeding of Allergenic Foods. J Lab & Clin Med 42:931-932, 1953. (Abstract)
95. Randolph TG. Food Allergy and Food Addiction. Interim Session, Amer Med Assn Ann Convention, St. Louis, MO: 1953. (Exhibit)
96. Randolph TG. Clinical Sensitivity to Petroleum, Coal, Gas, Pine and their Derivatives. J Allergy 25:81-82, 1954. (Abstract)
98. Randolph TG. Sensory Aspects of Cerebral Allergy. J Lab & Clin Med 44:910, 1954. (Abstract)

99. Randolph TG. Allergic Type Reactions to Industrial Solvents and Liquid Fuels. J Lab and Clin Med 44:910-911, 1954. (Abstract)
100. Randolph TG. Allergic Type Reactions to Mosquito Abatement Fogs and Mists. J Lab & Clin Med 44:911-912, 1954. (Abstract)
101. Randolph TG. Allergic Type Reactions to Motor Exhausts. J Lab & Clin Med 44:912, 1954. (Abstract)
102. Randolph TG. Allergic Type Reactions to Indoor Utility Gas and Oil Fumes. J Lab & Clin Med 44:913, 1954. (Abstract)
103. Randolph TG. Allergic Type Reactions to Chemical Additives of Foods and Drugs. J Lab & Clin Med 44:913-914, 1954. (Abstract)
105. Randolph TG, Clark HG. Sodium Bicarbonate in the Treatment of Allergic Conditions. J Lab & Clin Med 44:915, 1954. (Abstract)
106. Randolph TG. Depressions Caused by Home Exposures to Gas and Combustion Products of Gas, Oil & Coal. J Lab & Clin Med 46:942, 1955. (Abstract)
108. Randolph TG, Ahroon CR, Clark HG, Frauenberger GS, Interlandi J, Mitchell DS, Roberts RC, Watterson RP, Zotter H. Specific Adaptive Illness. 112th Ann Meeting of the Amer Psych Assn, Chicago: 1956. (Exhibit)
109. Randolph TG, Ahroon CR, Clark HG, Frauenberger GS, Interlandi J, Mitchell DS, Roberts RC, Watterson RP, Zotter H. Specific Adaptive Illness. 105th Ann Meeting Amer Med Assn, Chicago: 1956. (Exhibit)
110. Randolph TG, Ahroon CR, Clark HG, Frauenberger GS, Interlandi J, Mitchell DS, Roberts RC, Watterson RP, Zotter H. Specific Adaptive Illness: An Effective Nonpsychiatric Approach to Mental and Related Ills (Without Drugs). Chicago: Human Ecology Research Foundation, 1956.
111. Randolph TG. Descriptive Features of Food Addiction; Addictive Eating and Drinking. Quart J Stud Alcohol 17:198-224, 1956.
113. Randolph TG. Specific Adaptation Syndrome. J Lab & Clin Med 48:934, 1956. (Abstract)
115. Ahroon CR, Frauenberger GS, Mitchell DS, Zotter H, Randolph TG. The Specific Adaptation Syndrome. 13th Ann Congress Amer College of Allergists, Chicago: 1957. (Exhibit)
116. Randolph TG. Food Addiction and Alcoholism. J Lab & Clin Med 50:940-941, 1957. (Abstract)
117. Randolph TG, Mitchell DS. Specific Ecology and Chronic Illness. J Lab & Clin Med 52:936-937, 1958. (Abstract)
119. Randolph TG. The Specific Adaptation Syndrome. IL State Med Soc Meeting, Chicago: 1959. (Exhibit)
121. Randolph TG. The Specific Adaptation Syndrome. IL Acad of Gen Practice, 12th Ann Sci Assembly, Chicago: 1959. (Exhibit)
122. Raldolph TG. Ecologic Mental Illness — Psychiatry Exteriorized. J Lab & Clin Med 54:936, Dec, 1959. (Abstract)
123. Randolph TG. Food Susceptibility (Food Allergy). Current Therapy (Conn H, Ed.), Philadelphia: Saunders, 1960, p. 418-423.
125. Randolph TG. A Third Dimension of the Medical Investigation. Clin Physiology 2:42-47, 1960.
126. Randolph TG. Human Ecology. Clin Physiology 2:1-5, 1960.
127. Randolph TG. Human Ecology and the Specific Adaptation Syndrome. Clin Physiology 2:41-43, 1960.
128. Randolph TG. Correlation of Sexuality with Susceptibility and Exposure to Environmental Substances. J Lab & Clin Med 56:939-940, 1960. (Abstract)
129. Randolph TG. Human Ecology and Susceptibility to the Chemical Environment, Parts I & II. Ann Allergy 19:518-540, 1961.

130. Randolph TG. An Acute Psychotic Episode Following Intubation of an Allergen (Beets). 3rd World Congress of Psych, Montreal, Canada: 1961. (Motion Picture)
131. Randolph TG. Ecologic Mental Illness — Levels of Central Nervous System Reactions. Proc 3rd World Congress of Psych, 2:379-384, Montreal, Canada: Univ Toronto Press, 1961.
132. Randolph TG. Human Ecology and Susceptibility to the Chemical Environment, Part III. Ann Allergy 19:657-677, 1961.
133. Randolph TG. Human Ecology and Susceptibility to the Chemical Environment, Parts IV & V. Ann Allergy 19:779-799, 1961.
134. Randolph TG. Human Ecology and Susceptibility to the Chemical Environment, Parts VI, VII and VIII. Ann Allergy 19:908-929, 1961.
135. Randolph TG. The Specific Adaptation Syndrome. 4th Internat Congress of Allergology. Excerpta Medica, Internat Med Abstracting Service, Internat Congress Series, 42:23-24, New York: Excerpta and Foundation, 1961.
137. Randolph TG. *Human Ecology and Susceptibility to the Chemical Environment.* Springfield, IL: Thomas, 1962.
140. Randolph TG, Mitchell DS, Brooks CR, Frauenberger GS, Mackarness R, MacLennan JG, Roberts RC. The Ecology of Mental Health; The Causative Role of the Nonpersonal Environment in Mental and Related Ills. Chicago: The Human Ecology Research Foundation, 1962.
141. Randolph TG. Natural Food From a Cooperative Farm. Clin Physiology 4:252-262, 1962.
145. Randolph TG. Alkali Therapy of Acute Allergic Reactions. Program, 19th Ann Meeting Amer Acad of Allergy, Montreal, Quebec, Canada: 1963. (Abstract)
147. Randolph TG. Air Pollution. Proc National Conference on Air Pollution, US Dept Health, Education and Welfare, Public Health Service Publication No. 1022, US Govt Printing Office, Washington, DC, 1963, p. 157. (Discussion)
149. Randolph TG. An Experimentally Induced Acute Psychotic Episode Following Intubation of a Food Allergen (Beets). Premier Congres International D'Allergie Alimentaire et Digestive, Vichy, France: 1963. (Motion Picture)
150. Randolph TG. Human Susceptibility to Pesticide Exposures. Interagency Coordination in Environmental Hazards (Pesticides); Hearings, Subcommittee on Reorganization and International Organizations of the Committee on Government Operations, US Senate, 88th Congress, Part II, 1963. (Testimony)
155. Randolph TG. The Ecologic Unit, Part I. Hospital Management 97:45-47, 1964.
156. Randolph TG. The Ecologic Unit, Part II. Hospital Management 97:46-48, 1964.
157. Randolph TG. The Health Aspects of Human Conservation. Clin Physiology 5:20, 1964.
162. Randolph TG. Food Addiction, Obesity and Alcoholism. Internat J Soc Psych, Special Edition 4:70, 1st Internat Congress of Soc Psych, 1964.
165. Randolph TG. The Provocative Hydrocarbon Test, Preliminary Report. J Lab & Clin Med 64:995, 1964. (Abstract)
166. Randolph TG. An Ecologic Orientation in Medicine: Comprehensive Environmental Control in Diagnosis and Therapy. Ann Allergy 23:7, 1965.
167. Randolph TG. Clinical Manifestations of Individual Susceptibility to Insecticides and Related Materials. Indus Med & Surg 34:132, 1965.
172. Randolph TG. Food Addiction and Alcoholism. Chicago: Human Ecology Research Foundation, 1965.
173. Randolph TG. The Toxic Environment. Interactions of Man and His Environment, Proc of the Northwestern Univ Conference (Jennings BH, Murphy JE, Ed.), New York: Plenum Press, 1966, p. 131.
175. Brooks CR, Spetz M, Frauenberger GS, Randolph TG, Forman J. Society for Clinical Ecology — Announcement of Formation. 28th Series, Letters of the Internat Correspondence Soc of Allergists, (Forman J, Ed.), Worthington, OH:1965.

177. Randolph TG. Ingredients of Peanut Butter. Testimony in the Matter of a Definition and Standard of Identity for Peanut Butter. Docket No FDC-76, Before the Administrator, Federal Security Agency, 1966, p. 6511.

179. Randolph TG. The Identification and Management of Ecologic Mental Illness. Program, 22nd Ann Congress Amer College of Allergists, Chicago: 1966. (Abstract)

182. Spetz M. Allergies: Ecological Approach. Proc 1st Ann Meeting, Soc for Clin Ecology. Science 153:903, 1966.

184. Randolph TG. Clinical Ecology As It Affects the Psychiatric Patient. Internat J Soc Psych 12:245, 1966.

188. Randolph TG, Mitchell DS. An Ecologic Approach in Medicine. Bulletin of the Human Ecology Research Foundation, Chicago: 1967.

189. Randolph TG. An Ecologic Orientation in Medicine. J Lab & Clin Med 70:859, 1967. (Abstract)

197. Randolph TG. Application of Ecologically Oriented Medicine. "Health News and Views," Newsletter, Chicago: Human Ecology Study Group, 1969.

200. Cox WL, Randolph TG. Seizures Induced by Specific Foods, Relieved by Alkali Therapy — A Motion Picture. Presented at the 4th Ann Meeting Soc for Clin Ecology, Washington, DC: 1969.

201. Randolph TG. Thirty-five Years of Clinical Investigation. Chicago: Human Ecology Research Foundation, 1969. (Chart)

206. Randolph TG. Domiciliary Chemical Air Pollution in the Etiology of Ecologic Mental Illness. Internat J Soc Psych 16:243, 1970.

220. Randolph TG. Food Addiction: A Factor in the Development of Other Addictive Phenomena. J Lab & Clin Med 76:991, 1970. (Abstract)

221. Randolph TG. Human Conservation and Health. Arch Clin Ecology 2:33, 1971.

224. Randolph TG. Domiciliary Chemical Air Pollution in the Etiology of Ecologic Mental Illness. Digest of Neurology & Psych. Hartford, CT: The Institute of Living, June-July, 1971: 260. (Review)

232. Randolph TG. Analytic Versus Biologic Dietetics. J Lab & Clin Med 78:999, 1971. (Abstract)

240. Randolph TG. Annual reports, 1954 to present. Human Ecology Research Foundation.

248. Randolph TG. Ecologic Mental Illness: II. Observed Togetherness: An Experimentally Demonstrable Approach in Studying Apparent Interpersonal Reactions. Kyoto Conference on Clinico-Biological Psych Program & Abstracts, Kyoto, Japan: 1973, p. 13. (Abstract)

251. Randolph TG. The History of Ecologic Mental Illness. Ann Review of Allergy (Frazier CA, Ed.), Flushing, NY: Med Examination Pub Co, 1973, p. 425.

252. Randolph TG. Adaptation to Specific Environmental Exposures Enhanced by Individual Susceptibility. *Clinical Ecology* (Dickey LD, Ed.), Springfield, IL: Thomas, 1976, p. 46.

254. Randolph TG. Stimulatory and Withdrawal Levels and the Alternation of Allergic Manifestations. *Clinical Ecology,* (Dickey LD, Ed.) Springfield, IL: Thomas, 1976, p. 156.

255. Randolph TG. The Role of Specific Alcoholic Beverages. *Clinical Ecology* (Dickey LD, Ed.), Springfield, IL: Thomas, 1976, p. 321.

256. Randolph TG. Ecologically Oriented History Taking. *Clinical Ecology* (Dickey LD, Ed.), Springfield, IL: Thomas, 1976, p. 381.

257. Randolph TG. Biologic Dietetics. *Clinical Ecology* (Dickey LD, Ed.), Springfield, IL: Thomas, 1976, p. 107.

258. Randolph TG. The Enzymatic, Acid, Hypoxia, Endocrine Concept of Allergic Inflammation. *Clinical Ecology* (Dickey LD, Ed.), Springfield, IL: Thomas, 1976, p. 577.

259. Randolph TG. Hospital Comprehensive Environmental Control Program. *Clinical Ecology* (Dickey LD, Ed.), Springfield, IL: Thomas, 1976, p. 70.

260. Randolph TG. Food Sources of Alcoholic Beverages. "Health News and Views," Newsletter, Chicago: Human Ecology Study Group, 1974.

262. Randolph TG. Ecologic Mental Illness — Observed Togetherness: An Experimentally Demonstrable Approach in Studying Apparent Interpersonal Reactions. *Biological Mechanism of Schizophrenia and Schizophrenia-like Psychoses,* (Mitsuda H, Fukuda T, Ed.), Tokyo: Iguku Shoin Ltd, 1974, p. 13.

270. Randolph TG. Food Addiction — The Adaptational Basis of Addictive Behavior. 3rd Internat Congress of Soc Psych, Zagreb, Yugoslavia: Institute for the Study and Control of Alcoholism, Summaries 1:124, 1970.

272. Randolph TG. Demonstrable Role of Foods and Environmental Chemicals in Mental Illness. Japanese J of Allergology 23(7):445, 1974.

275. Randolph TG. Dynamics, Diagnosis and Treatment of Food Allergy. Otolaryng Clin North Amer 7:617, 1974.

276. Randolph TG. Restless Legs — Brain-fag Syndrome. Clin Research 2:644A, 1974. (Abstract)

284. Randolph TG. Ecologically Oriented Myalgia and Related Musculoskeletal Painful Syndromes. *Clinical Ecology* (Dickey LD, Ed.), Springfield, IL: Thomas, 1976, p. 213.

288. Randolph TG. The Role of Specific Sugars. *Clinical Ecology* (Dickey LD, Ed.), Springfield, IL: Thomas, 1976, p. 310.

289. Randolph TG. Historical Development of Clinical Ecology. *Clinical Ecology* (Dickey LD, Ed.), Springfield, IL: Thomas, 1976, p. 9.

290. Randolph TG. Principal Clinical Features of Various Stimulatory and Withdrawal Levels of Ecologic Disturbances. Clinical Ecology: Office Procedures Manual, Fort Collins, CO: LD Dickey, MD, 109 West Olive St, 80524, 1976. (Chart)

291. Randolph TG. Ecologically Oriented Rheumatoid Arthritis. *Clinical Ecology* (Dickey LD, Ed.), Springfield, IL: Thomas, 1976, p. 201.

305. Randolph TG. Allergenicity of Crystalline Sugars and Alcoholic Distillates. WY Postgraduate Course in Allergy and Immunology, Cheyenne, WY: RI Williams, MD, Director, 414 East 23rd St, 82001, 1977. (Abstract)

316. Randolph TG. Specific Adaptation. Ann Allergy 40:333, 1978.

321. Randolph TG. Biographical Sketch of Herbert J. Rinkel, MD, Emphasizing His Medical Contributions. Allergy: Including IgE in Diagnosis and Treatment, (Johnson F, Ed.), Miami: Symposia Specialists, 1979, p. 1.

323. Randolph TG. Francis Hare and "The Food Factor in Disease." 12th Advanced Seminar in Clin Ecology, 1978, p. 30. (Abstract)

325. Randolph TG, Kroker GF. Allergic Thrombocyteopenia. 12th Advanced Seminar in Clin Ecology, 1978, p. 103. (Abstract)

327. Randolph TG. The Scope of Food & Chemical Allergy/Addiction. Continuing Education for the Family Physician, 11:63, 1979.

328. Randolph TG. Food Addiction (Chronic Food Allergy) — The Apparent Basis of Other Addictions. Clin Research 26:700A, 1978. (Abstract)

340. Randolph TG, Buchholz IK. Comprehensive Environmental Control Unit Nursing Instructional Manual. Chicago: Human Ecology Research Foundation, 1982.

341. Randolph TG and Buchholz IK. Comprehensive Environmental Control Unit Patient Instructional Manual. Chicago: Human Ecology Research Foundation, 1982.

344. Randolph TG, Moss RW. *An Alternative Approach to Allergies. The New Field of Clinical Ecology Unravels the Environmental Causes of Mental and Physical Ills.* New York: Lippincott, 1980; Bantam Books, Inc., 1982.

355. Randolph TG. Demonstrable Role of Foods and Environmental Chemicals in Mental Illness. Diet Related to Killer Diseases, V, Nutrition and Mental Health, Hearing before the Select Committee on Nutrition and Human Needs of the US Senate, 95th Congress, First Session, Appendix: 219, Washington, DC: US Government Printing Office, 1977.

369. Randolph TG. Food Addiction: The Apparent Basis of Other Addictive Phenomena. (Abstract) Ann Meeting Program, Anaheim, CA: Amer Soc Opthalmol Otolaryng Allergy, 1980.

372. Kroker GF, Marshall RT, Stroud RM, Rea WJ, Carroll FM, Greenberg M, Bullock TM, Smiley RE, Randolph TG. Comprehensive Environmental Control and Its Effect on Rheumatoid Arthritis. Multiunit Arthritis Study I. FASTING. (Abstract). 14th Advanced Seminar in Clinical Ecology, Pine Mountain, GA: Society for Clin Ecology, 1980.

373. Kroker GF, Marshall RT, Stroud RM, Rea WJ, Carroll FM, Greenberg M, Bullock TM, Smiley RE, Randolph TG. Comprehensive Environmental Control and Its Effect on Rheumatoid Arthritis. Multiunit Arthritis Study II. FOOD AND CHEMICAL CHANGES. (Abstract). 14th Advanced Seminar in Clinical Ecology, Pine Mountain, GA: Society for Clin Ecology, 1980.

379. Randolph TG. The Ecologically Oriented Medical History. Clin Ecology 1:1, 1982.

380. Kroker GF, Marshall RT, Randolph TG. Acrylic denture intolerance in multiple food and chemical sensitivity. Clin Ecology 1:48, 1982.

384. Randolph TG. Emergence of the Speciality of Clinical Ecology. Clin Ecology 1:84, 1983.

386. Randolph TG. Graphic Representation of Clinical Ecology. (Scientific Exhibit) Clin Ecology 2:27, 1983.

387. Randolph TG. The Medical Contributions of Francis Hare. Clin Ecology 2:74, 1984.

388. Kroker GF, Stroud RM, Marshall RT, Bullock T, Carroll FM, Greenberg M, Randolph TG, Rea WJ, Smiley RE. Arthritis Study: Fasting and Rheumatoid Arthritis: A multicenter study. Clin Ecology 2:137, 1984.

389. Marshall RT, Stroud RM, Kroker GF, Bullock T, Carroll FM,. Greenberg M, Randolph TG, Rea WJ, Smiley R. Food Challenge Effects on Fasted Rheumatoid Arthritis Patients: A multicenter study. Clin Ecology 2:181, 1984.

391. Randolph TG. The Development of Ecologically Focused Medical Care. Clin Ecology 3:6-16, 1985.

392. Randolph TG. Distinctions Between Conventional Medical Care and Ecologically Focused Medical Care. Clin Ecology 3:117-121, 1986.

393. Randolph TG. Historical Background — Coca, Rowe and Rinkel Contributions. Clin Ecology 3:183-188, 1986.

394. Randolph TG and Crayton JW. "Brain-fag" From Foods and Chemicals. In Nutrition and Brain Dysfunction, Basel, Switzerland: S Karger AG. (in press)

395. Randolph TG. Food Addiction. In Rippere EV ed. Diet and Mental Disorders, London: Croom Helm. (in press)

396. Randolph TG. Aging and Clinical Ecology. (submitted for publication)

397. Randolph TG. Classification of Stimulatory and Withdrawal Manifestations of Clinical Ecology. (submitted for publication)

Index

Index

Prepared by Randolph Jorgen

Alphabetical order is letter-by-letter. Book and journal titles are italicized. *i* indicates illustration; *n* indicates note; *q* indicates quoted material.

ACE (adrenal cortex extract), 55
acid-anoxia-endocrine theory of allergy, 52-3, 120, 137-8, 268, 275, 277
ACTH. *see* adrenocorticotropic hormone
adaptation (*see also* general adaptation syndrome; individual susceptibility; specific adaptation)
    brain as major organ of, 258
    defined, 263
    and disease, 224
    in mental illness, ecologic, 196
    Randolph, T. G., early studies on, 9
addiction (*see also* alcoholism; food addiction; specific adaptation)
    family patterns, 255
    food and drug combinations, 253-6
    food vs. opiate, 251-2
    pyramid, 251-6
    response pattern, 109-10
    to tobacco, 109
    withdrawal, 104, 110
adrenal cortex extract (ACE), 55
adrenocorticotropic hormone (ACTH), 51, 55
    endogenous release, 52
Ahroon, C. Richard, 110, 113
air pollution (*see also* air pollution, indoor)
    as cause of disease, 224
    and chemical susceptibility, 144
    controls eased by Reagan administration, 227-8
    constituent analysis vs. overall health impact, 266
    National Conference on (1962), 146
    patterns in Chicago, 74, 75*i*
air pollution, indoor *(see also specific pollution sources and chemicals)*
    and chemical susceptibility, 78, 138, 144, 147-8, 166
    in mental illness, ecologic, 195-6
    predominant illness-related sources, 235
Akron, Ohio, 98
Alberts, Lee Winfield, 80
alcohol (*see also* alcoholism)
    allergenic food constituents, 40-3, 45-7, 185, 254
    and asthma, 83
    and enhanced food absorption, 43, 45

alcohol (cont.)
    hangover caused by food constituents, 256-7
    ingredient labeling, 239-40
    synthetic ethyl, in chemical susceptibility testing, 155-6
    synthetic ethyl and methyl, 47
    tolerance, 45, 85-6
    withdrawal, 86-7
alcoholic inebriation, simulated, 111
    in food allergy, 32
Alcoholics Anonymous, 42
alcoholism (*see also* alcohol)
    and alcohol abstention caused by failure of stimulatory response, 257
    and allergic tendencies, 42-3
    and ecologically focused medical care, 276
    and food addiction, 106-7, 121-4, 184, 254
    and food allergy to alcohol constituents, 41-6, 251
    Hare, Francis, research in, 85-9
    and house dust allergy, 122-3
    incipient, 43-4
    and influenza, 87
    psychosis and allergy, alternation with, 87
    as specific adaptation, 88
    as stimulatory response, 36, 193, 248, 254, 272
    and sugar, 86-8, 184
aldehydes, 232
Alimentary Anaphylaxis, 94
alkali salts
    in allergy treatment, 52-3, 120, 268, 275
    in chemical susceptibility testing, 137
allergic syndromes, 36
allergic toxemia, 27, 29
Allergies, Your Hidden Enemy [An Alternative Approach to Allergies] [Alternative Konsepte] (T.G. Randolph, R. Moss), 215
allergy (*see also* allergy, inhalant; chemical susceptibility; food allergy; individual susceptibility; mental illness, ecologic)
    acid-anoxia-endocrine theory, 52-3, 120, 268, 275, 277

allergy (cont.)
   alternative mechanisms of, 49-51
   vs. anaphylaxis, 91n
   clinical ecology, overlap with, 211
   and double blind studies, 221
   drug treatment, 172
   ecologically focused medical care for, 276
   and hyperactivity, 30
   immunologically defined, 13-4, 17, 49, 77, 90-1, 126-7, 219-20, 265, 277, 286, 288-9
   immunologic mechanisms, 102
   immunology, dissociation from, 168
   inductive vs. deductive concepts, 264-5
   instructional courses, 97-8, 150-1
   otorhinolaryngology, overlap with, 101-3
   psychosis, alternation with, 85-7
   systemic cerebral manifestations, 32-6, 244
   systemic physical manifestations, 27-31, 244
   traditional, vs. clinical ecology, 107, 216-7, 219, 222, 226-7, 261-2, 269-70, 277-9, 284-94, 301-4
Allergy (E. Urbach, P.M. Gottlieb), 14, 17, 90
allergy, inhalant (*see also* allergy)
   intradermal tests, 240
   lessened by air conditioning, 240
   low dosage treatment, 102
   optimum dosage therapy, 57
   Rinkel titration technique, 48, 151, 268
   serial dilution treatment, 1:5 vs. 1:10, 48
Allergy Discussion Group, 91
Alexander, Harry, 28
An Alternative Approach to Allergies [Allergies, Your Hidden Enemy] [Alternative Konsepte] (T.G. Randolph, R. Moss), 206, 215
Alternative Konsepte [An Alternative Approach to Allergies] [Allergies, Your Hidden Enemy] (T.G. Randolph, R. Moss), 206, 215
American Academy of Allergy and Immunology, 37, 66, 94
   clinical ecology criticisms, 217
   and drug industry, 292
   Position Paper on Clinical Ecology, 301-7
American Academy of Environmental Medicine — The Discipline of Clinical Ecology. *see* Society for Clinical Ecology
American Association of Immunologists, 13
American Board of Allergy and Immunology, 151

American Board of Family Practice, 266
American Board of Internal Medicine, 151
American Board of Pediatrics, 151
American College of Allergists, 54
   Allergy of the Nervous System, 127
   chemical susceptibility, attitudes toward, 77
   Food Allergy Committee report, 302
   formation of Society for Clinical Ecology, 159
   mental illness, ecologic, taught, 195
American College of Physicians, 20, 97
   Randolph, T.G., fellowship prevented, 79
American Medical Association, 110, 114, 245
American Psychiatric Association, 110, 245
   on somatoform disorders, 247
American Society of Ophthalmologic and Otolaryngologic Allergy, 102, 151, 97
Amnesia
   in food allergy, 32, 34, 35
   as withdrawal manifestation, 111, 193
amphetamines, 255
anaphylaxis, 126
   vs. allergy, 91n
Anderson, Jeffrey, 287q, 293q
anemia, 13
animal dander, 138
Annals of Allergy, 151
Ann Arbor, Michigan, 14
Annual Review of Allergy, 188
antibiotics, as cause of candidiasis, 267
antigen-antibody-mediator mechanisms, in redefinition of allergy, 13-4, 91, 126-7, 277, 286
antihistamines, and chemical susceptibility, 267
anxiety, 193, 248
aphasia, 193
apples
   in chemical susceptibility testing, 137
   sulfur dioxide contamination, 233
apricots, dried, sulfur dioxide contamination, 233
Archives for Clinical Ecology, 170-1
Armour and Company, 55
arrhythmia, 248
arteritis, 193
arthralgia, 244, 247
   in chemical susceptibility, 137, 143
   and food allergy, 30, 302
   as withdrawal manifestation, 193, 248
arthritis, 244

arthritis (cont.)
  in chemical susceptibility, 137, 143, 147, 234
  and food allergy, 30, 85
  rheumatoid, in fasting and food allergy, 212-4
  steroid use unnecessary, 267
  therapies delayed by tomato effect, 226
  as withdrawal manifestation, 193, 248, 257
aseptic technique, 56
asparagus, sulfur dioxide contamination of, 233
aspirin
  in arthritis treatment, 226
  sensitivity, 13, 49, 191
asthma
  accentuation with alkalinization, 53
  and alcohol, 83
  alternation with insanity, 188
  and blood histamine levels, 15
  in chemical susceptibility, 73, 147, 165
  and food allergy, double blind studies, 302
  IgE levels in, 289
  oxygen treatment for, 50
  steroid use unnecessary, 267
  as withdrawal response, 193, 248, 257
Asthma and Hay Fever in Theory and Practice (Coca, Walzer, Thommen), 14
Asthma; Its Pathology and Treatment (H.H. Salter), xii
Attleboro, Massachusetts, 195
automobile exhausts, chemical susceptibility to, 76, 92
Autumnal Catarrh (Wyman), 126

bacon, corn sugar content, 95
barley, food allergy to in alcohol drinkers, 254
Barry, Mr. & Mrs. Thomas, 145
beets. *see* beet sugar
beet sugar
  addictive potential, 253
  food allergy to, 37-8, 58-9
  in food ingredient labeling and geography of production, 239, 253
  in obesity, 40
  physical properties, 39
  psychotic reaction to, motion picture of, 33, 54, 68, 127-8
  sulfur dioxide contamination, 233
beef
  gelatin, in food allergy, 36
  in migraine, 182

behavioral reactions. *see* mental illness, ecologic
behaviorism, 169
  Randolph, T.G., early studies in, 9,11
  and Pavlov, I.P., 189
Beijing, China, 215
Bell, Iris, 292
Ber, A., 235
Berner, Eta S., 225
Bernstein, Leonard, 301
bicarbonate salts. *see* alkali salts
Binkley, E.L., 195
bipolar intermittent adapted-maladapted reactions, 32, 256-7
blackouts, 193
blueberries, canned, in chemical susceptibility testing, 137
Blue Cross, reluctance to pay comprehensive environmental control unit claims, 163, 214, 233
Boon, Donald J., 297
Boorstein, Seymour, 287q
Boris, M., 303
Bostock, J., xii
Boston Allergy Society (Sneeze, Wheeze and Itch Club), 90
bourbon whisky, in food allergy, 41, 43, 44, 45, 104, 198, 256
bowel changes, in ulcerative colitis, 29
Bowman, John L., 297
Boyles, John, 215
Brain Allergies, the Psycho-Nutrient Connection (W.H. Philpott, D.K. Kalita), 246
brain dysfunction, 111, 287
brain-fag
  in chemical susceptibility, 143, 147
  in food allergy, 36, 105, 250
  merges with physical symptoms, 258
  origin of term, 103-4
  as withdrawal manifestation, 111, 248, 257, 272
brandy, 43-4
  corn sugar content, 46-7, 69
Brisbane, Queensland, Australia, 81, 82
British Medical Association, 113, 128
British Society for Clinical Ecology, 128
broccoli, frozen, in chemical susceptibility testing, 137
bronchitis, 248
  in chemical susceptibility to indoor air pollution, 147
Brown, Dor, 150
Brown, Ethan Allen, 73, 159, 168, 220

Brown, T.R., 182
Buchholz, Ilene K., 214, 297
buckwheat, 182, 245
Budetti, Peter, 284
bulimia, and food addiction, 256
Buscher, David S., 297
Bush, George, 227-8

California Medical Association, 292, 293
calorimetry, 180
Campbell, Gary, 297
Canadian Medical Association, 113, 128
cancer, chemical exposure implicated in, 125-6
candidiasis, 63
   and antibiotic use as cause, 267
   prevalence of, 205
candy
   as alcohol substitute, 43, 184
   corn sugar content of, 39, 255
cane sugar
   food allergy to, 37, 38
   food allergy to, in alcohol drinkers, 42, 254
   in food ingredient labeling and geography of production, 239, 253
   in mental illness, ecologic 195
   in obesity and food addiction, 40
   physical properties, 39
   sulfur dioxide contamination, 233
can linings, in chemical susceptibility, 137
cardiac problems, and food allergy, double blind studies, 302
cardiovascular medicine, 276
cardiovascular system, and food allergy, 31
carpets
   in mental illness, ecologic, 195
   as formaldehyde source, 235
Carter administration
   chemical regulation order cancelled by Reagan administration, 227
   environmental stalemate in, 134, 206-7
case reports, value of, 199, 226, 276
cashews, 245
cauliflower, frozen, in chemical susceptibility testing, 137
celery, in chemical susceptibility testing, 137
cereal grains (see also corn; wheat)
   in arthritis, 213
   avoidance, in candidiasis, 205
   food allergy to, in alcoholics, 42, 184
   in migraine, 182
   as pulse decelerator, 155

cerebral dysfunction. see allergy, systemic cerebral manifestations of; brain fag; mental illness, ecologic
Chase, M.W., 91
chemicals (see also pesticides)
   detoxification program, 236
   endogenous storage of, in chemical susceptibility, 235-7, 270
   environmental, and public health, 227
   exogenous storage of, in chemical susceptibility, 232-5
   lawn, as indoor air pollution, 235
   testing and government regulation of, 166-8
chemical susceptibility. 279-84 (see also specific chemicals and symptoms; stimulatory-withdrawal responses)
   antihistamines as cause, 267
   attitudes toward, in medical establishment, 77, 78-9, 125-6, 195
   attitudes toward, of H.J. Rinkel, M. Zeller, 69
   from air pollution, indoor, 78, 138, 144, 147-8, 166, 235
   and cancer, 125
   cause and symptom preponderance, 143-4
   vs. chemical sensitivity, 77
   clearing of symptoms in chemical avoidance, 110
   Coca, A.F., reported by, 92-3
   and comprehensive environmental control, 114-5, 136-8, 155, 166, 233-4
   death caused by, 113
   in dietetics, biologic, 181, 182-3
   with exogenous exposures, 232-5
   and formaldehyde, 232
   "get well hotel" for treatment of, 237-8
   "hardening" in, with specific adaptation, 109
   and high degree of individual susceptibility, 78
   as injury, 144
   and mental illness, ecologic, 195, 196, 205-6, 245, 286-8
   non-immunologically mediated, 77
   and organically grown produce, 134, 144-5
   and phenolic food compounds, 235
   prevention, 276-7
   provocation-neutralization treatment for, 195
   provocative testing with synthetic ethyl alcohol, 155-6

chemical susceptibility (cont.)
  to psychiatric drugs, 173
    Randolph, T.G., and first case seen of, 73-4, 76
  vs. toxicity, 163-8, 172-3, 270
Chemical Victims (Human Ecology Study Group), 134, 145, 175
cherries, canned, in chemical susceptibility testing, 137
chest pain, 193
Chicago, Illinois, 48, 53, 102, 110, 127
  air pollution patterns, 74, 75*i*
  chosen by Randolph, T.G., for private practice, 23
Chicago Allergy Society, 79
Chicago Society of Medicine Physicians Review Committee, 234
chicken, in chemical susceptibility, 145
China, 215-6
chipboard, as formaldehyde source, 232, 238
chocolate
  addiction, 251, 253, 255
  corn sugar content, 39, 255
Chong-qing, China, 215
chronic illness, 135-6, 170, 231 (see also stimulatory-withdrawal responses)
  and air pollution, indoor, 147
  changed to acute illness by avoidance and reexposure, 25, 108-9, 264, 274
  in chemical susceptibility, 136, 165-6, 167-8
  clinical ecology treatments for, vs. traditional allergy, 227
  failure of traditional medicine to treat, 277-80
  multi-factorial origins, 223, 224
  rate doubles in young children, 284
cigarette smoking, 251 (see also tobacco)
  and food addiction, 253
Cincinnati, Ohio, 98
citrus peel, candied, sulfur dioxide contamination, 233
Clark, Harry G., 49, 51-2, 113
Clean Air Act, weakened by Reagan administration, 228
cleaning fluid, and chemical susceptibility, 283, 285
Cleveland, Ohio, 98
Clinical Allergy (F.K. Hansel), 102
Clinical Allergy Due to Foods, Inhalants, Contaminants, Fungi, Bacteria and Other Causes — Manifestations, Diagnosis and Treatment (A.H. Rowe), 95

clinical ecology, 41, 128 (see also allergy; individual susceptibility; medicine, traditional *and* environmental; stimulatory-withdrawal responses)
  analysis vs. synthesis in, 204-5, 263, 268
  books written for general audience, 205-6, 215
  chronic reactions in, 231
  clinical observations in, vs. immunologic mechanisms, 102
  as integrative discipline, 211-2, 262
  legal suit, 292
  medical history in, patient, 24, 273-4
  vs. medicine, traditional, 216-27
  onset of present illness in, 248, 249-50, 272
  origin of term, 107
  overlap with toxicology, 270
  Position Paper of American Academy of Allergy and Immunology, 301-4
  and specific adaptation, 263, 268-9
  vs. specific etiology doctrine, 262, 269-70
  training fellowships in, 199, 297
Clinical Ecology (L.D. Dickey), 160, 161, 199-200
Clinical Ecology — Archives for Human Ecology in Health and Disease, 212
Clinical Ecology Study Club, 160, 161
Coca, Arthur Fernandez, 89*i*, 126
  on alcoholism, 121-4*q*
  bibliography, 311-5
  on chemical susceptibility, 92-3, 191
  conflict with medical associates, 64, 91-2
  Familial Nonreaginic Food Allergy, 49, 63, 66*q*
  and food allergy, 91-2, 191
  Grove, Ella, marriage to, 91
  on hypersensitiveness, 17, 90
  on immunologic interpretation of allergy, 13-4, 17, 49, 90-91, 126-7, 220
  meeting with A.H. Rowe, H.J. Rinkel, T.G. Randolph, J. Interlandi, 66-8
  professional history, 90, 91
  on pulse test, 66-7*q*, 92, 119-21, 122
  terminal illness, 124-5
  views contrast with T.G. Randolph's, 119-21
  on writing for general public, 279
cocaine addiction, 251, 252
cocoa, food allergy to, 33
coffee
  and addiction, 110, 250, 253

coffee (cont.)
   food allergy to, among alcoholics, 42, 43, 184
   and mental illness, ecologic, 195
cognac, 43, 46
Cohen, S.G., 92
cola drinks, and food addiction, 251, 253
colchicine, in arthritis treatment, 226
colitis, 193, 247
   and food allergy, double blind studies, 302
collagen disease, 257
Columbia Hospital, Milwaukee, 16
Commonweal, 293
comprehensive environmental control, 113, 239, 261, 269
   addiction treatment, 109
   arthritis study, 213
   in chemical susceptibility testing, 114-5, 155, 166, 233-4
   and fellowship in clinical ecology, 199
   in food ingestion tests, 183, 240, 274
   at Lutheran General Hospital, 128, 163
   in observed togetherness, 198
   procedures, 136-9
   and provocative testing, 222-3
   services termed "not usual and customary" by Physicians Review Committee, 234
   unit closure at Herontin Hospital forced by insurance companies, 214
conjunctivitis
   in chemical susceptibility, 165
   and food allergy, double blind studies, 302
constipation, 247
construction materials recommendations, in chemical susceptibility, 237-8
Consumer Products Safety Commission, 228
convulsions, 193
   as stimulatory response, 193, 248
corn (see also corn sugar)
   in alcohol, 44, 46, 47, 185, 198, 254, 256-7
   in alcoholism, 42, 184, 254
   in arthritis, 213
   as cornstarch, as excipient, 37, 255
   in cornstarch-sized food containers, 37
   diet sources of, 44
   food addiction to, 252, 255-6
   food allergy to, 53, 54-5, 58, 65, 104, 113, 245

corn (cont.)
   food allergy to, disputed, 55, 58
   food allergy to, in alcohol drinkers, 254
   in mental illness, ecologic, 195, 198
   sulfur dioxide contamination, 233
corneal ulcers, and food allergy, 29
Cornell University Medical Center, 90
Corn Products Research Foundation, 54-5
corn sugar
   addictive potential, 40, 253, 255-6
   in alcohol, 46-7, 69
   as alcohol substitute, 43, 184
   in bacon, 65, 95
   in brandy, 46-7, 69
   as drug excipient, 65
   in ham, 65
   food allergy to, 58
   in food ingredient labeling and geography of production, 239, 253
   in intravenous alimentation, 37-8, 39-40, 65, 113
   in obesity, 40
   physical properties, 38-9
   sulfur dioxide contamination, 233
cortisone, 51, 55
Corwin, Alsoph, 90, 152-5q, 160
cosmetics
   and chemical susceptibility, 238
   in mental illness, ecologic, 195
cotton, formaldehyde treated, 232, 238
cottonseed oil, and food allergy, 37
Crayton, John W., 297
Criep, L.H., 92
Crook, William G., 65i, 205
Crowe, W.R., 191
crying, in food allergy, 35

Dart, Paul E., 297
Darwin, Charles, 107, 189
Davison, Hal McCluny, bibliography, 317-8
DDT, 235
Dees, S.C., 43
dementia, 191, 247, 258
deodorants, 238
depression
   accentuation with alkalinization, 53
   vs. brain-fag, 103
   in chemical susceptibility, 137, 143, 147, 156, 165
   and food allergy, 32-6, 84-5, 302
   merges with physical symptoms, 258
   oxygenation and exercise treatment for, 50

depression (cont.)
  and specific adaptation, 111
  as withdrawal response, 248, 250, 272
dermatitis, occupational, and chemical susceptibility, 109
dermatology, 276
Des Plaines, Iowa, 11
detoxification program, 236
Diabetic Diet Calculator, Squibb, 15
diagnostic related groups (DRGs), 228
Diamantina Hospital, Brisbane, 82, 191
diarrhea, 193, 247
Dickey, Lawrence D., 11
  China visit, 215
  Clinical Ecology, 160, 161, 162q, 199-200
  on non-specialization of clinical ecology, 212
  on provocative testing, 195
Dienes, Louis, 16
diesel fumes, and chemical susceptibility, 283
dietetics
  analytic, 88
  analytic vs. biologic, 179-87, 266
diet pills, 255
Disease Concept of Alcoholism, The (E.M. Jellinek), 107
disinfectants, as indoor air pollution, 147, 238
Dix, Irmgard, 297
dizziness, 247
Doerr, Robert, 17, 90
Dorfman, W., 190
Dorn, H., 195
double-blind studies
  impossibility of, in allergy, 221, 222
  need for disputed, in clinical ecology, 217
drug allergy, mechanisms unknown, 49
drug addiction
  allergy related to, 124
  and ecologically focused medical care, 276
  and food addiction, 251, 253-5
  as stimulatory response, 248, 255, 257-8
drugs, 164
  chemical additives in, 76
  chemical susceptibility, as cause of, 144, 166, 173-4, 275
  chronic illness, use in, 135
  in clinical ecology, use limited, 212
  corn excipients in, 65
  dyes in, FD & C, 73
  and eosinophilia, 18

drugs (cont.)
  in mental illness, ecologic, 195, 236, 268, 275
  psychiatric use, 173-4, 190, 195, 196, 197, 268
  safety regulation, governmental, 167
  and symptom suppression, 134, 173-4, 190
Dublin, Ireland, 82
Dubos, Rene, 218, 224q
Duke, William Waddell, 13, 95, 191
  bibliography, 319-23
  on food allergy specificity, 88
Dukes-Dubos, Francis N., 152-5q
DuNouy, L., 179-80
dust, house, 138
  allergy titration for, 48
Dust-Seal, 66, 122, 123
  chemical susceptibility to, 68, 124
dyes, 164
  chemical susceptibility to, 73, 144
dysentery, 83
dyslexia, in chemical susceptibility, 287
dyspepsia, 83

Earth Day, 176
ecologically focused medical care, 216-8, 223-4, 226, 261-80 (*see also* clinical ecology; comprehensive environmental control; medicine, traditional vs. environmental)
ecologic mental illness. *see* mental illness, ecologic
ecology (*see also* clinical ecology)
  first use of term in medicine by T.G. Randolph, 127
  popularization of term, 133
  vs. reductionism in science, 203-4
Ecolony, 169-70
eczema, 97
  in chemical susceptibility, 165
  as withdrawal manifestation, 193
Edelman, Gerald M., 204
edema, 248
Edinburgh, Scotland, 113
Efron, Vera, 106-7
eggplant, related to tobacco, 253
eggs
  food allergy to, 32, 36, 96, 105, 245
  in mental illness, ecologic, 195
electrolytes, in food allergy mechanisms, 51-2
elimination diets, 268

elimination diets (cont.)
    in food allergy diagnosis, 182, 274
    in food allergy diagnosis and treatment, 28-30, 95
    and Rowe, Albert, 95
    vs. skin tests, 19-20, 28
Elliotson, J., xii
Enkema, L. Neil, 297
environmental control. see comprehensive environmental control
environmental deterioration (see also environmental regulation, governmental)
    medical profession responsible for, 205
    in Reagan-Bush administration, 215
environmentalism, popularization of, 133-4
Environmental Policy Act of 1969, 134, 206
Environmental Protection Agency (EPA), 200
    agency analytical bias, 266-7
    creation, 176
    regulatory powers curtailed by Reagan Administration, 227-8
environmental regulation, governmental, 146, 176
    analytical bias, 266-7
    backlash in 1970s, 200, 206-7
    failure of, in pollution control, 262
    relaxed and politicized by Reagan administration, 227-8
eosinophilia, 15
    in appendicitis, 44
    in drug allergy, 18
eosinophils
    and blood histamine levels, 15
    counting technique, 17, 55
    in food and drug allergy, 49-50
    in otolaryngology, 102
epilepsy, 85
    ecologically focused medical care for, 276
    in food allergy, 29, 302
erythrocyte agglutination, in allergy, 51, 52
ethanol. see alcohol, synthetic ethyl
Etiologic Theory in America Prior to the Civil War (P. Allen), 7
exercise
    in allergy treatment, 50
    in chemical detoxification, 236

fabric, synthetic, 164
    as formaldehyde source, 232
Fadal, Richard, 151
Familial Nonreaginic Food Allergy (A.F. Coca), 49, 63, 66q, 92

fasting
    and arthritis, rheumatoid, 212-3
    in comprehensive environmental control, 136-7
    in food allergy relief, 83, 110
    in food ingestion tests, 79, 114-5
    and insanity, 84
fat deposits, chemical storage in, 231, 235-7
fatigue, 244, 247
    vs. brain-fag, 103
    in chemical susceptibility, 137, 143, 156, 165
    in food allergy, 27-30, 36
    oxygen treatment for, 50
Fauci, A.S., 220
Federal Pesticide Act of 1978, 206
Feinberg, Samuel M., 53
Fettes College, Edinburgh, Scotland, 82
Fife, Roger L., 297
Filice, Robert C., 297
Findley, Thomas, 11
fluorescent lights, 235, 237
fluorides, 77
food (see also dietetics)
    component analysis, 180
    ingredient labeling, 54-5, 163, 239-40
    phenolic compounds, 235
food addiction, 54, 79, 81, 89 (see also food allergy; stimulatory-withdrawal responses)
    and alcoholism, 88, 106-7, 121-4, 184, 254
    and drug addiction combined, 253-4, 255
    vs. food allergy, 104-7, 244-5, 251-2
    in obesity, 40, 251, 252
    opiate addiction comparison, 104, 251-2
    to potato, 109
    Rinkel, H.J., attitude toward, 68
    withdrawal effects, 110
food allergy (see also specific symptoms and foods; alcohol; alcoholism; allergy; elimination diets; food addiction; food ingestion tests; mental illness, ecologic; pulse test; rotary diversified diet; stimulatory-withdrawal responses)
    alcohol, reaction to, indicated by, 254
    and alcoholic inebriation, simulated, 32
    allergen persistence in blood, 28
    amnesia in, 32, 34, 35
    attitudes toward, in medical establishment, 53-5, 56, 57-8, 79, 186
    and brain-fag, 103, 104

food allergy (cont.)
  and candidiasis, 205
  chemical susceptibility, coexistence with, 137
  vs. chemical susceptibility, in food ingestion tests, 183
  chronic, 51
  Coca, A.F., research on, 92-3
  and corneal ulcers, 29
  death, implicated in, 113
  and depression, 32-6, 84-5
  vs. dietetics, analytic, 181
  and eosinophilia, 18
  and epilepsy, 29, 302
  fixed, 182, 244-5
  fixed vs. cyclic, 25
  vs. food addiction, 104-7, 244-5, 251-2
  food fractions in, 36-40
  and food inhalation, 30
  Hare, Francis, research in, 83-5
  and hyperactivity, 32, 40, 88, 302
  vs. idioblapsis, 93, 119-20, 121
  and immunologic interpretation of allergy, 17, 49
  intradermal tests, 240
  lymphocyte function in, 50
  manifestations, 175
  masked, 20, 24, 25, 40, 49, 96, 97, 107, 192
  masked vs. unmasked, 244-5, 271-2
  and myalgia, 30, 302
  opiate addiction, compared with, 104, 251-2
  predominant foods as cause, 26
  prevalence, in environmental illness, 239
  provocation-neutralization technique, 148, 183, 268
  and Rowe, Albert, 94-5
  sensitivity base in, 26
  skin tests, 19-20, 23, 28, 30, 49, 240
  skin tests vs. elimination diets, 19-20, 28
  sublingual tests, 240
  sugars predominating in, 88
  syndromes, 36
  systemic cerebral manifestations, 32-6
  systemic physical manifestations, 27-31
  and ulcerative colitis, 29, 175
  and unconsciousness, 96
Food Allergy (H.J. Rinkel, T.G. Randolph, M. Zeller), 31, 47, 63, 68, 93, 103, 246
Food Allergy (A.H. Rowe), 80, 94q
Food Allergy; It's Manifestations and Control, and the Elimination Diets — A Compendium (A.H. Rowe, H.J. Rinkel), 176

Food and Drug Administration
  bread standard identity hearings, 54
  peanut butter hearings, 163
food contaminants and additives, 236
  in arthritis, 212-3
  in dietetics, biologic, 181
  in chemical susceptibility testing, 137
  chemical susceptibility to, 76
  in food ingestion tests, 182
  in mental illness, ecologic, 195
Food Factor in Disease, The (F.W.E. Hare), 41, 81, 83, 85
food ingestion tests, individual, 105, 268, 274
  chemical susceptibility, differentiated from, 182
  and food allergy, masked, 97, 271-2
  in observed togetherness, 198
  and preliminary avoidance, 79, 108-9, 183
  procedures, 136-7
  vs. skin tests, for food allergy diagnosis, 19-20, 240
  symptom delay, in arthritis, 213
formaldehyde, 283
  as air pollution, indoor, 232, 235, 238
  from kerosene combustion, 73
Forman, Jonathan, 160
  award established, 161
  death, 170
Fox, Ruth, 121
Frauenberger, George S., 11, 65i, 113, 149, 159, 161
Frauenberger, Helen, 65i
Frazier, Claude A., 188
Freud, Sigmund, 190
Freund's adjuvant, 16
fruit *(see also specific fruits)*
  dried, sulfur dioxide contamination, 233
  food allergy to, and fatigue, 28
fuel oil, chemical susceptibility to, 76, 78, 144
furnishings recommendations, in chemical susceptibility, 237-8
furniture polish, 238

Galtman, A.M., 191
Gantt, W.H., 189
Gardner, Robert, 235
gasoline, in chemical susceptibility, 92, 191
gastrointestinal distress, 247
  as withdrawal response, 165, 193, 248
gastrointestinal medicine, 276

gas, utility
  as air pollution, indoor, 147, 235
  and arthritis, 234
  chemical susceptibility to, 76, 77, 78, 92, 138, 143, 166
  government regulation, 167
  "hardening" to, in chemical susceptibility, 109
  in mental illness, ecologic, 195
gelatin, in food allergy, 36-7
general adaptation syndrome, 15, 108 (see also individual susceptibility; specific adaptation)
  Coca, A.F., views on, 119, 120-1
  and environmental exposures, 135
  and food allergy, masked, 244
generally regarded as safe (GRAS)
  and chemical susceptibility, 144
  vs. individual susceptibility, 78, 270, 278
Gerdes, Kendall, 297
germ theory of disease, 7-8, 126, 127, 172
  and specific etiology doctrine, 218, 223, 266
Gerson, Max, 125-6
Gibson, Elizabeth, 18
Glaser, Jerome, 168, 220
"globaloney", 248
glue
  as air pollution, indoor, 235, 238
  sniffing, and drug addiction, 255
Godlowski, Z.Z. (Bob), 149
gold, in arthritis treatment, 226
Goodwin, James S., 225
Goodwin, Jean M., 225
Gorsuch, Anne M., 228
Gottlieb, P.M., 90
grains. see cereal grains
grapes, food allergy to in wine drinkers, 254
Green, Leonard, 66, 67, 124
Grove, Ella (second wife of A.F. Coca), 91
  and pulse test, 92
Guangzhou, China, 215
Guillet, Roberto, 297

Haeckel, Ernst, 107
hallucinations, 193
Hamburger, Maravene, 152-5q
Hansel, French Keller, 57, 97
  bibliography, 325-8
  and ophthalmology, 101-2
haptens, 231
"hardening", in chemical susceptibility, 109
Hare, Francis Washington Everard, 41, 80i, 81-2, 182

Hare, Francis Washington Everard (cont.)
  on alcoholism, 85-9
  bibliography, 329-31
  on food allergy, 83-5, 88-9
  Food Factor in Disease, The, 41, 81, 83, 85
  on mental illness, ecologic, 191
Harvard Medical School, 15
Hayden, Helen C., 65i
hay fever, 126
headache, 97, 193, 244, 247 (see also migraine)
  in chemical susceptibility, 137, 143, 147, 156, 165
  in food allergy, 19, 27, 30, 83, 175
  oxygen treatment, 50
  as withdrawal response, 248, 257
Hennecke, Thea, 160
Hensel, Hilda M., 65i
heroin addiction, 251, 252
Herontin Hospital, Chicago, 163, 199
  comprehensive environmental control unit, 214
  denies experimental nature of comprehensive environmental control, 234
hexachlorobenzene, 236
Hillsdale College, 8
Hippocratic ecological period of medicine, 218-9
Hippocrates, 223
histamine determinations, 15, 16
hives, 247
  as withdrawal response, 165, 193
Hoffer, Abram, 169
Holism and Evolution (J.C. Smuts), 223
homosexuality, 124
hoof and mouth disease quarantine, 3
Hosen, H., 195
housedust, 138
  allergy titration, 48
Houston, Texas, 98
How to Survive in America the Poisoned (L. Regenstein), 227
human conservation, 172, 174
human ecology, 167, 170, 173, 224 (see also clinical ecology)
Human Ecology and Susceptibility to the Chemical Environment (T.G. Randolph), 76, 77, 162, 215
Human Ecology Research Foundation (Rockwell M. Kempton Medical Research Fund), 113, 134, 159, 161
  formation, 80
  name change, 107

Human Ecology Study Group (Chemical Victims), 134, 145, 175
Humiston, Karl E., 297
Humphrey, Hubert, 145
hyperactivity, 244, 247
   in allergy, 30, 32-5, 36
   in chemical susceptibility, 156, 287
   ecologically focused medical care for, 276
   in food allergy, 32, 40, 88, 302
   and food coloring, 303
   as stimulatory response, 111, 193, 248, 272
hyperpyraemia, 83
hypersensitiveness, 17, 91
hypertension, 92, 247
   as withdrawal response, 248, 257
hypochondriacs
   and medical history recording, 24
   physician attitudes toward, 11-2
hypotension, 247

ice cream, corn sugar content, 184, 255
idioblapsis, 67 (*see also* food allergy)
   vs. specific adaptation, 119-20, 121
   term coined by A.F. Coca, 93
immunoglobulins
   IgE, 168, 220, 265, 277, 286, 289
   measurement in inhalant allergy, 151
immunology
   and allergic inflammation, 102
   and allergy, definition, 13-4, 17, 49, 77, 90-1, 126-7, 219-20, 265, 277, 289
   and allergy, dissociation from, 168
individual food tests. *see* food ingestion tests
individual susceptibility, 38 (*see also* food allergy; stimulatory-withdrawal responses)
   and addiction, 252
   and allergy, immunologic definition of, 126-7
   changes with time, 231
   degree indicated by avoidance and re-exposure, 244, 264, 275
   described by A.F. Coca, 220
   in dietetics, biologic, 181-2, 183
   in double-blind studies, 221
   hereditary basis disputed, 121
   relative predominance of substances responsible for, 78, 88, 232
   and specific adaptation, 109-12, 135, 170, 263, 271, 278

individual susceptibility (cont.)
   vs. term sensitivity, 107
   vs. toxicity, 163-8
influenza, and alcoholism, 87
insanity
   alternation with asthma, 188
   and fasting, 84
Insecticide, Fungicide and Rodenticide Act of 1972, 206
insecticides. *see* pesticides
insomnia, 193, 247
   as stimulatory response, 248
instructional courses, allergy, 97-8, 150-1
insurance companies, resistance to comprehensive environmental control claims payment, 214-5, 233-4, 285
Interactions of Man and His Environment, Proceedings of the Northwestern University Conference, 163-8q
Interlandi, Joseph, 66, 67, 113, 170
International Congress of Food and Digestive Allergy, 95-6, 150
International Correspondence Society of Allergists, 160
intradermal tests, 240, 261
intravenous alimentation
   and corn sugar allergy, 39-40, 65, 113
   sugar hazards in, 38
Ishizaka, L., 220
itching, 247
   as withdrawal response, 165, 193

JAMA, 30
janitorial supplies, 167
Jellinek, E.M., 107
Jerome, Michigan, 3
Johansson, H.G.O., 220
Jonathan Forman Gold Medal Award
   creation, 161
   to Frauenberger, G., 11
   to Randolph, T.G., 162
   to Rowe, A.H., 175
Journal of Allergy, 37, 96, 108
   rejects food allergy presentation of T.G. Randolph, 217
   rejects masked food allergy report of H.J. Rinkel, 97
   "symposium" on food allergy, 58
Journal of Immunology, 13, 90, 127
Journal of the American Dietetic Association, 37
Journal of the American Medical Association, 30

Journal of the Kansas Medical Society, 97
Journal of the Missouri Medical Society, 97
Juetersonke, George, 297
junk food, and corn addiction, 255-6

Kailin, Eloise W., 160, 195
Kansas City, Missouri, 97
Kaufman, W.
    on arthritis and arthralgia, 30-1
    on food allergy syndromes, 36
Kempton, Rena Bagley (mother of T.G. Randolph), 3, 5, 6, 7, 8
Kempton, Rockwell M. (uncle of T.G. Randolph), 5, 80
Kennedy, John F., 146
kerosene, 73
Kessler, Fred, 106
Kindwall, Josef, 79
King, David, 302
kinins, 277
kleptomania, 124
Klotz, Solomon, 169, 246
Kraepelin, E., 192, 197, 245
Kroker, George F., 297
Kruse, Marilyn, 297
Kuhn, Thomas, 225

Laseter, J.L., 236
Las Vegas, Nevada, 159
laughter, pathological, in food allergy, 36, 193
Lederle Laboratories, 91
Lee, Carleton, 148
Lee, James A., 224
Lempert, Julius, 57
Lerner, Michael, 293-4q
Lerner, Steve, 293
lettuce
    in chemical susceptibility testing, 137
    as pulse accelerator, 155
leukopenia
    in allergy, 50
    in food allergy, 16
Levin, Alan, 291-2q
Lewis, Helen, 297
Lockey, S.D., Sr., 73
Loveless, Mary, 58
Ludwig, Laurine P., 297
Lusk, F., 180
Lutheran General Hospital, Park Ridge, Illinois, 127-8, 145
    Randolph, T. G., termination of appointment to, 163

Lutheran Institute for Human Ecology, 163
Lyman, Richard B., 297
lymphocytes
    in allergy, function of, 50
    in allergy vs. infectious mononucleosis, 18

McGovern, Joseph J., 235, 284, 287
Mackarness, Richard, 113, 128, 149
Maclennan, John G., 195
McNamara, Steve, 284
Madison, Fred, 16
maggots, in osteomyelitis wounds, 14
Mandell, Marshall, 169, 195, 246
mania, 193, 248, 257, 272 (*see also* manic-depressive psychosis)
manic-depressive psychosis, 192, 193, 197, 245, 246, 256
Maniulit, Myrna T., 297
marihuana, 254, 256
marmalade, sulfur dioxide contamination, 233
Marquette University Medical School, Milwaukee, 16
Marshall, Robert T., 297
Massachusetts General Hospital, 15, 47
medical education, environmental, 267, 280
medical establishment
    allergy, attitudes toward, 126-7
    chemical susceptibility, attitudes toward, 77, 78-9, 125-6
    failure to apply concept of adaptation, 267
    food allergy, attitudes toward, 53-8, 79, 186
    mental illness, ecologic, attitudes toward, 168-9, 173-4, 195, 245-6
    resistance to change, 56-7, 66, 151, 216-7, 219-23, 234, 277-8
    responsible for environmental degradation, 205, 216
medical history, patient
    alcohol preferences, in relation to food allergy, 45
    in chemical susceptibility, 73-4
    direct recording, 24, 273-4
    ecologically focused, 273-4
Medicare, resistance to comprehensive environmental control claims payment, 214, 233, 285
medicine (*see also* clinical ecology)
    and drug-related symptomatic therapy, 134, 267-8, 275

medicine (cont.)
  environmental, 127, 170
  environmental, originated by Hippocrates, 223
  environmental vs. traditional, 127, 134-5, 138-9, 167, 170, 172-4, 211-2, 216-24, 243, 261-2
  federal regulatory interference, 228, 262, 265, 266, 279
  patient participation in, 138-9
  and problem solving, 225-6
  specialization and specific etiology in, 266
  synthesis vs. analysis in, 77, 135, 167, 179-82, 183-4, 186, 204-5, 263, 265-6, 277-8
  "third dimension" in, 138
  traditional, and failure in treatment of chronic illness, 277-80
Medicine, 30
melons, sulfur dioxide contamination, 233
Meniere's syndrome, as withdrawal response, 248, 257
mental illness, ecologic, 244 (*see also specific symptoms;* stimulatory-withdrawal responses)
  and chemical susceptibility, 195, 196, 205-6, 245, 286-8
  common environmental excitants of, 195-6
  development of concept, 245-7
  drug use in, 268
  etiology demonstrable, vs. psychiatry, 196-7
  history, 188-92
  need for new terminology, 103
  not distinct from physical illness, 258
  and observed togetherness, 197-9
  vs. psychiatry, traditional, 168-9, 173-4
  and specific adaptation, 111, 165, 168
Merck, Sharp and Dome, 55
metabisulfites, from food preservatives, 223
migraine
  and chemical fumes, 73
  in food allergy, 83-4, 182, 302
milk, 19
  and brain-fag, 104
  erythematous reaction to, 28
  and fatigue, in allergic toxemia, 27
  food allergy to, masked, 245
  in mental illness, ecologic, 195
  in migraine, 182
  pseudoalcoholic reaction to, 35

Mill, John Stuart, 221
Miller, Joseph B., 151, 195
Milwaukee, Wisconsin, 16, 47
Milwaukee Children's Hospital, 16
mimeograph fumes, and chemical susceptibility, 283
mineral oil, and chemical susceptibility, 191
Mirage of Health (R. Dubos), 218
Missing Diagnosis, The (O. Truss), 205
Mitchell, Donald S., 113, 114, 127, 128, 160
Monilia, 63
mononucleosis, infectious, and eosinophilia, 18
Morris, David, 73, 232
mosquito abatement fogs and mists, and chemical susceptibility, 76
Moss, Ralph W., 206, 215
motion picture, on food allergy and psychotic reaction to beets, 33, 54, 68, 127-8
motor exhaust, chemical susceptibility to, 76, 92
myalgia, 244, 247
  vs. brain-fag, 103
  in chemical susceptibility, 137, 143, 147
  and food allergy, 30, 302
  as withdrawal response, 193, 248
myositis, 193

Nalebuff, Donald J., 151
nasal secretion staining, 102
National Council on Alcoholism, 121
National Environmental Policy Act of 1969, 134, 206
National Institute of Mental Health, 168, 245
natural gas. *see* gas, utility
nectarines, dried, sulfur dioxide contamination, 233
Nesset, Norris, 127
neuralgia, 193, 247
  in chemical susceptibility to indoor air pollution, 147
neurasthenia, 84
neurosis (*see also* mental illness, ecologic)
  and allergy, 30
  causes sought by Sigmund Freud, 190
  and food allergy, 89
neutralization treatment. *see* provocation-neutralization
Newacheck, Paul, 284
Newbert, Louis H., 15
niacin supplementation, in sauna treatment, 236

nightshade family, 226, 253
Noon, L., 288
North Attleboro, Massachusetts, 169
Northwestern University Medical School, Chicago, 23, 53, 54, 96
    termination of T.G. Randolph, as "pernicious influence," 55-6, 113
Norwood Sanitarium, Beckenham, England, 85
Numoline, 69
nutrition (see also dietetics)
    overlap with clinical ecology, 211

Oakland, California, 94
obesity, 81, 244
    diet treatment for, 83
    ecologically focused medical care for, 276
    and food addiction, 40, 251, 252
    as stimulatory response, 36, 193, 248, 257, 272
observed togetherness, 197-9, 268
Odum, Eugene P., 203-4q
Of the Catarrhus Aestivus or Summer Catarrh (J. Bostock), xii
Oidiomycin, 63
Omaha, Nebraska, 20, 97
On Alcoholism (F.W.E. Hare), 41, 81-2
Ophthalmologic and Otolaryngologic Society of Allergy, 101
ophthalmology, 276
opiate addiction, 104
    compared with food addiction, 251-2
Oradell, New Jersey, 93
oranges, in ecologic mental illness, 195
Orr, William M., 297
OSHA (Occupational Safety and Health Act and Administration), 176, 200
    analytical bias of agency, 266-7
    regulatory powers curtailed by Reagan administration, 227
Otolaryngology
    certification denied by American Board of Allergy and Immunology, 151
    and ecologically focused medical care, 276
    instructional courses, 151
    overlap with allergy and ecologically oriented medicine, 101-2
    and surgical treatments, 102-3
oxygenation, tissue (see also acid-anoxia-endocrine theory of allergy)
    in allergy treatment, 50-1, 52, 275

pain, in allergic pain syndrome, 36
paint
    as air pollution, indoor, 147
    "hardening to" in chemical susceptibility, 109
Pan American Allergy Society, 292
paranoia, 193
particle board, as formaldehyde source, 232, 235, 238
Pasteur, Louis, 223
Pathology and Treatment of Asthma, The (H.H. Salter), 73
Pavlov, Ivan Petrovich, 189q, 190
peaches,
    canned, in chemical susceptibility testing, 137
    sulfur dioxide contamination, 233
pears, sulfur dioxide contamination, 233
pediatrics, 276
Peking Union Medical School, 215
peppers, 253
perception changes, in ecologic mental illness, 194
perfume, chemical susceptibility to, 92, 138, 235, 237
pesticides, 73, 164
    body fat storage, 231, 235-7
    chemical susceptibility to, 138, 143, 144, 166, 283
    government regulation relaxed by Reagan administration, 227
    in homes, 78, 235
Pfeiffer, Guy O., 170, 195
phenolic food compounds, and chemical susceptibility, 235
phenols, in chemical susceptibility, 283
Philosophy of Scientific Method (J.S. Mill), 220
Philpott, William H., 169, 195, 246
photophobia, in food allergy, 29
physical examination, in ecologically focused medical care, 274
placebo effect, 225-6
plastics, 164
    chemical susceptibility to, 138, 166
    as formaldehyde source, 232
plywood, indoor, as formaldehyde source, 232, 238
pollen allergy, as cause of ulcerative colitis, 29
pollinosis, 48, 126
pollution. see air pollution; food contaminants and additives; chemicals

pork, 36-7
potatoes
   food addiction to, 109, 252
   French fried, sulfur dioxide contamination, 233
   in mental illness, ecologic, 195
   related to tobacco, 253
Power, Bhaskar D., 297
Pretila, Edna B., 297
produce, organically grown, 134, 144-5
   testing by chemically susceptible persons, 145
provocation-neutralization technique
   in chemical susceptibility testing, 156
   and comprehensive environmental control, 221-3
   developed by C. Lee, H.J. Rinkel, 148, 150, 268
   double blind studies of, 302
   in food allergy testing, 183
   and symptom-suppressive state, 275
prunes, sulfur dioxide contamination, 233
psoriasis, 247
psychiatry (*see also* mental illness, ecologic)
   allergy, overlap with, 34-5, 101
   and brain-fag, 103
   drug use in, 173-4, 190, 196, 197, 268
   and ecologically focused medical care, 276
   and mental illness, ecologic, 168-9, 188-91, 195-7
   and observed togetherness, 197-9
psychoneurosis. *see* neurosis
psychosis (*see also* mental illness, ecologic)
   and allergy, alternation with, 85
   in chemical susceptibility, 147, 165
   in food allergy, 92
   manic-depressive, 192, 193, 197, 245, 246
psychosomatic medicine, 190, 276
puerperal sepsis, 56
pulmonary medicine, 276
pulse test, 66-7, 119
   in alcoholism treatment, 122
   analyzed, 152-5
   development, 92
   limitations, 120, 121
Pulse Test, The (A.F. Coca, L. Stuart), 120
purpura, as food and drug reaction, 16
Puza, Renee Sannes, 297

Quarterly Journal of Studies on Alcohol, 106

Rackeman, Francis M., 15, 16, 108
raisins, sulfur dioxide contamination, 233
ragweed pollinosis titration, 48, 126
Randolph, Fred Emerson (father of T.G. Randolph), 3, 6
   attitudes toward hygiene, 5, 7, 8
Randolph, Theron G. (*see also* Randolph, Theron G., publications cited)
   adaptation studies, 9
   alcoholism, incipient, 43-4
   on alcoholism and food addiction, 106-7, 123-4
   alcoholism research, 41-3
   alkali salts use in allergy, 268
   allergy, alternative mechanisms research, 49-53
   allergy, inhalant, research in, 47-8
   allergy and psychiatry connection discovered, 34-5, 289-90
   allergy meeting, first, 13
   allergy specialization in medical school, 14
   American Academy of Allergy and Immunology rejects food allergy presentation, 217
   American College of Allergists Merit Award, 215
   American College of Physicians fellowship prevented, 79
   at American International Hospital, 199
   appendicitis, 44
   on arthritis and artharlgia, 30-1
   bibliography, 349-57
   board certifications, 23
   boyhood and adolescence, 3, 4*i*, 5-9, 10*i*, 11
   brain-fag, origination of term, 103-4
   cereal grain sensitivity diagnosis, 27, 44
   chemical environment, first exposure to, 3
   chemical susceptibility research, 73-9
   Chicago, Illinois, private practice, 23-59
   Chicago Allergy Society, president of, 79
   China visit, 215
   clinical ecology, origination of term, 107
   Coca, A.F., views contrast, 119-21
   comprehensive environmental control, 114-5, 239, 268
   and comprehensive environmental control, legal suit over, 234
   on controversy between allergists and clinical ecologists, 289, 292
   corn sensitivity, 45-6

Randolph, Theron G. (cont.)
  death of first patient, 11
  Diabetic Diet Calculator, Squibb, 15
  divorce from Janet Sibley, 76
  draft status, 17
  Earth Day celebration, 176
  environmentally focused medical care concept formed, 243-7
  Environmental Quality Award received, 200
  eosinophil counting technique discovered, 17-8
  Evanston, Illinois, move to, 80
  fasting patients for food ingestion tests, 79, 114-5
  food allergy studies, 24-7, 30-47
  on food allergy vs. food addiction, 106-7, 123-4
  germ theory of disease, early experience of, 7
  Hare, Francis, research on, 80-2
  Harvard Medical School fellowship in allergy and immunology, 15-6
  at Herontin Hospital, 163, 214
  Hillsdale College, college education, 8-9
  on idioblapsis, 121
  immunologic theory of allergy teaching, 17
  internship and residency training (University of Michigan), 14-5
  Jonathan Forman Gold Medal Award received, 162
  Jonathan (son) born, 17
  at Lutheran General Hospital, 127, 163
  manuscripts refused for publication, 54
  marriage to Mitchell, Janet (Tudy), 77, 80
  marriage to Sibley, Janet, 17
  medical establishment and food industry conflict, over food allergy, 53-6, 57-8
  on medical history recording, 24, 273-4
  medical school education, University of Michigan, 9, 11, 13
  meeting with A.F. Coca, A.H. Rowe, H.J. Rinkel, J. Interlandi, 66-8
  meeting with G. Frauenberger, Z.Z. Godlowski, H.J. Rinkel, R. Mackarness, M. Spetz, 149
  mercury amalgam dental fillings replacement, 7
  milk avoidance, 5-6
  Milwaukee, Wisconsin, private practice, 16-7
  at Northwestern University Medical School, terminated as "pernicious influence," 55-6, 113

Randolph, Theron G. (cont.)
  on observed togetherness, 197-9, 268
  oxygenation research, 50-1
  pleurisy, 8, 14
  pneumothorax, 17
  psychology and psychiatry, early studies on, 9, 11
  on pulse test, 120, 121
  reluctance with new diagnoses and treatments, 63-4
  at St. Francis Hospital, 79, 80, 127
  scientific exhibit, Edinburgh, Scotland, 128
  Semmelweis, Ignaz, compared with, 56-7
  skin tests for food allergy diagnosis abandoned, 19-20
  Society for Clinical Ecology, formation of, 159-61
  on specific adaptation and individual susceptibility, 108, 114
  on steroid use, 267-8
  "stress" abandoned as term, 108
  at Swedish Covenant Hospital, Chicago, 127, 163
  tuberculosis, 13-4
  at University of Michigan Medical School Allergy Clinic, 17-20
  on writing for general public, 205-6, 279
Randolph, Theron G., publications cited
  acid-anoxia-endocrine theory of allergy, 52, 120, 137, 138
  adrenal cortex extract, 55
  adrenocorticotropic hormone, 55
  air pollution, 146-7q, 195
  alcoholism, 244
  alcoholism and allergy, 43, 107, 184-5
  allergy, immunologic definition of, 220
  allergy, inhalant, 48
  allergy, overlap with otolaryngology, psychiatry, 101
  An Alternative Approach to Allergies, 206, 215
  arthritis, rheumatoid, and fasting, 213, 234
  arthritis and arthralgia, 244
  beet sugar, specific allergy to, 58
  chemical susceptibility, 76, 143, 144, 155-6, 163-8q, 172-4, 182, 183, 233, 235, 245
  clinical ecology, 211-2, 245
  Clinical Ecology (L.D. Dickey), contributions to, 199
  Coca, A.F., A.H. Rowe, H.J. Rinkel, 89

Randolph, Theron G., publications cited (cont.)
 comprehensive environmental control, 109, 134-9q, 155, 163, 166, 173, 183, 195, 198, 200, 222, 234, 240, 246
 cortisone, 55
 cornstarch sized food containers, 37, 217
 cottonseed allergenicity, 37
 Diebetic Diet Chart, Squibb, 15, 31
 dietetics, analytic vs. biologic, 179, 181
 ecology, first use of term in medicine by Randolph, 127
 eosinophil changes in food and drug allergy, 49-50
 eosinophil counting techniques, 18
 excipients, allergenic, 37, 65
 fatigue, 30, 244
 FDA bread identity hearings, 55
 food addiction, 106, 107, 184, 244
 food allergy, 24, 27, 31, 32, 34, 58, 184, 192, 244
 Food Allergy (Randolph, H.J. Rinkel, M. Zeller), 31, 47, 63, 68, 192, 244, 246, 250
 food allergy diagnosis in alcohol drinkers, 45
 food allergy to food distillates, 47
 food ingestion tests, 20
 gelatin allergenicity, 37
 Hare, Francis, 82
 headache, allergic, 14, 27, 244
 histamine in allergy, 16
 "Human Conservation and Health," 171-5q
 Human Ecology and Susceptibility to the Chemical Environment, 143, 144, 156, 164, 166, 173, 182, 183, 200, 232, 233
 hyperactivity, 30, 40, 244
 leukopenia in allergy, 49-50
 lymphocytes in allergy, 18, 49-50
 medical history recording, ecological, 248
 medicine, environmental, 134, 138q, 170, 172-5q
 mental illness, ecologic, 34, 147, 168-9, 173, 188-97q, 244, 245
 mold allergy, 16
 myalgia, 30, 244
 natural food farm, 145
 obesity, in food allergy, 40, 244
 observed togetherness, 198-9
 peanut butter hearings, 163
 pesticide hearings, 145-6, 235

Randolph, Theron G., publications cited (cont.)
 "Provocative hydrocarbon test," 156q, 173
 psychotic reaction to beets (motion picture), 33, 128, 168, 245
 rheumatology, 101
 Rinkel, H.J., A.H. Rowe, A.F. Coca, 89
 Society for Clinical Ecology formation, 217
 specific adaptation, 110-2, 170, 195, 245
 stimulatory and withdrawal phases, 170, 192
 sugar allergenicity, 37, 253
 sugar in intravenous alimentation, 38, 40, 65
 sulfur dioxide, metabisulfites, sulfites, in chemical susceptibility, 233
randomized clinical trials, in clinical ecology, 217, 221, 222
Rapp, Doris, 220, 301-4q
RAST (radio allergo-sorbent test)
 for allergy, inhalant, 240
 in IgE identification, 168, 220, 265
Rea, Vera, 215
Rea, William J., 215, 287, 288
reaction levels, 129
Reagan-Bush administration
 environmental and medical policies, 134, 265
 environmental regulations relaxed and politicized, 227-8
 and government control of medicine, 228
 responsible for environmental degradation, 216
reductionism in science, vs. ecology, 203-4
refrigerants, in ecological mental illness, 195
Regenstein, Lewis, 227q, 228q
rheumatism, in chemical susceptibility, 147
rheumatology
 allergy, overlap with, 101
 and ecologically focused medical care, 276
rhinitis, 32-3, 41, 248
rhubarb, 182
Ribicoff, Abraham, 145
rice, food allergy to, 44
Richet, Charles, xii
Rinkel, Herbert John, 11q, 20, 57, 58, 63, 183
 allergy, inhalant, treatment for, 48, 268
 allergy contributions, 152

Rinkel, Herbert John (cont.)
  bibliography, 333-5
  on chemical susceptibility, 68, 69
  and corn allergy, 44
  on fatigue in food allergy, 27
  on food addiction, 69
  Food Allergy, (Rinkel, T.G. Randolph, M. Zeller), 31
  on food allergy, immunologic interpretation of, 49
  on food allergy, masked, 96, 97, 108-9
  on food allergy specificity, 88
  instructional courses, 97-8, 150-1
  meeting with A.F. Coca, J. Interlandi, A.H. Rowe, T.G. Randolph, 66-8
  meeting with G.S. Frauenberger, Z.Z. Godlowski, R. Mackarness, T.G. Randolph, M. Spetz, 149
  and Ophthalmologic and Otolaryngologic Society of Allergy, 101
  provocation-neutralization technique, 148, 268
  on rotary diversified diet, 45, 97, 268
  on Rowe, Albert, and corn sensitivity, 65q
  terminal illness, 148-50
Rinkel, Mrs. Herbert J., 149-50
Roberts, Ralph C., 113
Rockwell M. Kempton Medical Research Foundation. *see* Human Ecology Research Foundation
Rogers, Sherry, 215
Rollins, John P., 297
Rosen, Samuel, 57
Rosenzweig, L., 195
rotary diversified diet
  in food allergy treatment, 275, 276
  necessity of four day rotation, 44-5
  reported by H.J. Rinkel, 97, 268
Rowe, Albert Holmes, 36, 65i, 94i, 95i, 150
  antagonism of contemporaries, 64
  bibliography, 337-42
  and corn sensitivity, 65
  criticism from medical establishment, 95
  on elimination diets, 28-9q, 268
  on fatigue in food allergy, 27-30q
  on food allergy, 94q, 95-6, 175
  on food allergy specificity, 88
  on immunologic interpretation of food allergy, 49
  Jonathan Forman Gold Medal Award, 175
  meeting with H.J. Rinkel, AF Coca, J. Interlandi, T.G. Randolph, 66-8

Rowe, Albert Holmes (cont.)
  professional affiliations, 94
  publications, 95
Rowe, Albert, Jr., 176
Rowe, Mildred (wife of Albert Rowe), 175
Royal College of Physicians, 128
rubber, sponge
  and air pollution, indoor, 147, 238
  chemical susceptibility to, 138, 166
  in comprehensive environmental control, 115
  in mental illness, ecologic, 195
Ruckelshaus, William D., 176
rum, grape content, 46

Saifer, Phyllis, 284
St. Francis Hospital, Evanston, Illinois, 79, 115, 127
salmon, canned, in chemical susceptibility testing, 137
Salter, Henry Hyde, xii, 73, 83
Sanders, Sam, 151
San Francisco, California, 94
Sargent, Frederick II, 160
sauna, in chemical detoxification, 236
Savage, George M., 85, 188, 192, 197
Schafer, Walter L., 65i
schizophrenia, 246
Schloss, Oscar M., 95, 182
Schnare, D.W., 236
Schoonmaker, Jeanette S., 297
Scotch whisky, corn content, 46
Sechenov, I.M., 189
seizures, 92
  as stimulatory response, 193, 248
Selfridge, Grant L., 94
Selye, Hans
  A.F. Coca remarks on, 119
  and general adaptation syndrome, 15, 108, 120-1, 135, 244, 263
  "stress" defined, 108
Semmelweis, Ignaz, vii, 56, 57
sensitivity, vs. term susceptibility, 107
sexuality, changes in ecologic mental illness, 194
Shambaugh, George E., Jr., 56-7q, 101, 102-3
Shanghai, China, 215
Shannon, W.R., 182, 191
Sheldon, John, 14, 17, 18-9
shrimp, 182, 245
Sibley, Janet (first wife of T.G. Randolph), 17, 76

Siebold, George, 31
Sigmund Freud — a Pavlovian Critique (H.K. Wells), 190
Silver, Francis, 148, 160
Simon, R., 233
Simpson Memorial Institute, 12*i*, 13
Sinaiko, Robert, 285, 286*q*, 289*q*
skin tests
  for allergy, inhalant, 48
  for food allergy, 19-20, 23, 28, 30, 49
  inadequate in ecologic illness, 277
sleepiness, in chemical susceptibility, 156
Small, B., 232
Smuts, J.C., 179, 223
Sneeze, Wheeze and Itch Club (Boston Allergy Society), 90
Society for Clinical Ecology (American Academy of Environmental Medicine), 11, 82, 113, 162, 171, 204, 266 (*see also* Jonathan Forman Gold Medal Award)
  Archives for Clinical Ecology, 170
  formation, 159-61, 217
  legal suit, 292
  name change, 212
Society for the Study of Asthma and Allied Conditions, 94
solvents, 164
  body fat storage, 231
  and chemical susceptibility, 76, 109, 144
  in construction materials, 238
  in mental illness, ecologic, 195
somatoform disorders, 247
specialization, in medicine, 219, 223
specific adaptation
  vs. adaptation, 108, 263
  in clinical ecology, 263, 268-9
  essential concept in treatment of chronic illness, 278
  failure of, in onset of present illness, 249-50
  and individual susceptibility, 108-11, 135, 170, 173, 263-4, 271, 278
specific avoidance, 25, 32 (*see also* food ingestion tests)
specific etiology doctrine, 258, 262, 266-7, 269-70
  vs. clinical ecology, 222-3
  and germ theory of disease, 218
  and medical specialization, 219, 223
specific sensitivity, as allergy, 123-4
Spetz, Marseille, 149, 159
spinach, frozen, in chemical susceptibility testing, 137

Spohn, Richard, 292, 293
Squibb Diabetic Diet Calculator, 15
Squier, Theodore L., 16
Stahl, Richard, 98
starch, 83
  and alcoholism, 87
  allergenicity, 37
sterile technique, 56
steroids, 55
  excessive use of, in allergy, 267
Stigler, Del, 215
stimulatory-withdrawal responses, 33-5, 36, 111, 170, 258
  in double-blind studies, 221
  in food addiction, 244
  in mental illness, ecologic, 245
  reaction classes, 192-4, 197, 247-58, 271, 272
Stockholm, Sweden, 200
Stockman, David, 228
stress, 108
Structure of Scientific Revolutions (T. Kuhn), 225
Stuart, Lyle, 120
sublingual testing, 261
  for alcohol, synthetic ethyl, 155-6
  for food allergy, 240
  and Food Allergy Committee of American College of Allergists, 302
  for phenolic food compounds, 235
sublingual therapy, low dosage, 102
sucrose. *see* beet sugar; cane sugar
sugar (*see also* beet sugar; cane sugar; corn sugar)
  addictive potential, 252-3, 255
  and alcoholism, 43, 86, 87-8, 184, 255
  and food allergy, 37-8, 39, 58-9, 84
  and food labeling and geography of production, 239-40, 253
  and individual susceptibility, predominance in, 88
  and migraine, 182
  physical properties, 38-9
  and obesity, 83
Sugar Research Foundation, 58
Sugar Trap and How to Avoid It, The (B.T. Hunter), 39
sulfathiazole, 50
sulfites, and chemical susceptibility, 233
sulfur dioxide, and chemical susceptibility, 233
Sulzberger, Marion, 13, 14
sweating, in chemical detoxification, 236

Swedish Covenant Hospital, Chicago, 127, 163

tachycardia, 120, 247 (see also pulse test)
   as withdrawal response, 248, 257
tea, and addiction, 250, 253
Terr, Abba, 291, 292
textiles
   and chemical susceptibility, 144, 238
   and mental illness, ecologic, 195
Third Line Medicine — Modern Treatment for Persistent Symptoms (M.R. Werbach), 218, 222, 223, 258
Thomas, Charles C., 31, 160
Thompson, Willard O., 79
Thorndike, E.L., 189
thyroid supplement, corn content, 37
Tice's Practice of Medicine, 90, 17
titration, for inhalant allergy, 48, 151, 268
tobacco
   and food addiction, 253
   "hardening" to, in chemical susceptibility, 109
   smoke, as air pollution, indoor, 235
   smoke, as formaldehyde source, 232
   smoking and addiction, 251
tomato effect, 225-6
tomatoes, related to tobacco, 253
Tomidokoro, Ryuzo, 297
toxicity
   vs. chemical susceptibility, 163-8
   chemical susceptibility, overlap with, 144
   vs. individual susceptibility, 270, 278
toxicology, overlap with clinical ecology, 211, 270
Toxic Substances Control Act, 134, 200
   analytical bias in enforcement, 267
   regulations changed by Reagan administration, 228
training fellowships, in clinical ecology, 199, 297
Trichophyton, 63
Truss, Orion, 205
Tuft, L., 91
Tulane University, 11, 151
tuna, canned, in chemical susceptibility testing, 137
typhoid fever, immersion therapy for, 82

ulcerative colitis, 29, 175
unipolar adapted reactions, 32, 249, 250-1
   sustained, 251-6
unipolar maladapted reactions, 249-50

unipolar maladapted reactions (cont.)
   sustained, 257-8
United Nations Conference on the Human Environment, 200
University of Alabama, 151
University of California Medical School, San Francisco, 94, 284
University of Florida, 151
University of Hawaii, 151
University of Michigan Medical School, Ann Arbor, 9, 11, 12i, 14
   allergy clinic, 17, 18, 19, 47-8
University of Texas, 151
Urbach, E., 14, 17, 90
urea formaldehyde foam insulation, 232
utility gas. see gas, utility

vasculitis, 257
Vaughan, Warren T., 67
   bibliography, 343-6
   on food reactions, 36
Vichy, France, 96, 150
Virchow, Rudolf, 56
vitamins, first described, 180
Von Helsheimer, G., 195

Waickman, Francis J., 195
Waldbott, George, 77
Walker, Janet Mitchell (Tudy) (second wife of T.G. Randolph), 65i, 81, 82
   and chemical susceptibility, 76
   marriage, 77, 80
Walter, Clyde K., 297
Watson, John B., 9, 189
Watterson, Robert P., 114
Wells, H.K., 190
Werbach, Melvyn R., 218, 222-3, 258
Wesley Memorial Hospital, Chicago, 23, 34, 54, 289
Western Society for the Study of Hay Fever, Asthma and Allergic Diseases, 94
whealing response, 240
wheat
   in alcoholism, 42
   in arthritis, 213
   and food addiction, 252
   food allergy to, 35, 113, 244
   food allergy to, in alcohol drinkers, 254
   in mental illness, ecologic, 42
Wilkinson, Richard S., 297
Williams, Roger, 223
Williams, Russell I., 98, 150
Willoughby, James, 148, 150

Wilson, Stephen N., 297
Withers, Orval, 97
Wittich, F.W., 73
Wuhan, China, 215
Wyman, M., 126
Wyoming Postgraduate Course in Allergy and Immunology, 98

Xian, China, 215

Yale University Medical School, 106
Ye, Shitai, 215
Yeast, 63 (*see also* candidiasis)
  food allergy to, in alcohol drinkers, 42, 254

Yeast (cont.)
  infections, 205
  in mental illness, ecologic, 195
Yeast Connection, The (W.G. Crook), 205

Zeller, Lucille, 150
Zeller, Michael, 23, 31
  bibliography, 347-8
  on chemical susceptibility, 68
  Food Allergy (Zeller, T.G. Randolph, H.J. Rinkel), 31
  studied with H.J. Rinkel, 97
Zindler, George A., 78-9, 160, 243
Zotter, Hugo, 114

EPILOGUE
LAWRENCE D. DICKEY, M.D.   1905-1987

In Memory and Appreciation

This is a true epilogue, as this account is being written three days after this manuscript was submitted for publication and the day following the sudden death of Lawrence D. Dickey, M.D., November 24, 1987.

In my opinion, Larry Dickey, editor of the compendium, **Clinical Ecology,** *published in 1976, and the guiding light in planning, publishing and following through with our journal,* **Clinical Ecology—Archives for Human Ecology in Health and Disease,** *was, more than anyone else, responsible for presenting the recent documentation of clinical ecology.*

As such, this long-standing personal friend and remarkably generous individual deserves to be listed in the same book with the other pioneers of this new field. Finally, this accolade had not been written earlier in keeping with the policy of this book of documenting the medical accomplishments of only deceased persons.

Theron G. Randolph, M.D.